# HIGH-ORDER HARMONIC GENERATION IN SOLIDS

# HIGH-ORDER HARMONIC GENERATION IN SOLIDS

Editors

**Marcelo Ciappina**
*Guangdong Technion –*
*Israel Institute of Technology (GTIIT), China*
*Technion – Israel Institute of Technology, Israel*

**Paraskevas Tzallas**
*Foundation for Research and Technology –*
*Hellas (FORTH), Greece*
*Extreme Light Infrastructure-*
*Attosecond Light Pulse Source (ELI-ALPS), Hungary*

Foreword by
**Shambhu Ghimire** and **David Reis**
*Stanford University and*
*SLAC National Accelerator Laboratory, USA*

World Scientific

NEW JERSEY • LONDON • SINGAPORE • BEIJING • SHANGHAI • HONG KONG • TAIPEI • CHENNAI • TOKYO

*Published by*

World Scientific Publishing Co. Pte. Ltd.
5 Toh Tuck Link, Singapore 596224
*USA office:* 27 Warren Street, Suite 401-402, Hackensack, NJ 07601
*UK office:* 57 Shelton Street, Covent Garden, London WC2H 9HE

Library of Congress Control Number: 2024009298

**British Library Cataloguing-in-Publication Data**
A catalogue record for this book is available from the British Library.

**HIGH-ORDER HARMONIC GENERATION IN SOLIDS**



ISBN 978-981-12-7955-3 (hardcover)
ISBN 978-981-12-7956-0 (ebook for institutions)
ISBN 978-981-12-7957-7 (ebook for individuals)

For any available supplementary material, please visit
https://www.worldscientific.com/worldscibooks/10.1142/13506#t=suppl

Desk Editor: Shaun Tan Yi Jie

Typeset by Stallion Press
Email: enquiries@stallionpress.com

https://doi.org/10.1142/9789811279560_fmatter

# Foreword

The first experiments on the observation of non-perturbative high-order harmonics from solid materials subjected to strong laser fields were conducted by Shambhu Ghimire, David A. Reis and their collaborators and the results reported in a *Nature Physics* letter in 2010. Before this, experimental studies of strong field-driven high-order harmonic generation (HHG) phenomena were focused on isolated atoms and molecules in the gas phase. The challenges of intense laser-matter interactions in condensed matter lie in plasma formation and subsequent sample damage. In Ghimire *et al.*'s experiments, these concerns were mitigated by the tight focusing of laser beam to the exit face of 0.5 mm thick zinc oxide (ZnO) crystals. The ZnO crystal has a band gap of 3.2 eV and the driving laser's photon energy was 0.3 eV, making the band gap to photon energy ratio ~10, which is the same as the ratio of the ionization potential of argon atom (15.6 eV), the mostly used gaseous target, to the photon energy of the most common laser (1.5 eV) used in those experiments. Obtaining such a large ratio was important to ensure that multi-photon absorption is suppressed, and the interaction is well into the strong-field tunneling regime. In these experiments, the samples did not get damaged and there was no sign of plasma formation up to the peak intensity $\sim 5 \times 10^{12}$ W/cm$^2$ for repetitive excitation at 1 kHz. High-order harmonics up to 25$^{\text{th}}$ orders were observed, and the spectrum included a plateau feature, and an abrupt high-energy cutoff, very much like in atomic and molecular HHG. However, the high-energy cutoff law $I_p + 3.17\,U_p$, did not hold in solid-state HHG. The cutoff scaled linearly with the peak field of the laser instead of the quadratic scaling that has been well established in gaseous media. This new scaling hinted at a novel microscopic mechanism in solids attributed to the fundamental role of high density and periodicity. This paper sparked interest in the attosecond science community with several open questions at the time: (i) what are the relative roles of inter-band and intra-band channels, (ii) is there a unique relation between high-harmonic spectra and electronic band structure of the source material, (iii) can solid-state HHG be a novel probe of the atomic scale structure and dynamics of the source material, and (iv) can solid-state HHG support coherent attosecond pulse generation? Now it is well over a decade since the first experiments and these questions are at least partially answered, as presented beautifully in this book, and with a lot of exciting experimental and theoretical works ahead of us.

Shambhu Ghimire
David A. Reis

**Dr. Shambhu Ghimire** is a Lead Scientist at Stanford PULSE Institute, SLAC National Accelerator Laboratory, Stanford University, USA. He is a Fellow of OPTICA (formerly Optical Society of America), and is known for his pioneering work in generation of isolated attosecond pulse using polarization gating, strong-field physics in bulk and 2D crystals, nonlinear X-ray optics, and solid-state high-harmonic generation.

**Dr. David Reis** is the Director of Stanford PULSE Institute and is a Professor of Photon Science and Applied Physics at SLAC National Accelerator Laboratory and Stanford University, USA. He is a Fellow of OPTICA (formerly Optical Society of America) and the American Physical Society, and is known for his pioneering work in ultrafast and strong-field science including non-equilibrium lattice dynamics measurements, nonlinear X-ray optics, and solid-state high-harmonic generation.

https://doi.org/10.1142/9789811279560_fmatter

# Preface

High-order harmonic generation (HHG) in solids represents a captivating phenomenon at the intersection of attosecond science and condensed matter physics. Unlike the more commonly studied HHG in gases, which involves the interaction of intense laser fields with individual atoms or molecules, HHG in solids occurs when strong laser pulses interact with a crystalline lattice, leading to the emission of high-energy photons at harmonics of the driving laser frequency.

The phenomenon of HHG in solids involves both the acceleration of electrons within the crystal lattice and the electron-hole dynamics between the conduction and valence bands, both processes driven by the intense laser field. As these electrons gain energy from the laser, they undergo rapid oscillations within the lattice structure or transitions between the conduction and valence bands, emitting high-energy photons. This results in the generation of an attosecond burst of coherent radiation, enabling scientists to probe ultrafast dynamics in solid-state systems with unprecedented temporal resolution, as well as to characterize the crystal structure with sub-angstrom spatial resolution.

The Nobel Prize in Physics in 2018 was awarded to Gérard Mourou and Donna Strickland for their work on high-intensity ultrashort laser pulses. The techniques and principles they developed for the generation of intense laser pulses have paved the way for advancements in HHG research, including applications in solid-state systems.

Last year, the 2023 Nobel Prize in Physics was awarded to three trailblazers of attosecond science: Pierre Agostini, Ferenc Krausz, and Anne L'Huillier. The Laureates' experiments have produced pulses of light so short that they are measured in attoseconds, thus demonstrating that these pulses can be used to provide real-time images of processes inside atoms and molecules, and, recently, solid crystals.

By pushing the boundaries of our understanding of light-matter interactions in solids, HHG continues to unlock new frontiers in ultrafast science, offering unprecedented insights into the dynamics of electrons in condensed matter systems and laying the groundwork for future advancements in fields ranging from materials science to quantum electronics. The intent of this book is to provide a glimpse of recent theoretical and experimental developments in this research area.

We owe a great debt of gratitude to all of the authors, who have made this endeavor possible. We are especially grateful to Shaun, from World Scientific Publishing Co. Without him, this book would not be a reality. We hope that the present book serves as a valuable reference resource for many years to come.

Sincerely,

Marcelo Ciappina
Paraskevas Tzallas
Editors

https://doi.org/10.1142/9789811279560_fmatter

# Acknowledgments

Marcelo F. Ciappina acknowledges financial support from the Guangdong Provincial Key Laboratory of Materials and Technologies for Energy Conversion (MATEC), the Guangdong Province Science and Technology Major Project (Future functional materials under extreme conditions - 2021B0301030005) and the Guangdong Natural Science Foundation (General Program project No. 2023A1515010871).

https://doi.org/10.1142/9789811279560_fmatter

# About the Editors

**DSc. Dr. Marcelo Ciappina** is Associate Professor at the Guangdong Technion–Israel Institute of Technology, China. He is also Guest Professor at Nanjing University of Science and Technology, China, and Associate Editor of *Ultrafast Science*, a *Science* partner journal. He has co-authored over 250 journal papers with an H-index of 34. He is regarded as an expert in theory and numerical simulations of nonlinear laser interactions with atoms, molecules and complex systems. In recent years he has been one of the pioneers in a novel and fascinating field that merges two relatively new areas of research: attosecond and nanoscale physics.

**Dr. Paraskevas Tzallas** is Research Director at the Foundation for Research and Technology — Hellas (FORTH), Heraklion (Crete), Greece, Head of the Attosecond Science and Technology activity at FORTH, and Scientific Advisor of secondary sources at Extreme Light Infrastructure-Attosecond Light Pulse Source (ELI-ALPS), Szeged, Hungary. His research interest is in the field of attosecond science and quantum light engineering for applications in quantum information science. His research achievements, among others, include the first direct observation of attosecond light bursts emitted in gas and solid-state media, as well as the first observation of optical Schrodinger "cat" states using intense laser-matter interactions.

https://doi.org/10.1142/9789811279560_fmatter

# Contents

https://doi.org/10.1142/9789811279560_0001

# Chapter 1

# Probing Topological Phase Transition Using High-order Harmonic Generation

Shambhu Ghimire

*Stanford PULSE Institute,*
*SLAC National Accelerator Laboratory,*
*Menlo Park, CA 94025, USA*

## 1.1 Introduction to Three-Dimensional Topological Insulators

Three-dimensional topological insulators are known to host a multitude of emergent electrical phenomena from supporting helical currents [1] to dissipationless transport [2, 3], with the possibility of being building blocks for future technologies such as spintronics, quantum sensing and quantum computations [4]. They are often characterized in their electronic band structure with the presence of conducting surface bands forming an odd number of Dirac cones within the band gap of the bulk of the material [5,6]. The large spin-orbit couplings give rise to the band inversion and subsequent crossing of the valence and conduction bands of the surface states at or near the Fermi level. Because of these unique features in the band structure including the spin-momentum texture of the surface bands, topological insulators are the new class of emerging materials drawing renewed interests in investigating the fundamentals of intense light-matter interactions [7–11].

Attributed to their relatively small band gap, the laser peak intensities that are required to reach the strong-field limit [12] would typically damage these materials. For example, a typical 800 nm laser excitation would be above the band gap leading to generation of significant amount of carriers in the conduction bands. With the advent of intense few-cycle mid-infrared [7, 13] and terahertz laser pulses [14] it is now possible to excite these materials, well below their bulk band gap. In this case, because of the use of long wavelength, the strong-field regime can be reached at moderate peak intensities. In this regime, we expect that both the driven polarization and driven intra-band current radiate in the form of high-harmonics [8], analogous to the high-order harmonic generation (HHG) in wide band gap dielectrics subjected to intense near-infrared laser fields [15]. Precursors to these experimental activities were the advances in theoretical studies in model systems [16–18], which included the predictions of markedly different HHG efficiencies in topologically non-trivial and topologically trivial electronic states. In this chapter, we focus on the recent experimental progress in probing topological properties including chemically induced phase transitions in one of the prototypical three-dimensional topological insulators ($Bi_2Se_3$) subjected to intense mid-infrared laser fields.

### 1.1.1 *Crystal Structure of* $Bi_2Se_3$

$Bi_2Se_3$ has a layered rhombohedral structure with 5 atoms per unit cell forming quintuple layers that are stacked along the trigonal axis and are held together by weak van der Waals interactions [19,20]. It belongs to the space group notations $R\overline{3}m$, $D_{3d}^5$ and the unique space group number 166. The real space crystal structure is depicted schematically in Fig. 1.1(a), the top view is shown in Fig. 1.1(b), along with its corresponding 3D Brillouin zone (BZ) and the 2D projection of the (111) surface BZ in the momentum space, with proper labeling of high-symmetry points. It can be seen in the top view that the y-axis constitutes a mirror plane but otherwise the inversion symmetry is necessarily broken. In the momentum space, there are also three equivalent mirror planes along ΓM, in the $k_y$ direction, as shown in (Fig. 1.1(c)).

The bulk of the crystal is inversion symmetric, therefore a typical multi-cycle laser pulse would produce only odd-order harmonics that are polarized along the polarization direction of the laser field. However, as shown in the top view in Fig. 1.1(b), a direct consequence of the breaking of inversion symmetry at the surface is the generation of both even- and odd-order harmonics. A particularly interesting situation is that when the laser field is perpendicular to one of the mirror planes, the even-order harmonics were found to be polarized in the direction that is perpendicular to the polarization direction of the laser field [8]. The polarization direction flipping of even-order harmonics, from the direction along a mirror plane to a perpendicular direction, as the crystal is rotated, has been investigated

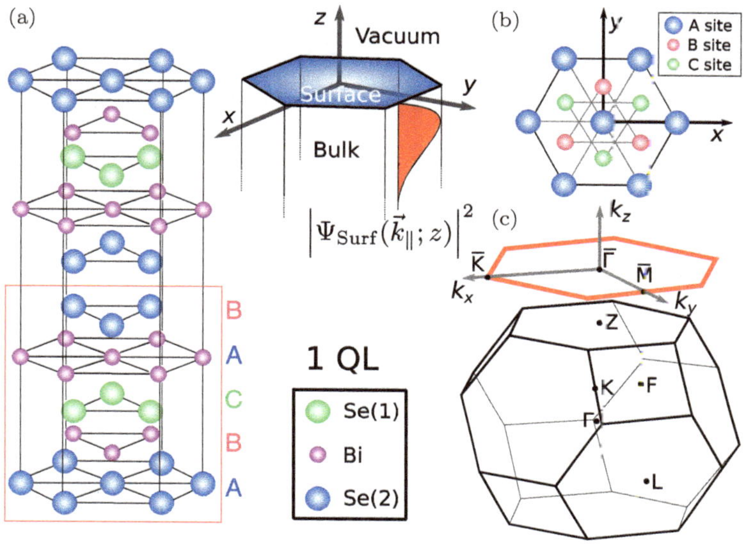


Fig. 1.1 (a) Crystal structure of $Bi_2Se_3$, which consists of alternating Bi and Se layers, stacked along the $z$-direction. Five consecutive layers form a quintuple layer (QL), which is the building block of the lattice (red rectangle). Each QL comprises five atoms: 2 equivalent Bi sites, 2 equivalent Se sites (Se(1)), and a third Se atom, Se(2), which assumes the role of an inversion center. The hexagonal lattice constants are $a = 4.14$ Å and $c = 28.70$ Å. (b) Schematic representation of the $C_{3v}$-symmetric $Bi_2Se_3$ (111) surface (rhombohedral convention), exposing a top Se layer and underlying Bi and Se layers. (c) Sketch of the 3D Brillouin zone of bulk $Bi_2Se_3$ (black) with the four time reversal-invariant points indicated $(\Gamma, L, F, Z)$. The projected 2D Brillouin zone of the (111) surface is shown as a red hexagon, with high-symmetry points labeled $\overline{\Gamma}, \overline{K}, \overline{M}$. Adapted with permission from [8].

in inversion symmetry-broken systems such as single-layer $MoSe_2$ [21]. Therefore, this is a general feature of inversion symmetry-broken systems [22–24], which does not necessarily represent the topologically non-trivial character of the surface state.

### 1.1.2 *Electronic Band Structure of $Bi_2Se_3$*

Electronic band structure of three-dimensional topological insulators are measured using angle-resolved photoemission spectroscopy (ARPES) [25], which is a surface-sensitive method. The left-most picture in Fig. 1.2 is one of such measurements of pristine $Bi_2Se_3$. The band structure contains a M-shaped valence bulk band, a well-separated conduction bulk band, and a

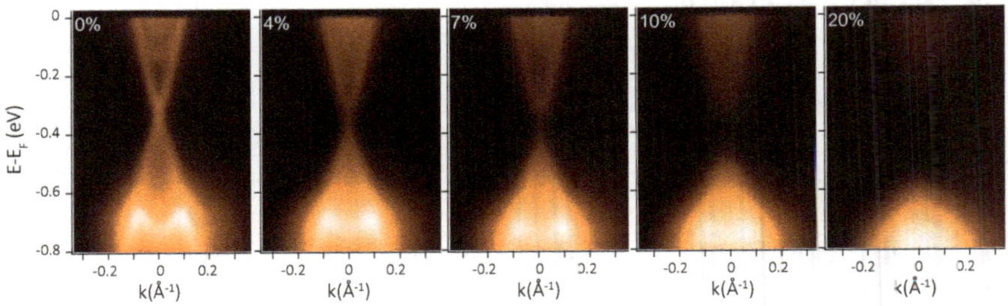


Fig. 1.2 ARPES measurement results. Measured electronic band structure of $(Bi_{1-x}In_x)_2Se_3$ for five different values of $x$, which represents fractional concentration of indium (in the plot, labeled as percentage). In pristine samples, topologically protected surface states and a single Dirac cone are clearly seen. The bulk valence band shows the expected M-shape, as a result of band inversion due to strong spin-orbit coupling. The bulk band gap is ~0.3 eV. As the concentration of doping increases from left to right the Dirac cone and linearly dispersive surface band begin to disappear. The weakening of the spin-orbit coupling and its subsequent effect in band inversion are prominent in the shape of the bulk valence band as it changes from M-shape to Λ-shape. For 20% doping the material turns into a band insulator. Adapted with permission from [27].

set of linearly dispersive surface valence and surface conduction bands that close within the bulk band gap energy at the crystal momentum around k = 0. The observed single Dirac cone is the signature of topologically non-trivial surface states. The subsequent data sets from left to right in Fig. 1.2 show measured band structures of $(Bi_{1-x}In_x)_2Se_3$, where $x$ is the fraction of indium concentration. The indium concentration increases from left to right in the figure. For higher concentrations, it can be seen that the bulk valence band changes from M-shape to a Λ-shape, which forms a minimum band gap at k = 0. Also, the surface bands start to disappear. For 20% doping the material turns completely into a band insulator. This is a signature of topological phase transition, which was originally reported in [26]. Results shown here are from measurements performed at the temperature of 20 K using XUV photon of photon energy ~35 eV. The chamber pressure was kept at less than $5 \times 10^{-11}$ Torr. Prior to the measurement, the protective selenium capping from the surface was removed by heating the samples to about 250°C while keeping the chamber pressure below $1 \times 10^{-9}$ Torr.

## 1.2 High-Harmonic Spectroscopy

The basic idea of high-harmonic spectroscopy relies on the implementation of the underlying microscopic mechanism of the HHG process to probe the atomic-scale structure of the source material. In the gas phase, this idea has been tested using aligned diatomic molecules [28, 29]. Molecular high-harmonic spectroscopy is largely based on the three-step recollision model [30, 31] comprising sub-cycle tunneling of electrons, their laser field-driven acceleration in the vacuum, and their subsequent returns with possible recombination to parent ions. The recombination is accompanied by the emission of high-order harmonics whose amplitude and phase depend on the recollision angle and emission time, endowing HHG with an atomic-scale spatial and temporal resolution [32]. The unprecedented temporal resolution is often obtained by taking advantage of recollisions corresponding to different harmonic orders that occur at different sub-cycle time scales.

Over the last decade, there has been a significant advancement in the investigation of the novel HHG mechanism in solid-state materials [33, 34], leading to the possibility of a similar spectroscopic approach for condensed matter systems. In the momentum space, underlying inter-band recollisions can probe the crystal momentum-dependent band gap [35] and intra-band Bloch oscillations can probe the shape of the conduction band [22, 36], whereas in the real space coherent recollisions can probe the valence charge density distributions [37, 38]. While these ideas are largely based on single active electron approximation they assume that the electronic band structure does not change during the excitation. At very high fields this approximation may not hold and particularly in materials that are known to host correlated electron dynamics the band gap can be modulated in the sub-cycle time scale [39, 40]. High-harmonics could probe dynamical band gap in these materials although a comprehensive theoretical approach is yet to emerge.

Recently, there has been efforts in scrutinizing spectroscopic capabilities of HHG to probe three-dimensional topological insulators [7, 13, 14]. The implications are three-fold: first, HHG could reveal the full potential of these novel materials for their technological applications; second, HHG could probe the driven non-equillibrium states including in ambient conditions; and finally, since these materials provide unique band structures such as the presence of Dirac cone and corresponding transition dipole moments with singularities, they can be useful to control the HHG process in a manner that is not possible in standard materials. For example, these materials are known to host large mean free paths for low-energy electrons and that could be useful for an efficient coherent driving of electrons.

To that aim, the outstanding initial questions are: (i) what is the signature of non-trivial topological states in HHG spectra, (ii) how do we separate the contributions of the surface and bulk state, and (iii) could HHG be used to probe topological phase transitions.

### 1.2.1 *Experimental Setup*

Because of its relatively large bulk band gap (∼0.3 meV) on this class of material and its well-studied band structure, $Bi_2Se_3$ is an excellent candidate for HHG driven by non-resonant excitation below the band gap, in the mid-infrared (MIR) and terahertz wavelength range. For ultrathin samples, the measurements can be performed using transmission geometry at normal incidence under ambient conditions as the harmonics are in the visible wavelength range. For bulk samples, they can be freshly exfoliated and measurements can be performed in reflection mode and at larger incident angles.

Intense femtosecond pulses in the MIR spectral range (5–10 $\mu$m) are generated in a setup based on a 1 kHz regenerative amplifier and optical parametric amplification followed by difference-frequency generation in GaSe. The output of the Ti:Sa amplifier (Coherent, 7 mJ, 45 fs, with a center wavelength of ∼790 nm) is used to pump a commercial optical parametric amplifier (TOPAS-HE, Light Conversion) in order to generate signal and idler pulses of ∼48–55 fs typical duration with central wavelengths tuned between 1415–1435 nm and 1715–1820 nm, respectively. Non-collinear type I phase-matched difference frequency mixing of the optical parametric amplifier output in a 500 $\mu$m thick, $z$-cut GaSe crystal (Eksma optics) yields MIR pulses tunable in the range 5–10 $\mu$m reaching typical energies of 80 $\mu$J at 7.5 $\mu$m. A 3 mm Ge filter is used to separate the MIR from the residual signal and idler beams. Then MIR laser is focused onto the sample either using a off-axis parabolic mirror or a lens. The polarization state of the driving field is varied with the aid of tunable half- ($\lambda/2$) and quarter-wave ($\lambda/4$) plates designed for the spectral region 1–19 $\mu$m (5 mm CdSe, Alphalas).

### 1.2.2 *Linearly Polarized Laser Field-Driven HHG*

Figure 1.3 shows measured and calculated crystal orientation dependence of HHG driven by linearly polarized MIR field. In panels a and b the polarizer in front of the detector has its axis fixed along and perpendicular to the MIR field direction, respectively. The three-fold crystal symmetry is reflected in the six-fold periodicity of the HHG signal as a function of the azimuthal crystal angle $\varphi$, as shown in the figure. Panels c and d of Fig. 1.3 display

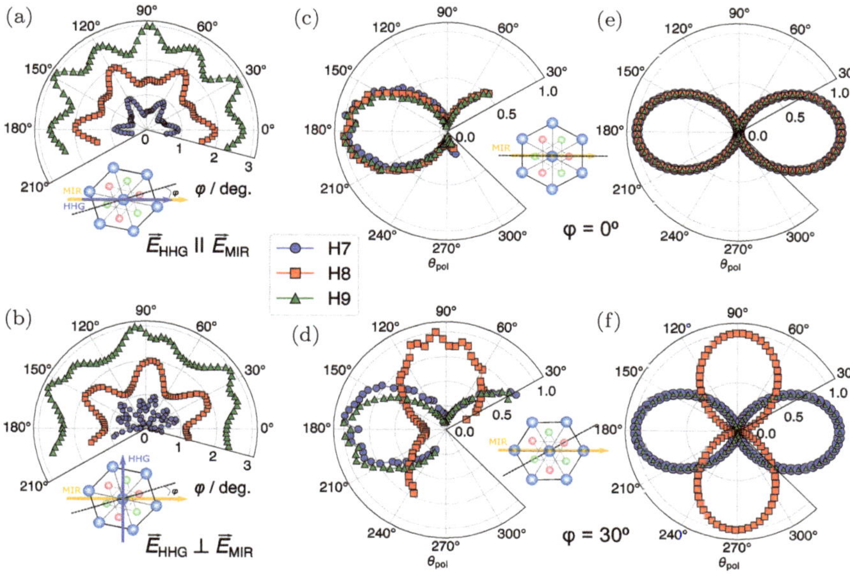


Fig. 1.3 Crystal orientation-dependent HHG and their polarization analysis. Measured crystal orientation-dependent yields for HOs 7, 8, and 9 for fixed polarization anlges, $\theta_{\rm pol}$, along (a) and perpendicular (b) to the polarization direction of MIR laser field. The relative orientation between the MIR laser polarization (horizontal) and the crystal's mirror axis (dashed line) is quantified by $\varphi$, as shown schematically on the left side in each panel. The curves for various harmonics are separated by an offset of 1 for clarity. Measured polarization-resolved yields of HOs 7, 8, and 9 for the MIR laser field along (c) and perpendicular (d) to the crystal's mirror axis. Panels (e) and (f) show calculated polarization-resolved harmonic yields of HOs 7, 8, and 9. Adapted with permission from [9].

measurements where both crystal and laser directions are fixed and the polarizer is rotated. They are polarization ellipses of the high orders (HOs) 7, 8 and 9 when the mirror axis of the crystal is along ($\varphi = 0°$) and perpendicular ($\varphi = 30°$) to the laser field, respectively. It is seen that in both cases, the polarization of the odd orders follows the driving field, whereas that of the even order (HO 8) depends on the orientation: it is parallel to the laser field when the latter is fixed at $\varphi = 0°$ (mirror plane), and perpendicular to it for the case $\varphi = 30°$. Panels e and f of Fig. 1.3 show theoretical polarization-resolved yields for the harmonic peaks HO 7, 8, and 9, derived from the surface-state calculation results [8]. While the general agreement is excellent, the non-vanishing signal for HO 8 at $\theta_{\rm pol} = 0$ in the measurement is likely due to spectral contamination from the strong HO 7 and HO 9.

Generation of even-order harmonics is a consequence of a broken inversion symmetry at the surface, but not necessarily a clear signature of non-trivial topology. A robust connection to topological properties can, however, be demonstrated through the response of the material to circularly polarized laser fields [8, 9].

### 1.2.3 *Circularly Polarized Laser Field-Driven HHG*

The HHG response to the elliptically polarized laser field is studied experimentally by varying the polarization state of the laser from right-circularly polarized ($\varepsilon = +1$) through linear ($\varepsilon = 0$) to left-circularly polarized ($\varepsilon = -1$) condition in small increments. In this process, the major polarization axis is fixed along $\varphi = 30°$. A remarkable result is that the intensity of HO 7 increases as the ellipticity increases from $\varepsilon = 0$ to $\varepsilon = \pm 1$, in stark contrast to the response of isotropic gas media or other solid-state systems [15, 24, 36], where the harmonic signal decays at high ellipticities.

These results are understood by calculating ellipticity dependence of HHG from a pair of surface and a pair of bulk bands. Calculation results are shown in Fig. 1.4(b). Generation efficiency of HO 7 from the bulk bands decreases monotonically as the laser ellipticity increases from $\varepsilon = 0$ to $\varepsilon = \pm 1$; however, the surface states show opposite behavior that is similar to the experimental results. With the aid of a detailed theory, as described in [8, 9], the enhancement of HHG from topological surface states for strongly elliptical fields can be attributed to some features in the electronic band structure. At relatively low peak intensities, the vortex-like pattern of transition dipole moment between valence and conduction bands are responsible for such behavior. At high peak intensities, the higher-order "warping" terms of the Hamiltonian, which are also responsible for the out-of-plane spin polarization of the surface bands, contribute to the enhancement with strongly elliptical or circularly polarized laser fields.

### 1.2.4 *Probing Topological Phase Transitions*

HHG response of pristine and 20% indium-doped $Bi_2Se_3$ is studied as they represent two extreme cases of topologically protected surface states, with the former being topologically non-trivial and the latter being topologically trivial. Experiments showed that for linearly polarized laser fields of peak

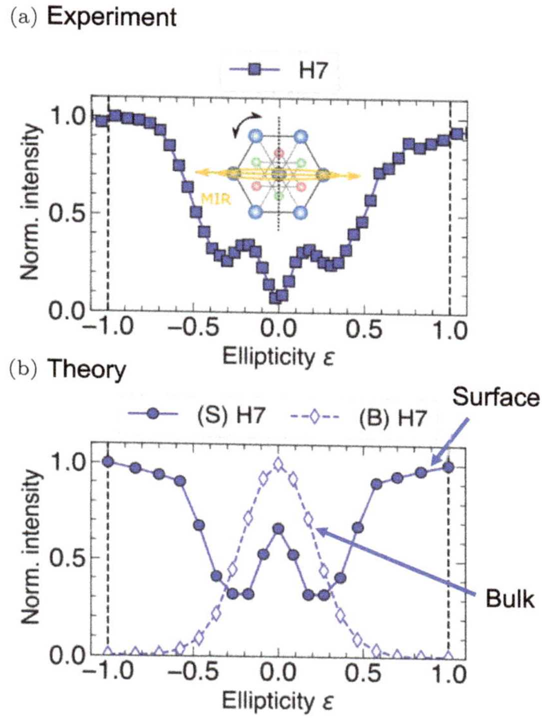


Fig. 1.4 Drive laser ellipticity dependence of HHG from $Bi_2Se_3$. (a) Experimental results; ellipticity-dependent response of HHG measured at normal incidence using an ultrathin sample. The inset shows the experimental geometry, where the major axis of the laser's polarization ellipse (indicated by the orange arrow) is fixed orthogonal to the crystal's mirror axis (dashed black line). (b) Calculated ellipticity-dependent response; the contributions from the bulk bands ("B", dashed lines, empty symbols) and from the surface bands ("S", solid lines, filled symbols) are shown separately. Experimental results look qualitatively similar to the dominant contributions from the surface states. Adapted with permission from [9].

field strengths around 0.2 $\text{Vnm}^{-1}$ and wavelength $\sim$4 $\mu$m, the pristine samples are about 8 times more efficient than doped samples.

The difference gets further enhanced for elliptically polarized laser fields. Figure 1.5 shows the intensity of a representative harmonic (HO5) for various MIR ellipticities, from right-circularly polarized ($\varepsilon = +1$) through linear ($\varepsilon = 0$) to left-circularly polarized ($\varepsilon = -1$) in small increments. For doped samples, the generation efficiency for HO 5 is maximum for $\varepsilon = 0$ (linear polarization), and drops monotonically as the ellipticity

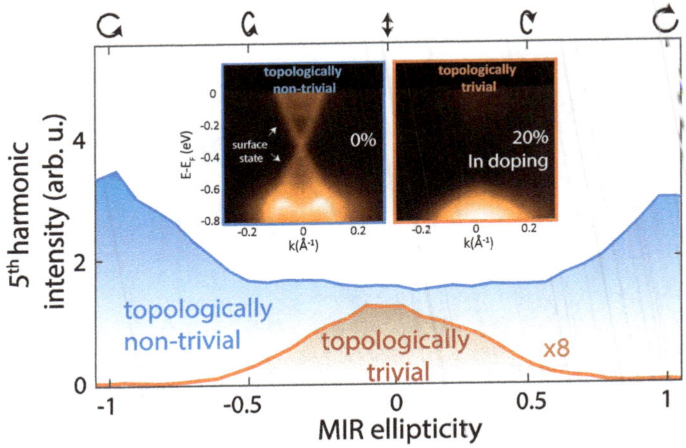


Fig. 1.5 Probing topological phase transitions using HHG. Inset: Measured electronic band structure, showing the topologically protected surface state for the pure $Bi_2Se_3$ sample (left). The surface state vanishes for high indium concentration (right) as the material turns into a band insulator. The intensity of the 5$^{th}$ harmonic as a function of MIR laser ellipticity is measured for both phases. Strong HHG enhancement is observed for the topologically non-trivial sample, especially for circular excitation. HHG and ARPES show one-to-one correspondence in terms of their ability to probe topological phase transition. Adapted with permission from [27].

increases to $\varepsilon = \pm 1$. For circular polarization, the sample does not produce detectable signal. In contrast, the pristine sample exhibits a strikingly different behaviour (blue line): higher absolute yield ($\sim$10), which grows with increasing laser ellipticity. For $\varepsilon = \pm 1$ the HHG yield is about twice as high as for $\varepsilon = 0$. The inset shows the measured electronic band structure of samples.

According to ARPES results the pristine $Bi_2Se_3$ has the non-trivial topological surface state as marked by the measured Dirac cone. The 20% doping fully destroys the Dirac cone, making the materials topologically trivial band insulators. While these are two extreme cases, HHG was studied in a range of doping concentrations.

#### 1.2.4.1 *Finding the Onset of Phase Transition*

In order to find the onset of topological phase transition, high-harmonics are generated from samples with a range of doping concentrations x = 0%,

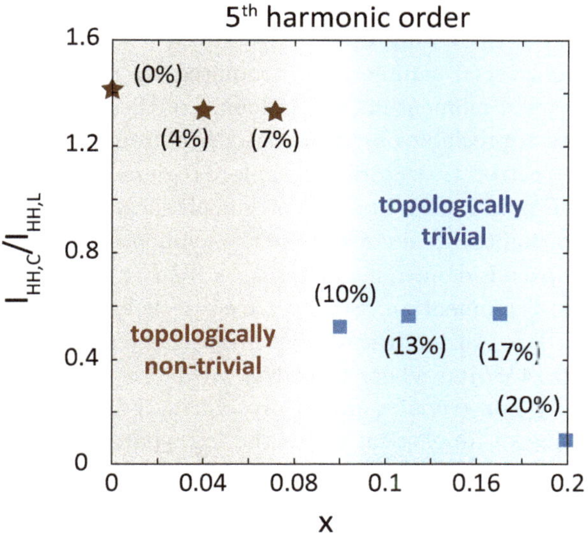


Fig. 1.6 Probing the onset of topological phase transition using high-harmonic spectroscopy. The ratio of harmonic yield for circularly polarized driver to linearly polarized driver is plotted as a function of the doping concentration x. There is a clear phase transition onset observed between 7% and 10% indium concentration. Adapted with permission from [27].

4%, 7%, 10%, and 13%. Figure 1.6 shows a ratio plot of HO5 generated by circularly polarized ($I_{\mathrm{HH,\,C}}$) versus linearly polarized ($I_{\mathrm{HH,\,L}}$) excitation. The ratio is higher than 1 for x = 0%, 4%, and 7%, and there is an abrupt change in the ratio to about 0.5 at x = 10% indicating an onset of topological phase transition. At about the same doping concentration x = 10% and 20%, ARPES data in Fig. 1.2 also show the disappearance of the Dirac cone and linearly dispersive surface state.

#### 1.2.4.2 *Laser Threshold and the Role of Spin-Orbit Coupling*

The ability to probe topological surface states through HHG relies on the fact that strong field-driven electrons probe sufficiently high momentum range of the BZ containing so-called hexagonal deformation of the band structure. This deformation feature was observed in photoemission experiments [41, 42], where it manifests itself as a "snowflake-like" shape. Mathematically, it corresponds to the higher-order ("warping") terms in the Hamiltonian. The exact analytical form of these terms is dictated by the combination of three-fold rotational symmetry ($C_3$), mirror symmetry as well as time-reversal symmetry, all imposed by the crystal structure. Being third-order in momentum, the influence of these terms on the HHG process becomes appreciable only when laser field intensities are sufficiently high to enable electron trajectories to explore regions of the BZ far enough away from the Γ point. Also, in terms of semiclassical electron trajectories, Bloch's acceleration theorem states that the momentum of the electrons in the reciprocal space is defined as: $k(t) = k_0 + \frac{e}{\hbar}\int E(\tau)d\tau$, with $k_0$ denoting the initial crystal momentum. Stronger electric fields can probe a larger fraction of the BZ. An empirically determined threshold field is around $E_{0,\text{threshold}} = 0.14\,\text{V/nm}$, where the HHG yields for circular excitation in the topologically non-trivial samples are equal. It means that for high-harmonic probing of topological states the laser parameters need to meet this threshold. The field threshold becomes lower for longer-wavelength drive. The detailed estimates can be found in [27].

## 1.3 Outlook

In recent years, there have been several reports regarding the use of strong light-field to induce topological phase transitions, including creation of novel non-equillibrium topological states that are not possible in equillibrium [43, 44]. For example, the laser field can drive a shear-mode phonon in $WTe_2$ and $MoTe_2$, and induce a structural phase transition from 1T' phase to $T_d$ phase. The crystal structure of 1T' phase is such that it breaks the inversion symmetry while that of $T_d$ phase recovers that symmetry; therefore, upon laser excitation $WTe_2$ undergoes a topological phase transition from type-II Weyl semi-metal to a trivial metal. Recently, this structural phase transition has been confirmed using MeV electron beam at SLAC [45]. The use of the HHG approach could reveal the underlying dynamics such as how the band structure evolves in the ultrafast time scale. Similarly, there are theoretical predictions that waveform-controlled intense laser fields can perform Floquet engineering of novel non-equillibrium topological states [46]. Depending on material systems these dynamics could either be only electronic or proceed through couplings to lattice motions and other degrees of freedom, making interpretations often difficult. Therefore, the HHG approach would be a novel table-top approach, which can complement conventional spectroscopic methods such as ARPES and ultrafast X-ray or electron diffraction.

## References

[1] J. W. McIver, D. Hsieh, H. Steinberg, P. Jarillo-Herrero and N. Gedik, Control over topological insulator photocurrents with light polarization, *Nature Nanotechnology* **7**, 2, pp. 96–100 (2012).
[2] P. Roushan, J. Seo, C. V. Parker, Y. S. Hor, D. Hsieh, D. Qian, A. Richardella, M. Z. Hasan, R. J. Cava and A. Yazdani, Topological surface states protected from backscattering by chiral spin texture, *Nature* **460**, 7259, pp. 1106–1109 (2009).
[3] S. OliaeiMotlagh, J. Wu, V. Apalkov and M. I. Stockman, Femtosecond valley polarization and topological resonances in transition metal dichalcogenides, *Physical Review B* **98**, 8, p. 081406(R) (2018).
[4] J. E. Moore, The birth of topological insulators, *Nature* **464**, 7286, pp. 194–198 (2010).
[5] M. Z. Hasan and C. L. Kane, Colloquium: Topological insulators, *Reviews of Modern Physics* **82**, 4, pp. 3045–3067 (2010).
[6] X. L. Qi and S. C. Zhang, Topological insulators and superconductors, *Reviews of Modern Physics* **83**, 4, p. 1057 (2011).
[7] Y. Bai, F. Fei, S. Wang, N. Li, X. Li, F. Song, R. Li, Z. Xu and P. Liu, High-harmonic generation from topological surface states, *Nature Physics* **17**, pp. 311–315 (2020).
[8] D. Baykusheva, A. Chacón, D. Kim, D. E. Kim, D. A. Reis and S. Ghimire, Strong-field physics in three-dimensional topological insulators, *Physical Review A* **103**, 2, p. 23101 (2021).
[9] D. Baykusheva, A. Chacón, J. Lu, T. P. Bailey, J. A. Sobota, H. Soifer, P. S. Kirchmann, C. Rotundu, C. Uher, T. F. Heinz, D. A. Reis and S. Ghimire, All-optical probe of three-dimensional topological insulators based on high-harmonic generation by circularly polarized laser fields, *Nano Letters* **21**, 21, pp. 8970–8978 (2021).
[10] C. Jürß and D. Bauer, Topological edge-state contributions to high-order harmonic generation in finite flakes, *Physical Review B* **106**, p. 054303 (2022).
[11] O. Neufeld, N. Tancogne-Dejean, H. Hübener, U. De Giovannini and A. Rubio, Are there universal signatures of topological phases in high-harmonic generation? probably not. *Physical Review X* **13**, p. 031011 (2023).
[12] S. Ghimire, G. Ndabashimiye, A. D. DiChiara, E. Sistrunk, M. I. Stockman, P. Agostini, L. F. DiMauro and D. A. Reis, Strong-field and attosecond physics in solids, *Journal of Physics B: Atomic, Molecular and Optical Physics* **47**, p. 204030 (2014).
[13] D. Baykusheva, J. Lu, J. A. Sobota, H. Soifer, C. R. Rotundu, P. S. Kirchmann, D. A. Reis and S. Ghimire, High-harmonic generation from topological insulators, in *2019 Conference on Lasers and Electro-Optics (CLEO)*, pp. 1–2 (2019), doi:10.1364/CLEO_QELS.2019.FM4M.6.
[14] C. P. Schmid, L. Weigl, P. Grössing, V. Junk, C. Gorini, S. Schlauderer, S. Ito, M. Meierhofer, N. Hofmann, D. Afanasiev, J. Crewse, K. A. Kokh, O. E. Tereshchenko, J. Güdde, F. Evers, J. Wilhelm, K. Richter, U. Höfer

and R. Huber, Generation in a topological insulator, *Nature* **593**, pp. 385–390 (2021).

[15] G. Ndabashimiye, S. Ghimire, M. Wu, D. A. Browne, K. J. Schafer, M. B. Gaarde and D. A. Reis, Solid-state harmonics beyond the atomic limit, *Nature* **534**, 7608, pp. 520–523 (2016).

[16] D. Bauer and K. K. Hansen, High-harmonic generation in solids with and without topological edge states, *Physical Review Letters* **120**, 17, p. 177401 (2018).

[17] A. Chacón, D. Kim, W. Zhu, S. P. Kelly, A. Dauphin, E. Pisanty, A. S. Maxwell, A. Picón, M. F. Ciappina, D. E. Kim, C. Ticknor, A. Saxena and M. Lewenstein, Circular dichroism in higher-order harmonic generation: Heralding topological phases and transitions in Chern insulators, *Physical Review B* **102**, p. 134115 (2020).

[18] R. E. Silva, Jiménez-Galán, B. Amorim, O. Smirnova and M. Ivanov, Topological strong-field physics on sub-laser-cycle timescale, *Nature Photonics* **13**, 12, pp. 849–854 (2019).

[19] H. Zhang, C.-X. Liu, X.-L. Qi, X. Dai, Z. Fang and S.-C. Zhang, Topological insulators in $Bi_2Se_3$, $Bi_2Te_3$ and $Sb_2Te_3$ with a single Dirac cone on the surface, *Nature Physics* **5**, 6, pp. 438–442 (2009).

[20] S. Mao, A. Yamakage and Y. Kuramoto, Tight-binding model for topological insulators: Analysis of helical surface modes over the whole Brillouin zone, *Physical Review B* **84**, 11, p. 115413 (2011).

[21] Y. Kobayashi, C. Heide, H. K. Kelardeh, F. Liu, T. F. Heinz and D. A. Reis, Polarization flipping of even-order harmonics in monolayer transition-metal dichalcogenides, *Ultrafast Science* **2021**, pp. 1–17 (2021).

[22] M. Garg, H. Y. Kim and E. Goulielmakis, Ultimate waveform reproducibility of extreme-ultraviolet pulses by high-harmonic generation in quartz, *Nature Photonics* **12**, pp. 291–296 (2018).

[23] Y. S. You, Y. Yin, Y. Wu, A. Chew, X. Ren, F. Zhuang, S. Gholam-Mirzaei, M. Chini, Z. Chang and S. Ghimire, High-harmonic generation in amorphous solids, *Nature Communications* **8**, p. 724 (2017).

[24] N. Liu, A. Chortos, T. Lei, L. Jin, T. R. Kim, W.-g. Bae, C. Zhu, S. Wang, R. Pfattner, X. Chen, R. Sinclair and Z. Bao, Ultratransparent and stretchable graphene electrodes, *Science Advances* **3**, p. e1700159 (2017).

[25] Y. Xia, D. Qian, D. Hsieh, L. Wray, A. Pal, H. Lin, A. Bansil, D. Grauer, Y. S. Hor, R. J. Cava and M. Z. Hasan, Observation of a large-gap topological-insulator class with a single Dirac cone on the surface, *Nature Physics* **5**, 6, pp. 398–402 (2009).

[26] M. Brahlek, N. Bansal, N. Koirala, S.-Y. Xu, M. Neupane, C. Liu, M. Z. Hasan and S. Oh, Topological-metal to band-insulator transition in $(Bi_{1-x}In_x)_2Se_3$ thin films, *Physical Review Letters* **109**, p. 186403 (2012)

[27] C. Heide, Y. Kobayashi, D. R. Baykusheva, D. Jain, J. A. Sobota, P. S. Kirchmann, T. F. Heinz, D. A. Reis and S. Ghimire, Probing topological phase transitions using high-harmonic generation, *Nature Photonics* **16**, pp. 620–624 (2022).

[28] J. Itatani, J. Levesque, D. Zeidler, H. Niikura, H. Pépin, J. C. Kieffer, P. B. Corkum and D. M. Villeneuve, Tomographic imaging of molecular orbitals, *Nature* **432**, 7019, pp. 867–871 (2004).

[29] P. Peng, C. Marceau and D. M. Villeneuve, Attosecond imaging of molecules using high harmonic spectroscopy, *Nature Reviews Physics* **1**, 2, pp. 144–155 (2019).

[30] K. C. Kulander, K. J. Schafer and J. L. Krause, Dynamics of short-pulse excitation, ionization and harmonic conversion, in B. Piraux, A. L'Huillier and K. Rzążewski (Eds.), *Super-Intense Laser-Atom Physics*. Plenum Press, pp. 95–110 (1993).

[31] P. B. Corkum, Plasma perspective on strong field multiphoton ionization, *Physical Review Letters* **71**, 13, pp. 1994–1997 (1993).

[32] A. J. Uzan, H. Soifer, O. Pedatzur, A. Clergerie, S. Larroque, B. D. Bruner, B. Pons, M. Ivanov, O. Smirnova and N. Dudovich, Spatial molecular interferometry via multidimensional high-harmonic spectroscopy, *Nature Photonics* **14**, 3, pp. 188–194 (2020).

[33] S. Ghimire and D. A. Reis, High-harmonic generation from solids, *Nature Physics* **15**, pp. 10–16 (2019).

[34] E. Goulielmakis and T. Brabec, High harmonic generation in condensed matter, *Nature Photonics* **16**, pp. 411–421 (2022).

[35] G. Vampa, T. J. Hammond, N. Thiré, B. E. Schmidt, F. Légaré, C. R. McDonald, T. Brabec and P. B. Corkum, Linking high harmonics from gases and solids, *Nature* **522**, 7557, pp. 462–464 (2015).

[36] S. Ghimire, A. D. Dichiara, E. Sistrunk, P. Agostini, L. F. Dimauro and D. A. Reis, Observation of high-order harmonic generation in a bulk crystal, *Nature Physics* **7**, 2, pp. 138–141 (2011).

[37] Y. S. You, D. A. Reis and S. Ghimire, Anisotropic high-harmonic generation in bulk crystals, *Nature Physics* **13**, 4, pp. 345–349 (2017).

[38] Y. S. You, E. Cunningham, D. A. Reis and S. Ghimire, Probing periodic potential of crystals via strong-field re-scattering, *Journal of Physics B: Atomic, Molecular and Optical Physics* **51**, 11 (2018).

[39] N. Tancogne-Dejean, M. A. Sentef and A. Rubio, Ultrafast modification of Hubbard U in a strongly correlated material: ab initio high-harmonic generation in NiO, *Physical Review Letters* **121**, 9, p. 097402 (2018).

[40] O. Grånäs, I. Vaskivskyi, X. Wang, P. Thunström, S. Ghimire, R. Knut, J. Söderström, L. Kjellsson, D. Turenne, R. Y. Engel, M. Beye, J. Lu, D. J. Higley, A. H. Reid, W. Schlotter, G. Coslovich, M. Hoffmann, G. Kolesov, C. Schüßler-Langeheine, A. Styervoyedov, N. Tancogne-Dejean, M. A. Sentef, D. A. Reis, A. Rubio, S. S. Parkin, O. Karis, J. E. Rubensson, O. Eriksson and H. A. Dürr, Ultrafast modification of the electronic structure of a correlated insulator, *Physical Review Research* **4**, 3, pp. 1–7 (2022).

[41] Y. L. Chen, J. G. Analytis, J.-H. Chu, Z. K. Liu, S.-K. Mo, X. L. Qi, H. J. Zhang, D. H. Lu, X. Dai, Z. Fang, S. C. Zhang, I. R. Fisher, Z. Hussain and Z.-X. Shen, Experimental realization of a three-dimensional topological insulator, $Bi_2Te_3$, *Science* **325**, 5937, pp. 178–181 (2009).

[42] D. Hsieh, Y. Xia, L. Wray, D. Qian, A. Pal, J. H. Dil, J. Osterwalder, F. Meier, G. Bihlmayer, C. L. Kane, Y. S. Hor, R. J. Cava and M. Z. Hasan, Observation of unconventional quantum spin textures in topological insulators, *Science* **323**, 5916, pp. 915–919 (2009).

[43] C. Bao, Light-induced emergent phenomena in 2D materials and topological materials, *Nature Reviews Physics* **4**, pp. 33–48 (2022).

[44] Jiménez-Galán, R. E. Silva, O. Smirnova and M. Ivanov, Lightwave control of topological properties in 2D materials for sub-cycle and non-resonant valley manipulation, *Nature Photonics* **14**, pp. 728–732 (2020).

[45] E. J. Sie, C. M. Nyby, C. D. Pemmaraju, S. J. Park, X. Shen, J. Yang, M. C. Hoffmann, B. K. Ofori-Okai, R. Li, A. H. Reid, S. Weathersby, E. Mannebach, N. Finney, D. Rhodes, D. Chenet, A. Antony, L. Balicas, J. Hone, T. P. Devereaux, T. F. Heinz, X. Wang and A. M. Lindenberg, An ultrafast symmetry switch in a Weyl semimetal, *Nature* **565**, 7737, pp. 61–66 (2019).

[46] M. S. Rudner and N. H. Lidner, Band structure engineering and non-equilibrium dynamics in Floquet topological insulators, *Nature Reviews Physics* **2**, pp. 229–244 (2020).

https://doi.org/10.1142/9789811279560_0002

# Chapter 2

# Manipulation of Quantum Properties in 2D Materials with Strong Tailored Fields

Eduardo B. Molinero[1], Rui E. F. Silva[1,2] and Álvaro Jiménez-Galán[1,2]

[1]*Instituto de Ciencia de Materiales de Madrid (ICMM), Consejo Superior de Investigaciones Científicas (CSIC), Sor Juana Inés de la Cruz 3, 28049, Madrid, Spain*
[2]*Max-Born-Institut, Max-Born Strasse 2A, D-12489, Berlin, Germany*

## 2.1 Introduction

Achieving on-demand electronic properties is one of the goals of modern condensed matter physics [1]. The synthesis of atom-thin monolayer materials marked one of the most important recent breakthroughs in that direction. Two-dimensional (2D) materials such as graphene and related transition metal dichalcogenides (TMDs) offer unique electronic properties [2–4], including record-high electron mobilities, topological physics [5–7], or spin- and valley-textures [8, 9]. The strong covalent bonds between the atoms of the monolayer and (weaker) interlayer van der Waals forces allow stacking of layers on top of each other and building of artificial materials with custom-design electronic properties [10].

At the same time, external electric fields are able to change material characteristics transiently during the time of the interaction, providing the opportunity to realize switches of on-demand electronic properties that operate on timescales shorter than the light cycle [11, 12]. Underpinning ultrafast control of electron motion is the feasibility to sculpt intense light with field strengths $E_0 \sim 1$ V/Å in different frequency ranges, from tens of THz to UV, at the level of individual oscillations; this includes control over the polarization, carrier-envelope phase, relative phases of multi-colour drivers or the orbital angular momentum [13–23]. As the light field strength changes from its peak value to zero in one quarter of its cycle, it produces a sudden voltage drop in the crystal that is comparable to the energy scales of electronic transitions, generating strong distortions of the charge distribution and a critical role of virtual charge carriers [24]. Furthermore, the electron motion is locked coherently to the light oscillations, allowing steering of electronic currents at the 0.1–10 fs scale with the sculpted light [25–31], i.e., lightwave electronics.

These timescales are particularly appealing for manipulating quantum electronic properties such as the valley degree of freedom or topological transport, which critically depend on fast relaxation and dephasing processes. For example, at cryogenic temperatures, the timescale at which dephasing occurs during electron transport in topological insulator films has been estimated to be on the order of few picoseconds [32] and the coherence between two valley states in a superposition is lost in around 100 fs [33]. Lightwave control of few-femtosecond electron motion in solids is key to exploit quantum properties in crystals, especially at room temperature. The question is how should one tailor the strong light waveform in order to control specific quantum properties of the crystal, and how to track these ultrafast changes during the laser-matter interaction.

In gases, the latter was made possible through the three-step model introduced by Corkum allowed mapping of the sub-cycle laser-driven electron dynamics to the amplitudes, phases and polarizations of the high harmonic radiation emitted by the oscillating dipole moment of the system [34–37]. In this way, the ultrafast laser-induced changes to the material are probed by the same fields that induce those changes, in an all-optical setup. Vampa *et al.* showed that an analogous three-step model can be applied to describe the generation of high harmonics in crystals, extending high harmonic spectroscopy to solids [38, 39]. This has allowed us to study the conduction band structure of wide-gap materials [40–43] as well as the strong laser-induced modifications to it [44], characterize topological invariants [45–48], time-resolve laser-driven electron motion in the conduction band [26, 49], and study phase transitions in strongly correlated systems [50–52].

In the next sections, we will review the use of tailored light waveforms to manipulate, on few-femtosecond timescales, two interconnected quantum properties: the valley degree of freedom and the Chern topological order of 2D semiconductors. Underlying the control of such quantum properties is control over two fundamental symmetries: spatial inversion and time reversal. We review the current efforts towards achieving few-femtosecond optical control of these symmetries with strong light fields in the route towards ultrafast manipulation and measurement of valley and topological degrees of freedom.

## 2.2 Symmetry Properties of 2D Materials

Ubiquitous synthesized monolayer materials include graphene, hexagonal boron nitride (hBN) or TMDs [53–55], all of which present a hexagonal arrangement of atoms in two triangular sub-lattices, as shown in Fig. 2.1(a). In graphene, each of the two sub-lattices host one carbon atom, preserving spatial inversion symmetry $\hat{\mathcal{P}}$. In absence of external fields or magnetic doping, monolayer materials preserve time-reversal symmetry $\hat{\mathcal{T}}$. The simultaneous conservation of $\hat{\mathcal{P}}$ and $\hat{\mathcal{T}}$ gives rise to the semi-metallic band structure of graphene, with the gap closing at the vertices of the Brillouin zone, that is, the $\mathbf{k} = \pm K$ crystal momenta (see Fig. 2.1(b)).

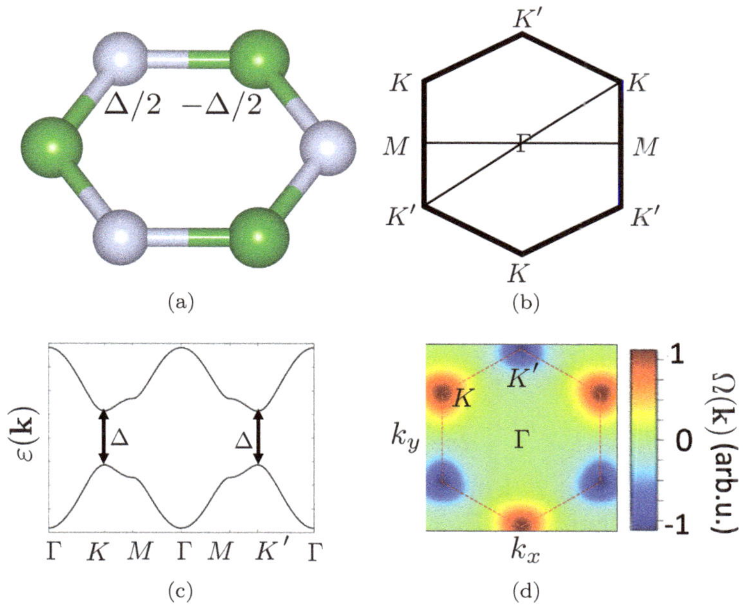


Fig. 2.1 Prototypical 2D hexagonal crystal. (a) Crystal lattice, made up of two triangular sub-lattices with on-site energies $\pm\Delta/2$. (b) Brillouin zone and high-symmetry points. (c) Band structure along the k-path shown by the lines in (b). (d) Berry curvature in the Brillouin zone.

In general, for a time reversal-invariant system, we have $\varepsilon(\mathbf{k}) = \varepsilon(-\mathbf{k})$ and $u^*_{n,-\mathbf{k}} = e^{i\varphi(\mathbf{k})}u_{n,\mathbf{k}}$, where $\varepsilon(\mathbf{k})$ is the band dispersion, $u_{n,\mathbf{k}}$ is the periodic part of the Bloch wavefunction of band $n$ at crystal momenta $\mathbf{k}$, and $\varphi(\mathbf{k})$ is an arbitrary phase reflecting the freedom of gauge. For a spatial inversion-invariant system, we have $\varepsilon(\mathbf{k}) = \varepsilon(-\mathbf{k})$ and $u_{n,-\mathbf{k}} = e^{i\varphi(\mathbf{k})}u_{n,\mathbf{k}}$, where we note the absence of the complex conjugate in the relation of the Bloch functions [56]. Thus, when both of these symmetries are present, $u_{n,\mathbf{k}}$ can be chosen real and $\pm K$ are indistinguishable.

In hBN and TMDs, the spatial inversion symmetry of the monolayer is broken by inequivalent atoms in the two sub-lattices. This opens a gap at both $\pm K$ Dirac points, which become energy-degenerate but inequivalent valleys in the band structure ($u^*_{n,-K} = e^{i\varphi(k)}u_{n,K}$, with $u_{n,k}$ complex), see Fig. 2.1(c). In such broken inversion-symmetric systems, we will refer to the $-K$ valley as $K'$. The Bloch electrons of these 2D gapped systems carry an orbital magnetic moment that arises from the self-rotation of the wavepacket along its center-of-mass motion [57, 58]. For the n-th band, it can be expressed as

$$\mathbf{m}_n(\mathbf{k}) = i\frac{\mu_B}{m}\sum_{j\neq n}\frac{\mathbf{P}_{n,j}(\mathbf{k})\times\mathbf{P}_{j,n}(\mathbf{k})}{\varepsilon_n(\mathbf{k})-\varepsilon_j(\mathbf{k})}, \tag{2.1}$$

where the Bohr magneton $\mu_B = \frac{e\hbar}{2m}$, $-e$ is the electron charge, $m$ is the electron mass, and the interband matrix element $P^{(\alpha)}_{n,j} = \langle u_{n,\mathbf{k}}|\hat{p}_\alpha|u_{j,\mathbf{k}}\rangle$, with $\alpha = x, y, z$. For the 2D monolayer, $\hat{p}_z = 0$ and the orbital magnetic moment is directed perpendicular to the plane of the monolayer, $m(\mathbf{k}) = \mathbf{m}(\mathbf{k})\cdot\hat{\mathbf{z}}$. When the dynamics in the system can be reduced to a two-band model, $m(\mathbf{k})$ will have the same value in both valence and conduction bands and can be written as

$$m(\mathbf{k}) = \frac{\mu_B}{2m}\frac{|P_-(\mathbf{k})|^2 - |P_+(\mathbf{k})|^2}{\Delta(\mathbf{k})}, \tag{2.2}$$

where $\Delta(\mathbf{k}) = \varepsilon_c(\mathbf{k}) - \varepsilon_v(\mathbf{k})$ is the band gap and $|P_\pm(\mathbf{k})|^2$ is the oscillator strength of $\sigma_\pm$ polarization, $P_\pm = P^{(x)}_{c,v} \pm i\,P^{(y)}_{c,v}$. For a system which is time reversal-invariant, $m(\mathbf{k}) = -m(-\mathbf{k})$, which results in circular dichroism for positive and negative crystal momenta according to Eq. (2.2). This circular dichroism becomes a perfect optical selection rule at the $K$ and $K'$ points: $\sigma_+$ ($\sigma_-$) circularly polarized light couples only to $K(K')$ [58], as shown in Fig. 2.2(a).

Because spatial symmetry is broken in the lattice, it gives rise to a permanent dipole. This permanent dipole can give rise to electron motion that is orthogonal to that of the driving field polarization. In the semiclassical equation of motion, where the electron is assumed to move only along one band, this orthogonal velocity appears explicitly through a term analogous to a momentum-space Lorentz force [59, 60],

$$\dot{\mathbf{r}}_n(\mathbf{k}) = \frac{1}{\hbar}\frac{\partial \varepsilon_n(\mathbf{k})}{\partial \mathbf{k}} - \dot{\mathbf{k}} \times \Omega_n(\mathbf{k}), \qquad \dot{\mathbf{k}} = -\frac{e}{\hbar}\mathbf{E}(t). \tag{2.3}$$

The first term in the right-hand side of the semiclassical velocity $\dot{\mathbf{r}}$ is the group velocity, which leads to the drift current, and is directed along the electric field polarization. The second (Lorentz-like) term is the so-called anomalous velocity, which contains the Berry curvature $\Omega_n(\mathbf{k})$ as a pseudo-magnetic field in momentum that is proportional to the magnetic moment,

$$\boldsymbol{\Omega}_n(\mathbf{k}) = i\frac{2\hbar\mu_B}{em}\sum_{j\neq n}\frac{\mathbf{P}_{n,j}(\mathbf{k}) \times \mathbf{P}_{j,n}(\mathbf{k})}{[\varepsilon_n(\mathbf{k}) - \varepsilon_j(\mathbf{k})]^2} = \frac{2\hbar}{e}\sum_{j\neq n}\frac{\mathbf{m}_n(\mathbf{k})}{\varepsilon_n(\mathbf{k}) - \varepsilon_j(\mathbf{k})}. \tag{2.4}$$

We note that this expression is analogous to that of the propensity field determining the forward-backward asymmetry in photoelectron circular dichroism of chiral molecules [61, 62]. As is clear from Eq. (2.4), the Berry curvature reflects the influence of the rest of the bands on the electron motion along one band. Yet, it vanishes locally at each $\mathbf{k}$ when all bands are considered,

$$\sum_n \boldsymbol{\Omega}_n(\mathbf{k}) = 0. \tag{2.5}$$

Like the orbital magnetic magnetic moment, the Berry curvature in the monolayer is directed along the surface normal, $\Omega(\mathbf{k}) = \boldsymbol{\Omega}(\mathbf{k}) \cdot \hat{\mathbf{e}}_\mathbf{z}$. It is shown in Fig. 2.1(d) for the prototypical gapped graphene system. The out-of-plane Berry curvature gives rise to an in-plane velocity that is orthogonal to the driving field. This generates an anomalous current $j_{\text{anom}}$ that is orthogonal to the drift current. For a 2D system and a linearly polarized field $\mathbf{E} = (E_x, 0)$,

$$j_{\text{anom},y} = -\frac{e}{(2\pi)^2}\left(\frac{-e}{\hbar}E_x\right)\sum_n \int_{BZ} d\mathbf{k}\, \rho_{n,n}(\mathbf{k})\, \Omega(\mathbf{k}), \tag{2.6}$$

where $\rho_{n,n}$ is the occupation of band $n$. From Eq. (2.6) we can define an anomalous Hall conductivity simply as,

$$\sigma_{xy} = \frac{e^2}{\hbar(2\pi)^2}\sum_n \int_{BZ} d\mathbf{k}\, \rho_{n,n}(\mathbf{k})\, \Omega_n(\mathbf{k}). \tag{2.7}$$

The Berry curvature satisfies the same symmetry relations as the magnetic dipole moment. When spatial inversion is conserved, $\Omega_n(\mathbf{k}) = \Omega_n(-\mathbf{k})$. When time reversal is conserved, $\Omega_n(\mathbf{k}) = -\Omega_n(-\mathbf{k})$. Hence, in the presence of both of these symmetries both $m(\mathbf{k})$ and $\Omega(\mathbf{k})$ vanish, and consequently the anomalous Hall conductivity also vanishes. This is the case of pristine graphene, bulk TMDs or TMDs with an even number of layers in which adjacent layers are rotated by 180° (2H stacking). For monolayers with broken inversion symmetry, the anomalous Hall conductivity is non-zero when $\rho_{n,n}(\mathbf{k}) \neq \rho_{n,n}(-\mathbf{k})$, e.g., when there is valley polarization.

For a 2D system in equilibrium, with fully occupied (empty) valence (conduction) bands,

$$\sigma_{xy}^{\text{eq.}} = \frac{e^2}{\hbar(2\pi)^2} \sum_n^{\text{occ}} \int_{BZ} d\mathbf{k}\, \Omega_n(\mathbf{k}) = \frac{e^2}{h} \sum_n^{\text{occ}} \mathcal{C}_n, \tag{2.8}$$

where $\mathcal{C} = \frac{1}{2\pi} \int_{BZ} d\mathbf{k}\, \Omega(\mathbf{k})$ is an integer called the Chern number that characterizes the static band topology [63]. If time-reversal symmetry is present, $\mathcal{C} = 0$ since $\Omega(\mathbf{k}) = -\Omega(-\mathbf{k})$. Hence, to have a non-zero Chern number in equilibrium, time-reversal symmetry must be broken, as first proposed by Haldane in his paradigmatic model of a Chern insulator [5] (see Section 2.4).

## 2.3 Valleytronics

While many materials present valleys in the band structure, the majority do not present any valley-contrasting quantity that couples to an external field. In contrast, hexagonal monolayers, by virtue of their broken spatial inversion symmetry, present valley-contrasting magnetic dipole moment and Berry curvature. The field of valleytronics looks for ways to initialize, control and read-out the valley state [8, 9], which is usually considered as a two-level system (K and K′) that can be represented as a Bloch sphere,

$$\psi_{\text{V}} = \cos(\theta/2)|K\rangle + e^{i\phi}\sin(\theta/2)|K'\rangle, \tag{2.9}$$

with $0 \leq \theta \leq \pi$ and $0 \leq \phi \leq 2\pi$. In this way, a valley index can be used as a new degree of freedom, additionally to the electron charge and spin, with potential applications in both classical and quantum computing (even if quantum entanglement between two valley qubits has not been realized yet).

Different methods have been used to generate valley polarization, including using external magnetic fields [64], doping [65], the magnetic proximity effect and strain [66], or external time-dependent electric fields [67–70]. We will focus on the latter, which allows for the fastest valley switches.

### 2.3.1 *Resonant Optical Valley Selection Rules*

In 2D TMDs, the residual interaction between the excited electron and the remaining hole is unusually strong due to the strong Coulomb interaction and weak dielectric screening, giving rise to exciton states with large binding energies (hundreds of meV) and oscillator strengths [72–74]. Such excitonic states retain the same valley-contrasting properties as that of the free carriers (Eqs. (2.2), (2.4)), and can be addressed by optical absorption from the valence band with visible frequencies.

The first strategies to generate valley polarization made use of circularly polarized fields resonant to the bright exciton states [67,75], creating valley-dependent circular dichroism $\rho_{n,n}(\mathbf{k}) \neq \rho_{n,n}(-\mathbf{k})$ according to Eq. (2.2). Consequently, this generates an anomalous Hall conductivity (Eq. (2.7)) that can be measured, e.g., with a Hall bar [76]. When the exciton is excited by linearly polarized (LP) light — an equal mix of $\sigma_+$ and $\sigma_-$ components — both $K$ and $K'$ valleys will populate equally. During the excitation process, the relative phase between the $K$ and $K'$ valleys remains fixed [77], a property referred to as valley coherence.

#### 2.3.1.1 *Valley Manipulation*

In order to make valleytronics feasible, the switching times of the valley state must be short [9]. For a classical operation, the manipulation of the valley state must be performed on timescales faster than the time it takes the carriers to scatter to the opposite valley (valley lifetime, $T_1^{(V)}$), which is on the order of a few picoseconds [8,9]. For a quantum operation, this manipulation should be faster than valley decoherence ($T_2^{(V)}$), which is on the order of a few hundred femtoseconds at temperatures below 100 K [33,78].

Langer *et al.* demonstrated valley population transfer using a circularly polarized pulse in combination with an overlapping strong, low-frequency driving pulse with wavelength $\lambda = 7.5$ $\mu$m [68]. The circular pulse injects population at one of the valleys, e.g., $K$, thanks to the optical valley selection rules. The low-frequency mid-infrared pulse is not sufficiently strong to inject significant population into the conduction band but drives the electron population that is injected by the circular pulse from one valley to the other following $\mathbf{k}(t) = \mathbf{k} + \frac{e}{\hbar}\mathbf{A}(t)$, where $\mathbf{A}(t) = -\int dt\mathbf{E}(t)$ is the vector potential of the field (see Eq. (2.3)). Valley population transfer occurs in one quarter-cycle of the mid-infrared field ($\approx$ 6.25 fs).

In the weak-field regime, the recent generation of trains of ultrashort pulses, down to the attosecond timescale [11], with a controllable phase and polarization relation between them [79–82] allows a coherent control protocol of valley switching. Silva *et al.* calculated that a sequence of identical linearly polarized ultrashort pulses with alternating polarization can turn on and off the valley state of 2D crystals on timescales shorter than 100 fs [71]. The scheme is illustrated in Fig. 2.2(b) for a gapped graphene system with band gap $\Delta = 6$ eV. First, a linearly polarized pulse injects population equally at both the $K$ and $K'$ valleys in the conduction band. A controlled time delay later, a replica of the first pulse but polarized in the orthogonal direction destroys the population at $K$ and enhances it in $K'$, creating valley polarization. After the second pulse, another time-delayed pulse replica polarized parallel to the first pulse can restore equal populations at $K$ and $K'$. Finally, a fourth replica polarized perpendicular to the first pulse can create again valley polarization. As long as the pulses are sufficiently short so that electron-hole dephasing, on the one hand, and valley lifetime ($T_1^{(V)} \sim 1$ ps), on the other, are longer than the duration of the four pulse sequence, the valley switch is realized.

Similarly, valley coherence can be controlled with a sequence of two delayed ultrashort pulses through the optical Stark effect using a slightly detuned circularly polarized field, as shown by Ye *et al.* [78].

### 2.3.2 *Non-Resonant, Strong-Field Valley Selection and Manipulation*

Above, we have discussed valley control with resonant pulses based on the optical valley selection rules (Eq. (2.2)). As we will now show, initializing and controlling the valley state with non-resonant, strong-driving fields offers some advantages, including a non-material-specific setup or valleytronics in inversion-symmetric materials.

Let us consider the population induced in the conduction band by a strong laser field with a frequency well below the material's band gap.

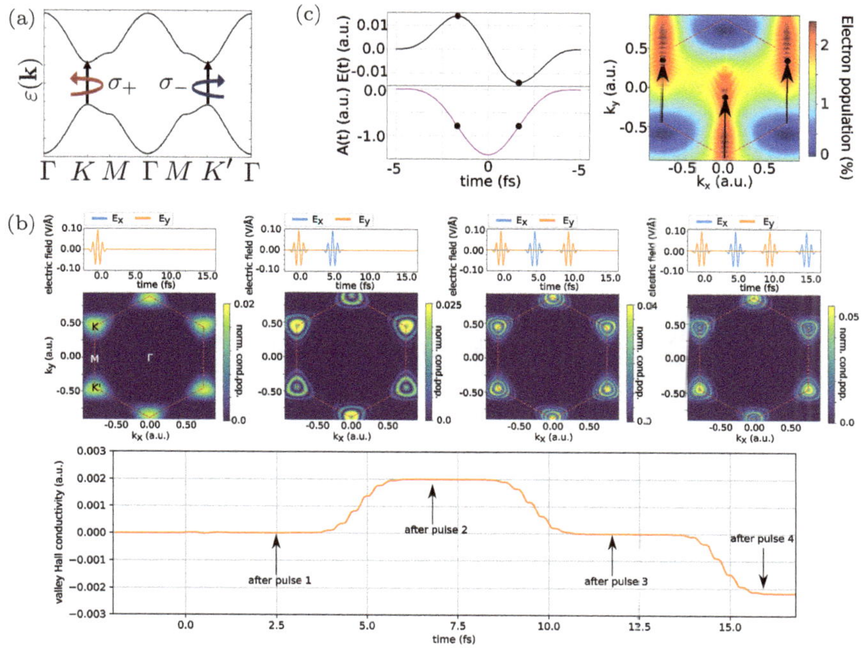


Fig. 2.2 (a) One-photon resonant valley-dependent optical selection rules. (b) Optical valley switch in gapped graphene using a train of ultrashort pulses. Each pulse (top panels) creates or destroys valley polarization. The middle panels correspond to the electron populations in the conduction band after the pulse sequence above. The bottom panel shows the change of the anomalous (valley) Hall conductivity (Eq. (2.7)) as a function of time for the four-pulse sequence in panel, demonstrating the optical valley switch. Reproduced with permission from [71]. (c) Valley polarization with a strong, single-cycle, linearly polarized pulse. Electric field (top left) and vector potential (bottom left) of a linearly polarized single-cycle pulse with $\lambda = 3$ $\mu$m and field maxima $F = 1.1$ V/Å. The black dots indicate the times when electron injection predominantly occurs, corresponding to a non-zero vector potential amplitude. The right panels are the conduction band electron populations (normalized at each k-point) after the pulse, showing valley polarization. The arrows indicate the streaking, from the minimum band gap crystal momenta $K$ and $K'$ to the final electron momenta according to the semiclassical equation of motion. We show a single-cycle pulse to convey the idea, but the same result applies using few-cycle pulses. Reproduced with permission from [70].

For clarity, we will focus on a two-band model of gapped graphene, but the results are applicable to real hexagonal monolayer systems such as hBN or TMDs. The electron population at the end of the pulse in the conduction band $\rho_{cc}(t, \mathbf{k})$ can be written in terms of the interband action $S_{cv}(t, t', \mathbf{k})$, see, e.g., [45]:

$$\rho_{cc}(t, \mathbf{k}) \simeq \left| e \int_{t_i}^{t_e} dt' e^{iS_{cv}(t_i, t', \mathbf{k})} \Delta\mathcal{A}(\mathbf{k}(t')) \cdot \mathbf{E}(t') \right|^2, \tag{2.10}$$

where $\mathbf{k}(t) = \mathbf{k} + \frac{e}{\hbar}\mathbf{A}(t)$, $t_e - t_i$ is the time of the injection process and $\Delta\mathcal{A}(\mathbf{k}) = \mathcal{A}_c(\mathbf{k}) - \mathcal{A}_v(\mathbf{k})$, with $\mathcal{A}_n(\mathbf{k}) = i\langle u_{n,\mathbf{k}}|\nabla_{\mathbf{k}}|u_{n,\mathbf{k}}\rangle$ the Berry connection of band $n$. The interband action $S_{cv}(t, t', \mathbf{k})$ is written in terms of a dynamical phase and a term proportional to the Berry connection, respectively,

$$S_{cv}(t_i, t, \mathbf{k}) = \int_{t_i}^{t} dt' \left[\Delta\varepsilon(\mathbf{k}(t')) + \mathbf{E}(t') \cdot \Delta\mathcal{A}(\mathbf{k}(t'))\right], \tag{2.11}$$

where $\Delta\varepsilon(\mathbf{k})$ is the energy difference between the conduction and valence bands. For strong fields, the injection step can be treated as a tunnelling step where the process occurs in imaginary time from a time of ionization $t_0 + i\tau$ to a time of tunnel exit $t_0$ [83],

$$\begin{aligned} &S_{cv}(t_0 + i\tau, t_0, \mathbf{k}) \\ &\quad = i \int_{\tau}^{0} d\tau' \left[\Delta\varepsilon(\mathbf{k}(t_0 + i\tau')) + \mathbf{E}(t_0 + i\tau') \cdot \Delta\mathcal{A}(\mathbf{k}(t_0 + i\tau')\right]. \end{aligned} \tag{2.12}$$

The integral of Eq. (2.10) is maximized when the imaginary part of the action is minimized with respect to $\mathbf{k}$, $\mathcal{I}\left\{\nabla_{\mathbf{k}} S_{cv}(t_0 + i\tau, t_0, \mathbf{k})\right\} = 0$,

$$\begin{aligned} \mathcal{R}\Bigg\{ &\int_{\tau}^{0} d\tau' \left[\Delta\varepsilon(\mathbf{k}(t_0 + i\tau')) + \mathbf{E}(t_0 + i\tau') \times \Delta\Omega(\mathbf{k}(t_0 + i\tau'))\right] \\ &+ \mathcal{A}(\mathbf{k}(t_0 + i\tau)) - \mathcal{A}(\mathbf{k}(t_0))\Bigg\} = 0, \end{aligned} \tag{2.13}$$

where we have used the fact that $\nabla_{\mathbf{k}}(\mathcal{A} \cdot \mathbf{E}) = (\mathbf{E} \times \Delta\Omega) + (\mathbf{E} \cdot \nabla_{\mathbf{k}})\mathcal{A}$, with $\Delta\Omega = \Omega_c - \Omega_v$ the difference in Berry curvatures between the conduction and valence bands. In the tunnelling limit, $\tau \to 0$ and the last two terms in Eq. (2.13) cancel. Assuming a doubly degenerate parabolic band dispersion close to the band gaps of both valleys with effective mass $m_{\text{eff}}$, we obtain the analytical expressions for the four saddle-point momenta (two for each valley)

$$k_x^{(K)} = K_x - \frac{e}{\hbar}A_x(t_0) - \frac{e}{\hbar}m_{\text{eff}}E_y(t_0)\Omega\left[k_x^{(K)} + \frac{e}{\hbar}A_x(t_0), k_y^{(K)} + \frac{e}{\hbar}A_y(t_0)\right],$$

$$k_y^{(K)} = K_y - \frac{e}{\hbar}A_y(t_0) + \frac{e}{\hbar}m_{\text{eff}}E_x(t_0)\Omega\left[k_x^{(K)} + \frac{e}{\hbar}A_x(t_0), k_y^{(K)} + \frac{e}{\hbar}A_y(t_0)\right],$$

$$
\begin{aligned}
k_x^{(K')} &= K_x - \frac{e}{\hbar}A_x(t_0) \\
&\quad - \frac{e}{\hbar}m_{\text{eff}}E_y(t_0)\Omega\left[k_x^{(K')} + \frac{e}{\hbar}A_x(t_0), k_y^{(K')} + \frac{e}{\hbar}A_y(t_0)\right], \\
k_y^{(K')} &= -K_y - \frac{e}{\hbar}A_y(t_0) \\
&\quad + \frac{e}{\hbar}m_{\text{eff}}E_x(t_0)\Omega\left[k_x^{(K')} + \frac{e}{\hbar}A_x(t_0), k_y^{(K')} + \frac{e}{\hbar}A_y(t_0)\right],
\end{aligned}
\tag{2.14}
$$

where the valleys $K = (K_x, K_y)$ and $K' = (K_x, -K_y)$.

In the low-frequency limit ($\omega \ll 1$), we have $E_i \ll A_i = E_i/\omega$ so that

$$
\begin{aligned}
k_x^{(K,0)} &= K_x - \frac{e}{\hbar}A_x(t_0), \\
k_y^{(K,0)} &= K_y - \frac{e}{\hbar}A_y(t_0), \\
k_x^{(K',0)} &= K_x - \frac{e}{\hbar}A_x(t_0), \\
k_y^{(K',0)} &= -K_y - \frac{e}{\hbar}A_y(t_0).
\end{aligned}
\tag{2.15}
$$

This is equivalent to the familiar streaking expression [84]. At the end of the pulse, conduction band electrons are accelerated in momentum space by a quantity proportional to the vector potential at the time of injection. This has been used in graphene to control the direction of residual currents [30].

While the valley-dependent optical selection rules do not allow generation of valley polarization with linearly polarized fields, Eq. (2.15) suggests that it is possible to do so in the strong-field regime, as demonstrated in [70,85,86]. In order to achieve valley polarization with a linearly polarized field, we need its vector potential magnitude at time $t_0$ to be equal to the momentum separation between the valleys. For example, $(A_x(t_0), A_y(t_0)) = \frac{\hbar}{e}(0, -2K_y)$. For graphene, this distance corresponds to $\approx$0.9 atomic units. Figure 2.2(c) shows the electron population after a single-cycle pulse of field strength $E = 1.1$ V/Å and wavelength $\lambda = 3$ $\mu$m is applied, corresponding to $A(t_0) \approx 0.9$ a.u., which clearly displays valley polarization. As suggested by Eq. (2.15), the population at the end of the pulse is predominantly localized at $\mathbf{k}^{(K,0)} = (K_x, 3K_y)$, corresponding to the $\Gamma$ point, and at $\mathbf{k}^{(K',0)} = (K_x, K_y)$, corresponding to $K$, as we see in Fig. 2.2(c). In order to achieve linearly polarized pulses in which the vector potential is non-zero at the maxima of the electric field $t_0$, as we see in Fig. 2.2(c), the pulses must be either very short [70,85] or a combination of two colors [86,87].

It is worth noting that since $K$ and $K'$ are time-reversal partners of one another, time-reversal symmetry must be broken in order to generate valley polarization. For example, valley switching using circularly polarized fields occurs when the helicity is changed from left to right, i.e., left and right circular fields are time-reversal partners of one another. In the case of the single-cycle field, the time-reversal partner of the electric field in Fig. 2.2(c), which will switch the valley polarization, corresponds to a change of the carrier-envelope phase of $\pi$. Thus, the carrier-envelope phase of the few-cycle pulse controls time-reversal symmetry breaking and valley polarization in the same way as the ellipticity/helicity of the circular drivers does for the one-photon valley selection rules. Yet, under the conditions of Eq. (2.15), valley polarization is achieved without influence of the Berry curvature or the magnetic moment and thus it can be generated in inversion-symmetric materials such as graphene [87], bilayer or bulk TMDs.

One can continue the expansion of Eq. (2.14) to the next order, which includes the contribution of the Berry curvature,

$$\begin{aligned} k_x^{(K,1)} &= K_x - \frac{e}{\hbar}A_x(t_0) - \frac{e}{\hbar}m_{\text{eff}}E_y(t_0)\Omega(K_x, K_y), \\ k_y^{(K,1)} &= K_y - \frac{e}{\hbar}A_y(t_0) + \frac{e}{\hbar}m_{\text{eff}}E_x(t_0)\Omega(K_x, K_y), \\ k_x^{(K',1)} &= K_x - \frac{e}{\hbar}A_x(t_0) - \frac{e}{\hbar}m_{\text{eff}}E_y(t_0)\Omega(K_x, -K_y), \\ k_y^{(K',1)} &= -K_y - \frac{e}{\hbar}A_y(t_0) + \frac{e}{\hbar}m_{\text{eff}}E_x(t_0)\Omega(K_x, -K_y). \end{aligned} \tag{2.16}$$

The Berry curvature is opposite in both valleys, so $\Omega(K_x, K_y) = -\Omega(K_x, -K_y) \equiv \Omega$. Let us assume the pulse is circularly polarized so that the vector potential is $\mathbf{A}(t) = A_0(-\sin(\omega t), \pm\cos(\omega t))$ and the electric field is $\mathbf{E}(t) = A_0\omega(\cos(\omega t), \pm\sin(\omega t))$, with $\pm$ corresponding to $\sigma_\pm$ light, respectively. The saddle point momenta are then

$$\begin{aligned} k_x^{(K,1)} &= K_x - \frac{e}{\hbar}A_0\sin(\omega t_0)\left[-1 \pm \omega\,\Omega\, m_{\text{eff}}\right], \\ k_y^{(K,1)} &= K_y - \frac{e}{\hbar}A_0\cos(\omega t_0)\left[\pm 1 - \omega\,\Omega\, m_{\text{eff}}\right], \\ k_x^{(K',1)} &= K_x - \frac{e}{\hbar}A_0\sin(\omega t_0)\left[-1 \mp \omega\,\Omega\, m_{\text{eff}}\right], \\ k_y^{(K',1)} &= -K_y - \frac{e}{\hbar}A_0\cos(\omega t_0)\left[\pm 1 + \omega\,\Omega\, m_{\text{eff}}\right]. \end{aligned} \tag{2.17}$$

When $\omega\,\Omega\,m_{\rm eff} \approx 1$, a $\sigma_+$ pulse will localize electrons in $K$ while a $\sigma_-$ pulse will localize electrons in $K'$. Valley polarization in this case arises due to phase interference between the dynamical phase and the phase proportional to the Berry connection, as first noted by Azar Oliaei Motlagh *et al.* [69].

Up to now, we have seen that switching of valley polarization, from $K$ to $K'$, is induced by time-reversing the electric field. In the next section, we will review a different method, where switching the valley population can be done without time-reversing the field. Instead, the strong field induces a controlled time-reversal symmetry breaking parameter in the Hamiltonian, connecting band structure topology with valley physics.

## 2.4 Light-Driven Topological Transitions

### 2.4.1 *Topological Insulators*

#### 2.4.1.1 *Haldane Model*

In his seminal article [5], Haldane devised a tight-binding model in which the Chern number of a 2D hexagonal system could be non-zero, leading to the appearance of an anomalous Hall current in equilibrium (see Eq. (2.8)) and gapless edge states [88]. Due to its relevance for the following, we briefly review it here. The Haldane Hamiltonian is a modification of the gapped-graphene Hamiltonian with real first neighbour hopping $t_1$ and complex couplings $t_2$ between second neighbours that break time-reversal symmetry (see Fig. 2.3(a)),

$$
\begin{aligned}
H = M \sum_i \left(a_i^\dagger\, a_i - b_i^\dagger\, b_i\right) - t_1 \sum_{\langle ij\rangle} \left(a_i^\dagger\, b_j + b_j^\dagger\, a_i\right) \\
- |t_2| \sum_{\langle\langle ij\rangle\rangle} \left(e^{i\phi} a_i^\dagger\, a_j + e^{-i\phi} b_i^\dagger\, b_j + \text{h.c.}\right),
\end{aligned}
\tag{2.18}
$$

where $\langle\rangle$ and $\langle\langle\rangle\rangle$ means sum over first neighbours and second neighbours, respectively. The operators $a$ and $b$ act on the $A$ and $B$ triangular sublattices, respectively, that have on-site energies $E_A = M$ and $E_B = -M$. The direct vectors of the lattice are

$$
\begin{aligned}
\mathbf{a}_1 &= \frac{r_0}{2}(3, \sqrt{3}) \\
\mathbf{a}_2 &= \frac{r_0}{2}(3, -\sqrt{3}),
\end{aligned}
\tag{2.19}
$$

where $r_0$ is the distance between the two nearest-neighbour atoms in the unit cell. The distance from atom $j$ to atom $i$ can be written in terms of $r_0$

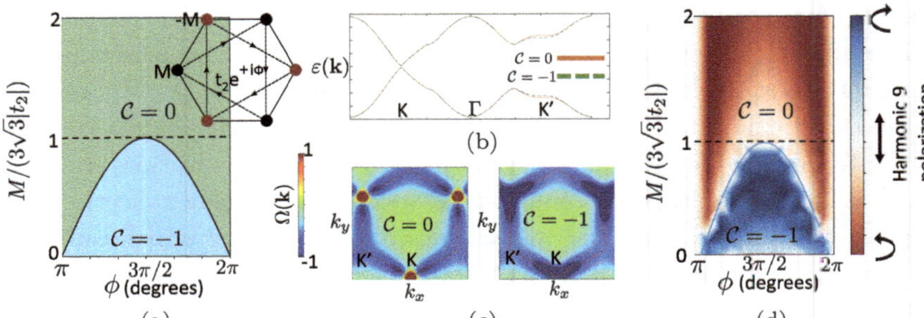


(a) (b) (c) (d)

Fig. 2.3 Haldane system and probing through polarization-resolved high-order harmonic generation. (a) Haldane phase diagram showing the trivial insulating state ($\mathcal{C} = 0$) and topological state ($\mathcal{C} = -1$) depending on the ratio between the on-site energy $M$ and the complex second neighbour hopping $|t_2|\, e^{i\phi}$. The inset shows the Haldane lattice with $t_2 e^{i\phi}$ breaking time-reversal symmetry. (b) Band structure along a representative k-path showing the lifting of valley degeneracy in the time symmetry-broken 2D crystal for two points in the phase diagram: black curve (trivial phase) and orange curve (topological phase). (c) Berry curvature in the Brillouin zone of the trivial and topological phases. (d) Helicity-resolved signal of the 9th harmonic of a strong, linearly polarized 3 μm pulse interacting with each of the systems contained in the Haldane phase diagram. The helicity of the harmonic changes sign as the system undergoes the topological phase transition due to the change of the Berry curvature at the valley that closes the band gap. Adapted with permission from [45].

and the angle $\alpha$ between those two atoms,

$$\mathbf{r} = [x_{ij}, y_{ij}] = (\sqrt{3})^m\, r_0\, [\cos\alpha_{ij}, \sin\alpha_{ij}], \tag{2.20}$$

where $m = 0, 1$ for nearest neighbours and next-nearest neighbours, respectively. In momentum space, the Haldane Hamiltonian can be written as

$$H(\mathbf{k}) = H_0(\mathbf{k})I + \mathbf{d}\cdot\sigma, \tag{2.21}$$

where $I$ is the identity matrix, $\sigma$ are the Pauli matrices and

$$\begin{aligned}
H_0(\mathbf{k}) &= -2|t_2|\cos\phi \sum_{\mathbf{d}} \cos(\mathbf{k}\cdot\mathbf{d}),\\
d_x(\mathbf{k}) &= -t_1 \sum_{\delta} \cos(\mathbf{k}\cdot\delta),\\
d_y(\mathbf{k}) &= -t_1 \sum_{\delta} \sin(\mathbf{k}\cdot\delta),\\
d_z(\mathbf{k}) &= M - 2|t_2|\sin\phi \sum_{\mathbf{d}} \sin(\mathbf{k}\cdot\mathbf{d}),
\end{aligned} \tag{2.22}$$

where $\delta$ and $\mathbf{d}$ are first and second neighbour vectors respectively. The gap is then,

$$\Delta(\mathbf{k}) = 2\sqrt{d_x(\mathbf{k})^2 + d_y(\mathbf{k})^2 + d_z(\mathbf{k})^2}. \tag{2.23}$$

At the $K$ and $K'$ valleys, $d_x = d_y = 0$ and

$$\begin{aligned} d_z(K') &= M - 3\sqrt{3}|t_2|\sin\phi, \\ d_z(K) &= M + 3\sqrt{3}|t_2|\sin\phi. \end{aligned} \tag{2.24}$$

The band gap is different at the $K$ and $K'$ valleys whenever $\phi \neq n\pi$, as shown in Fig. 2.3(b). If $M > 3\sqrt{3}|t_2|\sin\phi$, $\text{sign}(d_z(K)) = \text{sign}(d_z(K'))$ and the system has zero Chern number — it is a trivial insulator. If $M < 3\sqrt{3}|t_2|\sin\phi$, $\text{sign}(d_z(K)) = -\text{sign}(d_z(K'))$, the system is a Chern insulator (see Fig. 2.3(a)), supporting chiral edge states, with $\mathcal{C} = \pm 1$, depending if the gap is closed at $K$ ($\phi = 3\pi/2$) or $K'$ ($\phi = \pi/2$).

In order to induce the complex $t_2$, Haldane proposed to break time-reversal symmetry through a staggered magnetic field within the lattice. The extreme difficulty in realizing such experiment was already noted by Haldane in his paper, who considered it a toy model. In 2014, Jotzu *et al.* reproduced the Haldane system with ultracold atoms in an optical lattice [89]. Recently, Mitra *et al.* realized the Haldane system in a light-driven hBN crystal using polarization-tailored fields [90], following the theoretical proposal in [91], that we detail in subsection 2.5.2.

#### 2.4.1.2 *Quantum Spin Hall Insulators*

For completeness, we mention some of the other models of topological insulators that followed that of Haldane. Kane and Mele proposed a time reversal-invariant model made of two copies of the (spinless) Haldane Hamiltonian with different spin projections and with a spin-orbit interaction term [6,92]. The spin-orbit splitting $\Delta_{\text{SO}}$ acts as the term $|t_2|\sin(\pm\pi/2)$ in the Haldane model, with an opposite sign for each copy. Since the system is time-reversal invariant, it preserves the valley degeneracy, but with opposite spins at each valley. When $M < 3\sqrt{3}\Delta_{\text{SO}}$, the system is a quantum spin Hall insulator characterized by the $Z_2$ topological invariant. Realization of the Kane-Mele model was first proposed in graphene ($M = 0$), where the spin-orbit coupling opens a topological gap. However, the small spin-orbit splitting of graphene makes this gap effectively zero even at cryogenic temperatures. Other Group 14 monolayers (silicene, germanene, stanene) are currently being studied as candidates to observe the quantum spin Hall state proposed by Kane and Mele [93, 94]. The first experimental realization of the quantum spin Hall effect was performed in HgTe quantum wells [95], and was then observed in 3D crystals following the theoretical prediction of Bernevig, Hughes and Zhang [96], who showed that a quantum spin Hall state occurs when the valence and conduction posses opposite parity and a parameter, such as the thickness in quantum wells or the spin-orbit interaction in bulk materials, induces band inversion [97]. The Bernevig-Hughes-Zhang model is the basis of essentially all experimental realizations of quantum spin Hall insulators, and has also been used to confirm the quantum anomalous Hall effect in a magnetically doped crystal [98].

### 2.4.2 *High-Order Harmonic Generation as a Probe of Topological Phases*

Despite topological indices being defined in equilibrium, recent works have showed that they can be retrieved, at least theoretically, in the highly out-of-quilibrium regime using high-order harmonic generation. Bauer and Hansen showed that when topological edge states are present in the Su–Schrieffer–Heeger chain, sub-bandgap harmonics are greatly enhanced [99]. Silva *et al.* later devised a method to reconstruct the Chern number of the Haldane system using polarization-resolved high harmonic generation [45]. In the Haldane system, the Berry curvature changes sign at one of the valleys upon closing of the gap (see Fig. 2.3(c)), inducing a $\pi$-shift in the anomalous component of the current (see Eq. (2.6)). Thus, the harmonics generated by a linearly polarized field will be elliptical, due to the non-zero anomalous current in Eq. (2.3), and with an opposite helicity between the trivial and topological phases, as shown in Fig. 2.3(d). Note that this result is valid only when time-reversal symmetry is broken and therefore only one of the valleys participates in the dynamics, namely, the one at which the band gap is closed. Similar results were found by Chacón *et al.* using circular dichroism [46]. Neufeld *et al.* recently used time-dependent density functional theory on two model materials, a quantum spin Hall insulator and a crystalline topological insulator [100], to show that such dichroic high-order harmonic generation measurements do not distinguish topology in these materials [101]. Indeed, both the quantum spin Hall insulator and crystalline topological insulator preserve time-reversal symmetry and have zero Chern number, so different approaches are required. It is worth noting, as pointed out by Neufeld [101], that in order to identify topological signatures in the high harmonic spectrum it is necessary to disentangle them from other non-topological effects, such as those coming from different bulk band structures. This has been achieved by theoretically comparing cases with the same bulk band structure and different topology [45, 46], and by identifying the effect of topological surface states, as in recent experiments performed in 3D quantum spin Hall insulators [47, 48, 102].

## 2.5 Lightwave Control of Topological Properties

### 2.5.1 *Floquet Topological Insulators*

In 2009, Oka and Aoki [103] used Floquet theory to show that a circularly polarized field opens a topological gap in graphene, turning the laser-driven graphene system into a Chern insulator with Chern number $\mathcal{C} = \pm 1$ depending on the light helicity. Their work provided the first example of light-driven topological phase transitions, and was recently confirmed experimentally [104]. Soon after the work by Oka and Aoki, Linder *et al.* showed a similar effect in semiconducting quantum wells [105], where the circular driver turned the initially topologically trivial state into a quantum spin Hall state. In monolayer $WS_2$, Sie *et al.* found that when $\hbar\omega > \Delta$, with $\Delta$ the band gap, the Floquet bands closed at one of the valleys [106].

These pioneering works marked the beginning of Floquet engineering and topological physics in synthetic dimensions [107–110]. In the Floquet formalism, a long, periodic field of frequency $\omega$ 'dresses' the band structure with a series of copies separated by $\hbar\omega$. Each energy level is replicated $n$ times, with $n$ the number of photons involved. The intensity and frequency of the driving field controls the energies and couplings of the dressed system, allowing modification of the overall properties [109].

Yet, when many photons are involved, the Floquet perspective does not capture essential dynamical information of the response [31]. A prime example is the failure of the Floquet perspective to characterize topological invariants when $\hbar\omega < W$. In this regime, edge states may exist even when the Chern number of the Floquet bands that result from the cycle-averaged Hamiltonian is zero [109,111,112]. These anomalous edge states do not obey the bulk-edge correspondence in a static framework, and their appearance is uniquely linked to the sub-cycle dynamics of the system [113]. They have now been observed in different setups [114–117].

### 2.5.2 *Lightwave Control of Topological Properties*

Floquet engineering relies on resonant or above-resonant photon absorption processes, where $\hbar\omega > \Delta$. The timescale at which the light-induced properties can be controlled and studied is thus limited by the duration of the pulse envelope.

The strong-field regime of light-induced topological transitions, i.e., $\hbar\omega < \Delta$, was shown in [91]. By matching the spatial symmetry of the pulse to the symmetry of the 2D crystal lattice, both inversion and time-reversal symmetries can be manipulated through rotation of the light waveform in space. The bicircular field, i.e., a combination of two counter-rotating circular fields of frequencies $\omega$ and $2\omega$, offers the three-fold symmetry necessary for this process, as shown in Fig. 2.4(a). Without loss of generality,

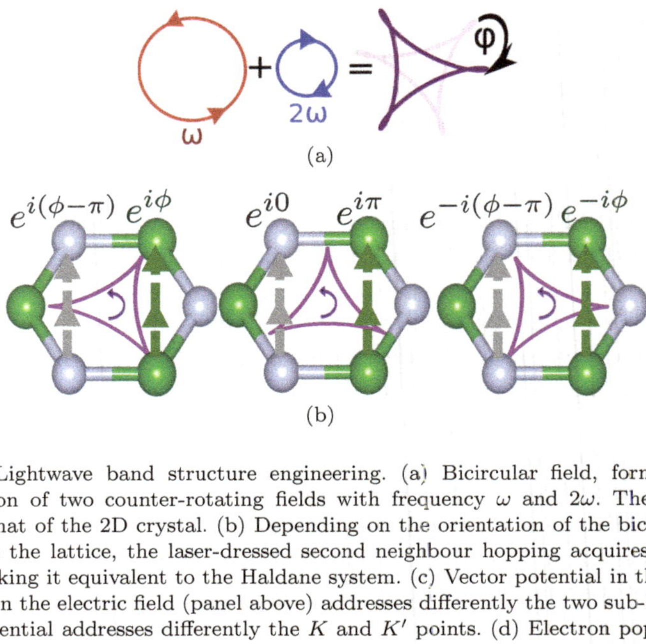


Fig. 2.4 Lightwave band structure engineering. (a) Bicircular field, formed by the combination of two counter-rotating fields with frequency $\omega$ and $2\omega$. The symmetry matches that of the 2D crystal. (b) Depending on the orientation of the bicircular field relative to the lattice, the laser-dressed second neighbour hopping acquires a complex phase, making it equivalent to the Haldane system. (c) Vector potential in the Brillouin zone. When the electric field (panel above) addresses differently the two sub-lattices, the vector potential addresses differently the $K$ and $K'$ points. (d) Electron populations in the first Brillouin zone after the interaction with the pulses above. Valley polarization happens when the phase of the laser-dressed second neighbour hopping $\phi \neq n\pi$ $(n = 0, 1, 2, \ldots)$. We note that there is not perfect valley polarization switching, i.e., the injected population is greater at the $K'$ valley (right panel). This is a manifestation of the fact that the fields are not time-reversal partners. (e) Band structures obtained using the laser-dressed corrected values of the first and second neighbour hopping. Valley polarization coincides with the valley at which the band gap is lower. The time-dependent perturbative model does show perfect switching of $K$ and $K'$ (compare left and right panels) since its derivation does not include the rotation of the field.

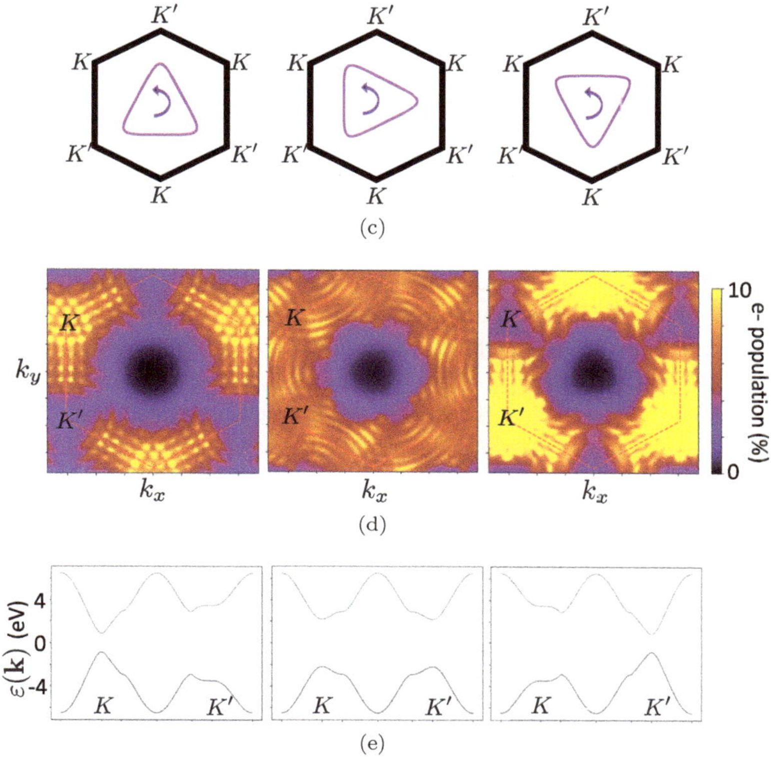


Fig. 2.4 (*Continued*)

we define the bicircular field as a clockwise rotating field with vector potential

$$\mathbf{A}(t) = [A_x(t), A_y(t)] = \left[\frac{F_\omega}{\omega}\cos(\omega t) - \frac{F_{2\omega}}{2\omega}\cos(2\omega t + \varphi),\right.$$
$$\left. -\frac{F_\omega}{\omega}\sin(\omega t) - \frac{F_{2\omega}}{2\omega}\sin(2\omega t + \varphi)\right]. \quad (2.25)$$

The field strengths of the fundamental and second harmonic fields are $F_\omega$ and $F_{2\omega}$, respectively, and $\varphi$ is the phase delay between the two fields. This phase delay can be controlled on timescales shorter than one optical cycle and determines the spatial orientation of the bicircular field. Depending on its orientation, the bicircular field addresses differently the two sub-lattices in real space and the vector potential addresses differently the two valleys in reciprocal space, as shown in Fig. 2.4(b, c).

Figure 2.4(d) shows the electron populations of the conduction band of a gapped graphene system after the interaction with the bicircular field in three different orientations. We observe clear valley polarization and switching. For valley polarization to arise, time-reversal symmetry must be broken. We note, however, that the helicity of both the $\omega$ and $2\omega$ fields is unchanged in Fig. 2.4, yet we see valley switching. Such effect has been explored theoretically in [91, 118], and has been recently observed both in hBN monolayers [90], and in bulk $MoS_2$ [119].

In order to understand how time-reversal symmetry is broken in this system, we start with the gapped-graphene tight binding Hamiltonian up to first neighbour hoppings, which is equivalent to the Haldane Hamiltonian in Eq. (2.18) with $t_2 = 0$. To study the interaction of this system with the bicircular field, we calculate the lowest order time-dependent correction to the field-free hoppings $t_1$ and $t_2$. We assume the latter is initially negligible. We note that a laser field modifies $t_1$ between an orbital in sub-lattice $A$ and an orbital in sub-lattice $B$ according to the Peierls substitution (atomic units are used throughout the derivation),

$$t_{1,AB}(t) = t_1 e^{-i\mathbf{r}_{AB}\cdot\mathbf{A}(t)}, \tag{2.26}$$

where $\mathbf{r}_{AB}$ is the distance from the site in sub-lattice $B$ to the site in sub-lattice $A$, and $\mathbf{A}(t)$ is the vector potential of the field. The laser-induced hopping term can be separated into one cycle-averaged term and one term that contains the dynamical corrections to this cycle average,

$$t_{1,AB}(t) = t_1 \langle e^{-i\mathbf{r}_{AB}\cdot\mathbf{A}(t)}\rangle + V_{AB}(t), \tag{2.27}$$

where $\langle\rangle$ is cycle-averaged. Therefore, we have

$$V_{AB}(t) = t_1 e^{-i\mathbf{r}_{AB}\cdot\mathbf{A}(t)} - \gamma_1 \langle e^{-i\mathbf{r}_{AB}\cdot\mathbf{A}(t)}\rangle. \tag{2.28}$$

The time-dependent perturbation $V_{AB}(t)$ can lead to transitions between next nearest neighbours ($A \leftarrow A'$) of the same sub-lattice through virtual nearest neighbours. To lowest order, the transition amplitude for such second-order process reads (atomic units are used below)

$$\mathcal{A}^{(2)}_{A\leftarrow A'}(t) = -\sum_B \int_{t_0}^{t} dt' \int_{t_0}^{t'} dt'' e^{i(E_{AB}t' + E_{BA'}t'')} V_{AB}(t') V_{BA'}(t''), \tag{2.29}$$

where the energy difference $E_{AB} = E_A - E_B = \Delta$, and $\Delta$ is the minimum band gap energy (we assume $E_A > E_B$). The summation in principle runs along all possible intermediate sites in sub-lattice $B$. However, we note that there is only one nearest neighbour site that can participate to lowest order, so we can remove the summation. Using Eq. (2.28),

$$\mathcal{A}^{(2)}_{A\leftarrow A'}(t) = -t_1^2 \int_{t_0}^{t} dt' e^{i\Delta t'} \left[e^{-i\mathbf{r}_{AB}\cdot\mathbf{A}(t')} - \langle e^{-i\mathbf{r}_{AB}\cdot\mathbf{A}(t)}\rangle\right] \times \int_{t_0}^{t'} dt'' e^{-i\Delta t''} \left[e^{-i\mathbf{r}_{BA'}\cdot\mathbf{A}(t'')} - \langle e^{-i\mathbf{r}_{BA'}\cdot\mathbf{A}(t)}\rangle\right]. \quad (2.30)$$

We perform a change of variables in the second integral, $u = -i[\Delta t'' + \mathbf{r}_{BA'}\cdot\mathbf{A}(t'')]$, $du = -i[\Delta - \mathbf{r}_{BA'}\cdot\mathbf{F}(t'')]dt''$. In the strong-field, low-frequency regime, we have $|\mathbf{F}(t)\cdot\mathbf{r}_{BA}| \ll |\Delta|$. For example, the gap in a typical TMD is $\Delta \simeq 0.06$ a.u., while the field strengths for strong-field processes are usually one order of magnitude smaller. Thus, we can neglect the second term in $du$ and write $du = -i\Delta dt''$. In this way, the second integral is easily computed as

$$\begin{aligned}\int_{t_0}^{t'} du \frac{e^u}{-i\Delta} &- \langle e^{-i\mathbf{r}_{BA'}\cdot\mathbf{A}(t)}\rangle \int_{t_0}^{t'} dt'' e^{-i\Delta t''} \\ &= \frac{i}{\Delta} e^{-i\Delta t'} \left[e^{-i\mathbf{r}_{BA'}\cdot\mathbf{A}(t')} - \langle e^{-i\mathbf{r}_{BA'}\cdot\mathbf{A}(t)}\rangle\right] \\ &\quad - \frac{i}{\Delta} e^{-i\Delta t_0} \left[1 - \langle e^{-i\mathbf{r}_{BA'}\cdot\mathbf{A}(t)}\rangle\right]. \end{aligned} \quad (2.31)$$

Substituting Eq. (2.31) into Eq. (2.30),

$$\begin{aligned}\mathcal{A}^{(2)}_{A\leftarrow A'}(t) = -\frac{i\gamma_1^2}{\Delta} \int_{t_0}^{t} dt' \Big\{ &\left[e^{-i\mathbf{r}_{AB}\cdot\mathbf{A}(t')} - \langle e^{-i\mathbf{r}_{AB}\cdot\mathbf{A}(t)}\rangle\right] \\ &\times \left[e^{-i\mathbf{r}_{BA'}\cdot\mathbf{A}(t')} - \langle e^{-i\mathbf{r}_{BA'}\cdot\mathbf{A}(t)}\rangle\right] \\ &- e^{-i\Delta(t_0 - t')}\left[1 - \langle e^{-i\mathbf{r}_{BA'}\cdot\mathbf{A}(t)}\rangle\right] \\ &\times \left[e^{-i\mathbf{r}_{AB}\cdot\mathbf{A}(t')} - \langle e^{-i\mathbf{r}_{AB}\cdot\mathbf{A}(t)}\rangle\right]\Big\}. \end{aligned} \quad (2.32)$$

The above expression is the lowest-order transition amplitude expression between two sites in sub-lattice $A$. From it, we can identify the transition matrix element between second neighbour sites $A'$ and $A$,

$$\begin{aligned} t_{2,AA'} = \frac{t_1^2}{\Delta}\Big\{ &\left[e^{-i\mathbf{r}_{AB}\cdot\mathbf{A}(t')} - \langle e^{-i\mathbf{r}_{AB}\cdot\mathbf{A}(t)}\rangle\right]\left[e^{-i\mathbf{r}_{BA'}\cdot\mathbf{A}(t')} - \langle e^{-i\mathbf{r}_{BA'}\cdot\mathbf{A}(t)}\rangle\right] \\ &- e^{-i\Delta(t_0 - t)}\left[1 - \langle e^{-i\mathbf{r}_{BA'}\cdot\mathbf{A}(t)}\rangle\right]\left[e^{-i\mathbf{r}_{AB}\cdot\mathbf{A}(t')} - \langle e^{-i\mathbf{r}_{AB}\cdot\mathbf{A}(t)}\rangle\right]\Big\}. \end{aligned} \quad (2.33)$$

Averaging now over one cycle,

$$\langle t_{2,AA'}\rangle = \frac{t_1^2}{\Delta}\{\langle e^{-i\mathbf{r}_{AA'}\cdot\mathbf{A}(t)}\rangle - \langle e^{-i\mathbf{r}_{AB}\cdot\mathbf{A}(t)}\rangle\langle e^{-i\mathbf{r}_{BA'}\cdot\mathbf{A}(t)}\rangle\}. \tag{2.34}$$

Using the Jacobi-Anger expansion, one can express the imaginary component of $\langle t_{2,AA'}\rangle$, to lowest order in the vector potential magnitude, as

$$\Im\langle t_{2,AA'}\rangle \propto J_2\left(\frac{\sqrt{3}\,r_0\,F_\omega}{\omega}\right) J_1\left(\frac{\sqrt{3}\,r_0\,F_{2\omega}}{2\omega}\right)\cos(\varphi), \tag{2.35}$$

where $J_{1,2}$ are the first- and second-order Bessel function of the first kind and $r_0$ is the nearest-neighbour distance. If the transition is between two next-nearest neighbours of the other sub-lattice ($B$), then we need to substitute $\Delta \to -\Delta$, and

$$\langle t_{2,BB'}\rangle = -\langle t_{2,AA'}\rangle. \tag{2.36}$$

Figure 2.3(b) shows the complex phases of $\langle t_{2,BB'}\rangle$ and $\langle t_{2,AA'}\rangle$ for different orientations. We note that the cycle-averaged Hamiltonian of the laser-driven system is written in the same form as Eq. (2.21), but with the changes

$$\begin{aligned} &H_0(\mathbf{k}) = 0,\\ &d_z(\mathbf{k}) = M - 2|t_2|\sin\phi\sum_{\mathbf{d}}\sin(\mathbf{k}\cdot\mathbf{d}) - 2|t_2|\cos\phi\sum_{\mathbf{d}}\cos(\mathbf{k}\cdot\mathbf{d}). \end{aligned} \tag{2.37}$$

Figure 2.3(e) shows the effective band structures obtained in the gapped graphene system with the laser-dressed $\langle t_2\rangle$ and $\langle t_1\rangle$ parameters. When the vector potential maxima points towards the $K$ or $K'$ valleys, we have $\phi = \pm\pi/2$, and the laser-driven crystal is the exact realization of the Haldane model with imaginary $t_2$, lifting the valley degeneracy depending on the orientation of the field relative to the lattice. When the vector potential maxima points towards the $M$ point, $t_2$ is real ($\phi = 0, \pi$). Here, the band gap is still reduced relative to the field-free case, albeit equally for both valleys.

## 2.6 Outlook

Lightwave-driven manipulation of electronic and topological properties of quantum materials holds many opportunities, and presents many exciting challenges for both theory and experiment. Generation of topological phase transitions on the timescale of the lightwave, and the prospect of controlling such transitions with the spatial properties of the field can lead the way to the dynamical counter-part of twistronics. Here, the composite system is the 2D crystal + polarization-tailored field, and new properties are also induced by rotating one respect to the other. In this case, however, the rotation is achieved on few femtoseconds. Two recent parallel experiments have provided the proof of concept of this method, demonstrating ultrafast selective modification of the crystal band structure by the lightwave [90, 119]. Added to the opportunity to control and characterize the temporal profile, spatial profile, spin, angular momentum and topology of the strong light field and its harmonics on ultrafast timescales [20, 120–130], the field of strong-field physics in solids holds unprecedented potential for on-demand, on-the-fly manipulation and measurement of transient electronic properties of materials [131, 132]. Exploration of the robustness of topological edge states or other exotic states, such as Majorana, to strong fields, as well as the real-time observation of their formation and evolution are only some of the enticing prospects, a few of which have started to be explored already theoretically [133]. In the same way, excitons dominate the optical absorption spectrum in 2D systems and have been a key ingredient in weak-field valleytronics. Yet, their effect during the strong laser-matter interaction has only recently started to be appreciated [134, 135]. Also, in order to achieve ultrafast switches of quantum properties, a more adequate description of dephasing times in strong-field driven processes is required. This may come from a real-space perspective of the laser-crystal interaction, as recently suggested by Brown *et al.* [136], opening the way to strong-field physics in solids on the atomic scale [137–143].

## Acknowledgements

The authors would like to thank Misha Ivanov for fruitful discussions and his continued support. A. J. G acknowledges financial support from the Comunidad de Madrid through TALENTO Grant 2022-T1/IND-24102. E. B. M. and R. E. F. S. acknowledge support from the fellowship LCF/BQ/PR21/11840008 from "La Caixa" Foundation (ID 100010434) and the Grant PID2021-122769NB-I00 funded by MCIN/AEI/10.13039/501100011033.

## References

[1] D. Basov, R. Averitt and D. Hsieh, Towards properties on demand in quantum materials, *Nature Materials* **16**, 11, pp. 1077–1088 (2017).

[2] A. C. Neto, F. Guinea, N. M. Peres, K. S. Novoselov and A. K. Geim, The electronic properties of graphene, *Reviews of Modern Physics* **81**, 1, p. 109 (2009).

[3] A. C. Ferrari, F. Bonaccorso, V. Fal'Ko, K. S. Novoselov, S. Roche, P. Bøggild, S. Borini, F. H. Koppens, V. Palermo, N. Pugno *et al.*, Science and technology roadmap for graphene, related two-dimensional crystals, and hybrid systems, *Nanoscale* **7**, 11, pp. 4598–4810 (2015).

[4] Y. Liu, X. Duan, H.-J. Shin, S. Park, Y. Huang and X. Duan, Promises and prospects of two-dimensional transistors, *Nature* **591**, 7848, pp. 43–53 (2021).

[5] F. D. M. Haldane, Model for a quantum hall effect without Landau levels: Condensed-matter realization of the "parity anomaly", *Physical Review Letters* **61**, 18, p. 2015 (1988).

[6] C. L. Kane and E. J. Mele, Z 2 topological order and the quantum spin hall effect, *Physical Review Letters* **95**, 14, p. 146802 (2005).

[7] M. Z. Hasan and C. L. Kane, Colloquium: topological insulators, *Reviews of Modern Physics* **82**, 4, p. 3045 (2010).

[8] J. R. Schaibley, H. Yu, G. Clark, P. Rivera, J. S. Ross, K. L. Seyler, W. Yao and X. Xu, Valleytronics in 2D materials, *Nature Reviews Materials* **1**, 11, pp. 1–15 (2016).

[9] S. A. Vitale, D. Nezich, J. O. Varghese, P. Kim, N. Gedik, P. Jarillo-Herrero, D. Xiao and M. Rothschild, Valleytronics: opportunities, challenges, and paths forward, *Small* **14**, 38, p. 1801483 (2018).

[10] A. K. Geim and I. V. Grigorieva, Van der Waals heterostructures, *Nature* **499**, 7459, pp. 419–425 (2013).

[11] F. Krausz and M. Ivanov, Attosecond physics, *Reviews of Modern Physics* **81**, 1, p. 163 (2009).

[12] E. Goulielmakis, V. S. Yakovlev, A. L. Cavalieri, M. Uiberacker, V. Pervak, A. Apolonski, R. Kienberger, U. Kleineberg and F. Krausz, Attosecond control and measurement: lightwave electronics, *Science* **317**, 5839, pp. 769–775 (2007).

[13] A. Wirth, M. T. Hassan, I. Grguraš, J. Gagnon, A. Moulet, T. T. Luu, S. Pabst, R. Santra, Z. Alahmed, A. Azzeer *et al.*, Synthesized light transients, *Science* **334**, 6053, pp. 195–200 (2011).

[14] M. Schultze, E. M. Bothschafter, A. Sommer, S. Holzner, W. Schweinberger, M. Fiess, M. Hofstetter, R. Kienberger, V. Apalkov, V. S. Yakovlev *et al.*, Controlling dielectrics with the electric field of light, *Nature* **493**, 7430, pp. 75–78 (2013).

[15] B. Wolter, M. G. Pullen, M. Baudisch, M. Sclafani, M. Hemmer, A. Senftleben, C. D. Schröter, J. Ullrich, R. Moshammer and J. Biegert, Strong-field physics with mid-IR fields, *Physical Review X* **5**, 2, p. 021034 (2015).

[16] P. Steinleitner, N. Nagl, M. Kowalczyk, J. Zhang, V. Pervak, C. Hofer, A. Hudzikowski, J. Sotor, A. Weigel, F. Krausz *et al.*, Single-cycle infrared waveform control, *Nature Photonics* **16**, 7, pp. 512–518 (2022).

[17] M. Khokhlova, M. Y. Emelin, M. Y. Ryabikin and V. Strelkov, Polarization control of quasimonochromatic XUV light produced via resonant high-order harmonic generation, *Physical Review A* **103**, 4, p. 043114 (2021).

[18] E. Pisanty and Á. Jiménez-Galán, Strong-field approximation in a rotating frame: High-order harmonic emission from $p$ states in bicircular fields, *Physical Review A* **96**, 6, p. 063401 (2017).

[19] D. Hui, H. Alqattan, S. Zhang, V. Pervak, E. Chowdhury and M. T. Hassan, Ultrafast optical switching and data encoding on synthesized light fields, *Science Advances* **9**, 8, p. eadf1015 (2023).

[20] E. Pisanty, G. J. Machado, V. Vicuña-Hernández, A. Picón, A. Celi, J. P. Torres and M. Lewenstein, Knotting fractional-order knots with the polarization state of light, *Nature Photonics* **13**, 8, pp. 569–574 (2019).

[21] L. Rego, K. M. Dorney, N. J. Brooks, Q. L. Nguyen, C.-T. Liao, J. San Román, D. E. Couch, A. Liu, E. Pisanty, M. Lewenstein *et al.*, Generation of extreme-ultraviolet beams with time-varying orbital angular momentum, *Science* **364**, 6447, p. eaaw9486 (2019).

[22] D. Ayuso, O. Neufeld, A. F. Ordonez, P. Decleva, G. Lerner, O. Cohen, M. Ivanov and O. Smirnova, Synthetic chiral light for efficient control of chiral light–matter interaction, *Nature Photonics* **13**, 12, pp. 866–871 (2019).

[23] L. Rego and D. Ayuso, Structuring the local handedness of synthetic chiral light: global chirality versus polarization of chirality, *New Journal of Physics* **25**, 9, p. 093005 (2023).

[24] T. Boolakee, C. Heide, A. Garzón-Ramírez, H. B. Weber, I. Franco and P. Hommelhoff, Light-field control of real and virtual charge carriers, *Nature* **605**, 7909, pp. 251–255 (2022).

[25] A. Schiffrin, T. Paasch-Colberg, N. Karpowicz, V. Apalkov, D. Gerster, S. Mühlbrandt, M. Korbman, J. Reichert, M. Schultze, S. Holzner *et al.*, Optical-field-induced current in dielectrics, *Nature* **493**, 7430, pp. 70–74 (2013).

[26] O. Schubert, M. Hohenleutner, F. Langer, B. Urbanek, C. Lange, U. Huttner, D. Golde, T. Meier, M. Kira, S. W. Koch *et al.*, Sub-cycle control of terahertz high-harmonic generation by dynamical bloch oscillations, *Nature Photonics* **8**, 2, pp. 119–123 (2014).

[27] T. Higuchi, C. Heide, K. Ullmann, H. B. Weber and P. Hommelhoff, Light-field-driven currents in graphene, *Nature* **550**, 7675, pp. 224–228 (2017).

[28] J. Reimann, S. Schlauderer, C. Schmid, F. Langer, S. Baierl, K. Kokh, O. Tereshchenko, A. Kimura, C. Lange, J. Güdde *et al.*, Subcycle observation of lightwave-driven Dirac currents in a topological surface band, *Nature* **562**, 7727, pp. 396–400 (2018).

[29] V. Hanus, V. Csajbók, Z. Pápa, J. Budai, Z. Márton, G. Z. Kiss, P. Sándor, P. Paul, A. Szeghalmi, Z. Wang *et al.*, Light-field-driven current control in

solids with pJ-level laser pulses at 80 MHz repetition rate, *Optica* **8**, 4, pp. 570–576 (2021).

[30] C. Heide, T. Boolakee, T. Eckstein and P. Hommelhoff, Optical current generation in graphene: Cep control vs. $\omega + 2\omega$ control, *Nanophotonics* **10**, 14, pp. 3701–3707 (2021).

[31] S. Ito, M. Schüler, M. Meierhofer, S. Schlauderer, J. Freudenstein, J. Reimann, D. Afanasiev, K. Kokh, O. Tereshchenko, J. Güdde *et al.*, Build-up and dephasing of Floquet–Bloch bands on subcycle timescales, *Nature* **616**, pp. 696–701 (2023).

[32] J. Liao, Y. Ou, H. Liu, K. He, X. Ma, Q.-K. Xue and Y. Li, Enhanced electron dephasing in three-dimensional topological insulators, *Nature Communications* **8**, 1, p. 16071 (2017).

[33] K. Hao, G. Moody, F. Wu, C. K. Dass, L. Xu, C.-H. Chen, L. Sun, M.-Y. Li, L.-J. Li, A. H. MacDonald *et al.*, Direct measurement of exciton valley coherence in monolayer $WSe_2$, *Nature Physics* **12**, 7, pp. 677–682 (2016).

[34] P. B. Corkum, Plasma perspective on strong-field multiphoton ionization, *Physical Review Letters* **71**, 13, pp. 1994–1997 (1993).

[35] M. Lewenstein, P. Balcou, M. Y. Ivanov, A. L'Huillier and P. B. Corkum, Theory of high-harmonic generation by low-frequency laser fields, *Physical Review A* **49**, pp. 2117–2132 (1994).

[36] J. L. Krause, K. J. Schafer and K. C. Kulander, High-order harmonic generation from atoms and ions in the high intensity regime, *Physical Review Letters* **68**, 24, p. 3535 (1992).

[37] O. Smirnova, Y. Mairesse, S. Patchkovskii, N. Dudovich, D. Villeneuve, P. Corkum and M. Y. Ivanov, High harmonic interferometry of multi-electron dynamics in molecules, *Nature* **460**, 7258, pp. 972–977 (2009).

[38] G. Vampa, C. McDonald, G. Orlando, D. Klug, P. Corkum and T. Brabec, Theoretical analysis of high-harmonic generation in solids, *Physical Review Letters* **113**, 7, p. 073901 (2014).

[39] G. Vampa, T. Hammond, N. Thiré, B. Schmidt, F. Légaré, C. McDonald, T. Brabec and P. Corkum, Linking high harmonics from gases and solids, *Nature* **522**, 7557, pp. 462–464 (2015).

[40] T. T. Luu, M. Garg, S. Y. Kruchinin, A. Moulet, M. T. Hassan and E. Goulielmakis, Extreme ultraviolet high-harmonic spectroscopy of solids, *Nature* **521**, 7553, pp. 498–502 (2015).

[41] G. Vampa, T. Hammond, N. Thiré, B. Schmidt, F. Légaré, C. McDonald, T. Brabec, D. Klug and P. Corkum, All-optical reconstruction of crystal band structure, *Physical Review Letters* **115**, 19, p. 193603 (2015).

[42] N. Tancogne-Dejean, O. D. Mücke, F. X. Kärtner and A. Rubio, Impact of the electronic band structure in high-harmonic generation spectra of solids, *Physical Review Letters* **118**, 8, p. 087403 (2017).

[43] A. J. Uzan, G. Orenstein, Á. Jiménez-Galán, C. McDonald, R. E. F. Silva, B. D. Bruner, N. D. Klimkin, V. Blanchet, T. Arusi-Parpar, M. Krüger *et al.*, Attosecond spectral singularities in solid-state high-harmonic generation, *Nature Photonics* **14**, 3, pp. 183–187 (2020).

[44] A. J. Uzan-Narovlansky, Á. Jiménez-Galán, G. Orenstein, R. E. F. Silva, T. Arusi-Parpar, S. Shames, B. D. Bruner, B. Yan, O. Smirnova, M. Ivanov *et al.*, Observation of light-driven band structure via multiband high-harmonic spectroscopy, *Nature Photonics* **16**, 6, pp. 428–432 (2022).

[45] R. E. F. Silva, Á. Jiménez-Galán, B. Amorim, O. Smirnova and M. Ivanov, Topological strong-field physics on sub-laser-cycle timescale, *Nature Photonics* **13**, 12, pp. 849–854 (2019).

[46] A. Chacón, D. Kim, W. Zhu, S. P. Kelly, A. Dauphin, E. Pisanty, A. S. Maxwell, A. Picón, M. F. Ciappina, D. E. Kim *et al.*, Circular dichroism in higher-order harmonic generation: Heralding topological phases and transitions in Chern insulators, *Physical Review B* **102**, 13, p. 134115 (2020).

[47] Y. Bai, F. Fei, S. Wang, N. Li, X. Li, F. Song, R. Li, Z. Xu and P. Liu, High-harmonic generation from topological surface states, *Nature Physics* **17**, 3, pp. 311–315 (2021).

[48] D. Baykusheva, A. Chacón, J. Lu, T. P. Bailey, J. A. Sobota, H. Soifer, P. S. Kirchmann, C. Rotundu, C. Uher, T. F. Heinz *et al.*, All-optical probe of three-dimensional topological insulators based on high-harmonic generation by circularly polarized laser fields, *Nano Letters* **21**, 21, pp. 8970–8978 (2021).

[49] M. Hohenleutner, F. Langer, O. Schubert, M. Knorr, U. Huttner, S. W. Koch, M. Kira and R. Huber, Real-time observation of interfering crystal electrons in high-harmonic generation, *Nature* **523**, 7562, pp. 572–575 (2015).

[50] R. E. F. Silva, I. V. Blinov, A. N. Rubtsov, O. Smirnova and M. Ivanov, High-harmonic spectroscopy of ultrafast many-body dynamics in strongly correlated systems, *Nature Photonics* **12**, 5, pp. 266–270 (2018).

[51] E. B. Molinero and R. E. F. Silva, Coherent control of the photoinduced transition in a strongly correlated material, *Physical Review Research* **4**, 4, p. 043198 (2022).

[52] J. Alcalà, U. Bhattacharya, J. Biegert, M. Ciappina, U. Elu, T. Graß, P. T. Grochowski, M. Lewenstein, A. Palau, T. P. Sidiropoulos *et al.*, High-harmonic spectroscopy of quantum phase transitions in a high-Tc superconductor, *Proceedings of the National Academy of Sciences* **119**, 40, p. e2207766119 (2022).

[53] K. S. Novoselov, A. K. Geim, S. V. Morozov, D.-E. Jiang, Y. Zhang, S. V. Dubonos, I. V. Grigorieva and A. A. Firsov, Electric field effect in atomically thin carbon films, *Science* **306**, 5696, pp. 666–669 (2004).

[54] C. Jin, F. Lin, K. Suenaga and S. Iijima, Fabrication of a freestanding boron nitride single layer and its defect assignments, *Physical Review Letters* **102**, 19, p. 195505 (2009).

[55] K. F. Mak, C. Lee, J. Hone, J. Shan and T. F. Heinz, Atomically thin $MoS_2$: a new direct-gap semiconductor, *Physical Review Letters* **105**, 13, p. 136805 (2010).

[56] R. M. Martin, *Electronic Structure: Basic Theory and Practical Methods.* Cambridge University Press (2004).

[57] D. Xiao, W. Yao and Q. Niu, Valley-contrasting physics in graphene: magnetic moment and topological transport, *Physical Review Letters* **99**, 23, p. 236809 (2007).

[58] W. Yao, D. Xiao and Q. Niu, Valley-dependent optoelectronics from inversion symmetry breaking, *Physical Review B* **77**, 23, p. 235406 (2008).

[59] X. Wang, J. R. Yates, I. Souza and D. Vanderbilt, Ab initio calculation of the anomalous Hall conductivity by Wannier interpolation, *Physical Review B* **74**, 19, p. 195118 (2006).

[60] D. Xiao, M.-C. Chang and Q. Niu, Berry phase effects on electronic properties, *Reviews of Modern Physics* **82**, 3, p. 1959 (2010).

[61] A. F. Ordonez and O. Smirnova, Propensity rules in photoelectron circular dichroism in chiral molecules. i. chiral hydrogen, *Physical Review A* **99**, 4, p. 043416 (2019).

[62] A. F. Ordonez and O. Smirnova, Propensity rules in photoelectron circular dichroism in chiral molecules. ii. general picture, *Physical Review A* **99**, 4, p. 043417 (2019).

[63] D. J. Thouless, M. Kohmoto, M. P. Nightingale and M. den Nijs, Quantized hall conductance in a two-dimensional periodic potential, *Physical Review Letters* **49**, 6, p. 405 (1982).

[64] Y. Li, J. Ludwig, T. Low, A. Chernikov, X. Cui, G. Arefe, Y. D. Kim, A. M. Van Der Zande, A. Rigosi, H. M. Hill *et al.*, Valley splitting and polarization by the Zeeman effect in monolayer $MoSe_2$, *Physical Review Letters* **113**, 26, p. 266804 (2014).

[65] Y. Cheng, Q. Zhang and U. Schwingenschlögl, Valley polarization in magnetically doped single-layer transition-metal dichalcogenides, *Physical Review B* **89**, 15, p. 155429 (2014).

[66] L. Xu, M. Yang, L. Shen, J. Zhou, T. Zhu and Y. P. Feng, Large valley splitting in monolayer $WS_2$ by proximity coupling to an insulating antiferromagnetic substrate, *Physical Review B* **97**, 4, p. 041405 (2018).

[67] K. F. Mak, K. He, J. Shan and T. F. Heinz, Control of valley polarization in monolayer $MoS_2$ by optical helicity, *Nature Nanotechnology* **7**, p. 494 (2012).

[68] F. Langer, C. P. Schmid, S. Schlauderer, M. Gmitra, J. Fabian, P. Nagler, C. Schüller, T. Korn, P. G. Hawkins, J. T. Steiner, U. Huttner, S. W. Koch, M. Kira and R. Huber, Lightwave valleytronics in a monolayer of tungsten diselenide, *Nature* **557**, 7703, pp. 76–80 (2018).

[69] S. A. O. Motlagh, F. Nematollahi, A. Mitra, A. J. Zafar, V. Apalkov and M. I. Stockman, Ultrafast optical currents in gapped graphene, *Journal of Physics: Condensed Matter* **32**, 6, p. 65305 (2019).

[70] Á. Jiménez-Galán, R. E. F. Silva, O. Smirnova and M. Ivanov, Sub-cycle valleytronics: control of valley polarization using few-cycle linearly polarized pulses, *Optica* **8**, 3, pp. 277–280 (2021).

[71] R. E. F. Silva, M. Ivanov and Á. Jiménez-Galán, All-optical valley switch and clock of electronic dephasing, *Optics Express* **30**, 17, pp. 30347–30355 (2022).

[72] A. Chernikov, T. C. Berkelbach, H. M. Hill, A. Rigosi, Y. Li, B. Aslan, D. R. Reichman, M. S. Hybertsen and T. F. Heinz, Exciton binding energy and nonhydrogenic Rydberg series in monolayer $WS_2$, *Physical Review Letters* **113**, 7, p. 076802 (2014).

[73] F. Wu, F. Qu and A. H. MacDonald, Exciton band structure of monolayer $MoS_2$, *Physical Review B* **91**, 7, p. 075310 (2015).

[74] J. Xiao, M. Zhao, Y. Wang and X. Zhang, Excitons in atomically thin 2D semiconductors and their applications, *Nanophotonics* **6**, 6, pp. 1309–1328 (2017).

[75] H. Zeng, J. Dai, W. Yao, D. Xiao and X. Cui, Valley polarization in $MoS_2$ monolayers by optical pumping, *Nature Nanotechnology* **7**, 8, pp. 490–493 (2012).

[76] K. F. Mak, K. L. McGill, J. Park and P. L. McEuen, The valley Hall effect in $MoS_2$ transistors, *Science* **344**, 6191, pp. 1489–1492 (2014).

[77] A. M. Jones, H. Yu, N. J. Ghimire, S. Wu, G. Aivazian, J. S. Ross, B. Zhao, J. Yan, D. G. Mandrus, D. Xiao *et al.*, Optical generation of excitonic valley coherence in monolayer $WSe_2$, *Nature Nanotechnology* **8**, 9, pp. 634–638 (2013).

[78] Z. Ye, D. Sun and T. F. Heinz, Optical manipulation of valley pseudospin, *Nature Physics* **13**, 1, pp. 26–29 (2017).

[79] K. M. Dorney, L. Rego, N. J. Brooks, J. San Román, C.-T. Liao, J. L. Ellis, D. Zusin, C. Gentry, Q. L. Nguyen, J. M. Shaw, A. Picón, L. Plaja, H. C. Kapteyn, M. M. Murnane and C. Hernández-García, Controlling the polarization and vortex charge of attosecond high-harmonic beams via simultaneous spin–orbit momentum conservation, *Nature Photonics* **13**, 2, pp. 123–130 (2019).

[80] I. Pupeza, D. Sánchez, J. Zhang, N. Lilienfein, M. Seidel, N. Karpowicz, T. Paasch-Colberg, I. Znakovskaya, M. Pescher, W. Schweinberger, V. Pervak, E. Fill, O. Pronin, Z. Wei, F. Krausz, A. Apolonski and J. Biegert, High-power sub-two-cycle mid-infrared pulses at 100 MHz repetition rate, *Nature Photonics* **9**, 11, pp. 721–724 (2015).

[81] V. Stummer, T. Flöry, G. Krizsán, G. Polónyi, E. Kaksis, A. Pugžlys, J. Hebling, J. A. Fülöp and A. Baltuška, Programmable generation of terahertz bursts in chirped-pulse laser amplification, *Optica* **7**, 12, pp. 1758–1763 (2020).

[82] L. Rego, J. San Román, L. Plaja and C. Hernández-García, Trains of attosecond pulses structured with time-ordered polarization states, *Optics Letters* **45**, 20, pp. 5636–5639 (2020).

[83] O. Smirnova and M. Ivanov, Multielectron high harmonic generation: simple man on a complex plane, in T. Schultz and M. Vrakking (Eds.), *Attosecond and XUV Physics: Ultrafast Dynamics and Spectroscopy*, John Wiley & Sons, pp. 201–256 (2014).

[84] J. Itatani, F. Quéré, G. L. Yudin, M. Y. Ivanov, F. Krausz and P. B. Corkum, Attosecond streak camera, *Physical Review Letters* **88**, 17, p. 173903 (2002).

[85] A. Hashmi, S. Yamada, A. Yamada, K. Yabana and T. Otobe, Valley polarization control in $WSe_2$ monolayer by a single-cycle laser pulse, *Physical Review B* **105**, 11, p. 115403 (2022).

[86] A. Hashmi, S. Yamada, K. Yabana and T. Otobe, Enhancement of valley-selective excitation by a linearly polarized two-color laser pulse, *Physical Review B* **107**, 23, p. 235403 (2023).

[87] M. Mrudul, Á. Jiménez-Galán, M. Ivanov and G. Dixit, Light-induced valleytronics in pristine graphene, *Optica* **8**, 3, pp. 422–427 (2021).

[88] X.-L. Qi, Y.-S. Wu and S.-C. Zhang, General theorem relating the bulk topological number to edge states in two-dimensional insulators, *Physical Review B* **74**, p. 045125 (2006).

[89] G. Jotzu, M. Messer, R. Desbuquois, M. Lebrat, T. Uehlinger, D. Greif and T. Esslinger, Experimental realization of the topological haldane model with ultracold fermions, *Nature* **515**, 7526, pp. 237–240 (2014).

[90] S. Mitra, Á. Jiménez-Galán, M. Neuhaus, R. E. F. Silva, V. Pervak, M. F. Kling and S. Biswas, Light-wave-controlled Haldane model in monolayer hexagonal boron nitride, *Nature* **628**, pp. 752–757 (2024).

[91] Á. Jiménez-Galán, R. E. F. Silva, O. Smirnova and M. Ivanov, Lightwave control of topological properties in 2D materials for sub-cycle and non-resonant valley manipulation, *Nature Photonics* **14**, 12, pp. 728–732 (2020).

[92] C. L. Kane and E. J. Mele, Quantum spin Hall effect in graphene, *Physical Review Letters* **95**, 22, p. 226801 (2005).

[93] M. Dávila, L. Xian, S. Cahangirov, A. Rubio and G. Le Lay, Germanene: a novel two-dimensional germanium allotrope akin to graphene and silicene, *New Journal of Physics* **16**, 9, p. 095002 (2014).

[94] F.-F. Zhu, W.-J. Chen, Y. Xu, C.-L. Gao, D.-D. Guan, C.-H. Liu, D. Qian, S.-C. Zhang and J.-F. Jia, Epitaxial growth of two-dimensional stanene, *Nature Materials* **14**, 10, pp. 1020–1025 (2015).

[95] M. Konig, S. Wiedmann, C. Brune, A. Roth, H. Buhmann, L. W. Molenkamp, X.-L. Qi and S.-C. Zhang, Quantum spin Hall insulator state in HgTe quantum wells, *Science* **318**, 5851, pp. 766–770 (2007).

[96] B. A. Bernevig, T. L. Hughes and S.-C. Zhang, Quantum spin Hall effect and topological phase transition in HgTe quantum wells, *Science* **314**, 5806, pp. 1757–1761 (2006).

[97] H. Zhang, C.-X. Liu, X.-L. Qi, X. Dai, Z. Fang and S.-C. Zhang, Topological insulators in $Bi_2Se_3$, $Bi_2Te_3$ and $Sb_2Te_3$ with a single Dirac cone on the surface, *Nature Physics* **5**, 6, pp. 438–442 (2009).

[98] C.-Z. Chang, J. Zhang, X. Feng, J. Shen, Z. Zhang, M. Guo, K. Li, Y. Ou, P. Wei, L.-L. Wang *et al.*, Experimental observation of the quantum anomalous Hall effect in a magnetic topological insulator, *Science* **340**, 6129, pp. 167–170 (2013).

[99] D. Bauer and K. K. Hansen, High-harmonic generation in solids with and without topological edge states, *Physical Review Letters* **120**, 17, p. 177401 (2018).

[100] L. Fu, Topological crystalline insulators, *Physical Review Letters* **106**, 10, p. 106802 (2011).
[101] O. Neufeld, N. Tancogne-Dejean, H. Hübener, U. De Giovannini and A. Rubio, Are there universal signatures of topological phases in high-harmonic generation? Probably not., *Physical Review X* **13**, p. 031011 (2023).
[102] C. P. Schmid, L. Weigl, P. Grössing, V. Junk, C. Gorini, S. Schlauderer, S. Ito, M. Meierhofer, N. Hofmann, D. Afanasiev *et al.*, Tunable non-integer high-harmonic generation in a topological insulator, *Nature* **593**, 7859, pp. 385–390 (2021).
[103] T. Oka and H. Aoki, Photovoltaic Hall effect in graphene, *Physical Review B* **79**, 8, p. 081406 (2009).
[104] J. W. McIver, B. Schulte, F.-U. Stein, T. Matsuyama, G. Jotzu, G. Meier and A. Cavalleri, Light-induced anomalous Hall effect in graphene, *Nature Physics* **16**, 1, pp. 38–41 (2020).
[105] N. H. Lindner, G. Refael and V. Galitski, Floquet topological insulator in semiconductor quantum wells, *Nature Physics* **7**, p. 490 (2011).
[106] E. J. Sie and E. J. Sie, Valley-selective optical stark effect in monolayer $WS_2$, in E. J. Sie (Ed.), *Coherent Light-Matter Interactions in Monolayer Transition-Metal Dichalcogenides*, Springer, pp. 37–57 (2018).
[107] A. Gómez-León and G. Platero, Floquet-Bloch theory and topology in periodically driven lattices, *Physical Review Letters* **110**, 20, p. 200403 (2013).
[108] A. G. Grushin, Á. Gómez-León and T. Neupert, Floquet fractional Chern insulators, *Physical Review Letters* **112**, 15, p. 156801 (2014).
[109] M. S. Rudner and N. H. Lindner, Band structure engineering and non-equilibrium dynamics in Floquet topological insulators, *Nature Reviews Physics* **2**, 5, pp. 229–244 (2020).
[110] T. Ozawa and H. M. Price, Topological quantum matter in synthetic dimensions, *Nature Reviews Physics* **1**, 5, pp. 349–357 (2019).
[111] T. Kitagawa, E. Berg, M. Rudner and E. Demler, Topological characterization of periodically driven quantum systems, *Physical Review B* **82**, 23, p. 235114 (2010).
[112] M. S. Rudner, N. H. Lindner, E. Berg and M. Levin, Anomalous edge states and the bulk-edge correspondence for periodically driven two-dimensional systems, *Physical Review X* **3**, 3, p. 031005 (2013).
[113] F. Nathan and M. S. Rudner, Topological singularities and the general classification of Floquet–Bloch systems, *New Journal of Physics* **17**, 12, p. 125014 (2015).
[114] M. A. Broome, A. Fedrizzi, B. P. Lanyon, I. Kassal, A. Aspuru-Guzik and A. G. White, Discrete single-photon quantum walks with tunable decoherence, *Physical Review Letters* **104**, 15, p. 153602 (2010).
[115] Q. Cheng, Y. Pan, H. Wang, C. Zhang, D. Yu, A. Gover, H. Zhang, T. Li, L. Zhou and S. Zhu, Observation of anomalous $\pi$ modes in photonic Floquet engineering, *Physical Review Letters* **122**, 17, p. 173901 (2019).
[116] S. Mukherjee, A. Spracklen, M. Valiente, E. Andersson, P. Öhberg, N. Goldman and R. R. Thomson, Experimental observation of anomalous

topological edge modes in a slowly driven photonic lattice, *Nature Communications* **8**, 1, p. 13918 (2017).

[117] L. J. Maczewsky, J. M. Zeuner, S. Nolte and A. Szameit, Observation of photonic anomalous Floquet topological insulators, *Nature Communications* **8**, 1, p. 13756 (2017).

[118] H. Cui, L. Li, T. Huang, J. Li, P. Lan and P. Lu, Valley-resolved interband excitation and emission in gapped graphene, *Physical Review A* **106**, 4, p. 043505 (2022).

[119] I. Tyulnev, Á. Jiménez-Galán, J. Poborska, L. Vamos, P. S. J. Russell, F. Tani, O. Smirnova, M. Ivanov, R. E. F. Silva and J. Biegert, Valleytronics in bulk $MoS_2$ with a topologic optical field, *Nature* **628**, pp. 746–751 (2024).

[120] A. Fleischer, O. Kfir, T. Diskin, P. Sidorenko and O. Cohen, Spin angular momentum and tunable polarization in high-harmonic generation, *Nature Photonics* **8**, 7, pp. 543–549 (2014).

[121] A. S. Wyatt, T. Witting, A. Schiavi, D. Fabris, P. Matia-Hernando, I. A. Walmsley, J. P. Marangos and J. W. Tisch, Attosecond sampling of arbitrary optical waveforms, *Optica* **3**, 3, pp. 303–310 (2016).

[122] M. F. Ciappina, J. A. Pérez-Hernández, A. S. Landsman, W. A. Okell, S. Zherebtsov, B. Förg, J. Schötz, L. Seiffert, T. Fennel, T. Shaaran *et al.*, Attosecond physics at the nanoscale, *Reports on Progress in Physics* **80**, 5, p. 054401 (2017).

[123] G. Vampa, H. Fattahi, J. Vučković and F. Krausz, Attosecond nanophotonics, *Nature Photonics* **11**, 4, pp. 210–212 (2017).

[124] Á. Jiménez-Galán, G. Dixit, S. Patchkovskii, O. Smirnova, F. Morales and M. Ivanov, Attosecond recorder of the polarization state of light, *Nature Communications* **9**, 1, p. 850 (2018).

[125] M. Garg, H. Kim and E. Goulielmakis, Ultimate waveform reproducibility of extreme-ultraviolet pulses by high-harmonic generation in quartz, *Nature Photonics* **12**, 5, pp. 291–296 (2018).

[126] Á. Jiménez-Galán, N. Zhavoronkov, D. Ayuso, F. Morales, S. Patchkovskii, M. Schloz, E. Pisanty, O. Smirnova and M. Ivanov, Control of attosecond light polarization in two-color bicircular fields, *Physical Review A* **97**, p. 023409 (2018).

[127] E. Pisanty, L. Rego, J. San Román, A. Picón, K. M. Dorney, H. C. Kapteyn, M. M. Murnane, L. Plaja, M. Lewenstein and C. Hernández-García, Conservation of torus-knot angular momentum in high-order harmonic generation, *Physical Review Letters* **122**, 20, p. 203201 (2019).

[128] P.-C. Huang, C. Hernández-García, J.-T. Huang, P.-Y. Huang, C.-H. Lu L. Rego, D. D. Hickstein, J. L. Ellis, A. Jaron-Becker, A. Becker *et al.*, Polarization control of isolated high-harmonic pulses, *Nature Photonics* **12**, 6, pp. 349–354 (2018).

[129] A. de las Heras, A. K. Pandey, J. San Román, J. Serrano, E. Baynard, G. Dovillaire, M. Pittman, C. G. Durfee, L. Plaja, S. Kazamias *et al.*, Extreme-ultraviolet vector-vortex beams from high harmonic generation, *Optica* **9**, 1, pp. 71–79 (2022).

[130] A. Korobenko, S. Rashid, A. Y. Naumov, D. M. Villeneuve, D. A. Reis, P. Berini, P. B. Corkum and G. Vampa, In situ high-harmonic microscopy of a nanostructured solid, *Optica* **10**, 5, pp. 642–649 (2023).

[131] J. Schoetz, Z. Wang, E. Pisanty, M. Lewenstein, M. Kling and M. Ciappina, Perspective on petahertz electronics and attosecond nanoscopy, *ACS Photonics*, **6**, 12, pp. 3057–3069 (2019).

[132] M. Borsch, M. Meierhofer, R. Huber and M. Kira, Lightwave electronics in condensed matter, *Nature Reviews Materials* **8**, 10, pp. 668–687 (2023).

[133] N. Baldelli, U. Bhattacharya, D. González-Cuadra, M. Lewenstein and T. Graß, Detecting Majorana zero modes via strong field dynamics, *ACS Omega* **7**, 50, pp. 47424–47430 (2022).

[134] J. Freudenstein, M. Borsch, M. Meierhofer, D. Afanasiev, C. P. Schmid, F. Sandner, M. Liebich, A. Girnghuber, M. Knorr, M. Kira *et al.*, Attosecond clocking of correlations between Bloch electrons, *Nature* **610**, 7931, pp. 290–295 (2022).

[135] E. B. Molinero, B. Amorim, M. Malakhov, G. Cistaro, Á. Jiménez-Galán, M. Ivanov, A. Picón, P. San-José and R. E. F. Silva, Formation, stability, and highly nonlinear optical response of excitons to intense light fields interacting with two-dimensional materials, *arXiv preprint arXiv:2307.16647* (2023).

[136] G. G. Brown, Á. Jiménez-Galán, R. E. F. Silva and M. Ivanov, A real-space perspective on dephasing in solid-state high harmonic generation, *arXiv preprint arXiv:2210.16889* (2022).

[137] H. Lakhotia, H. Kim, M. Zhan, S. Hu, S. Meng and E. Goulielmakis, Laser picoscopy of valence electrons in solids, *Nature* **583**, 7814, pp. 55–59 (2020).

[138] E. N. Osika, A. Chacón, L. Ortmann, N. Suárez, J. A. Pérez-Hernández, B. Szafran, M. F. Ciappina, F. Sols, A. S. Landsman and M. Lewenstein, Wannier-Bloch approach to localization in high-harmonics generation in solids, *Physical Review X* **7**, 2, p. 021017 (2017).

[139] A. Parks, G. Ernotte, A. Thorpe, C. McDonald, P. Corkum, M. Taucer and T. Brabec, Wannier quasi-classical approach to high harmonic generation in semiconductors, *Optica* **7**, 12, p. 1764 (2020).

[140] Á. Jiménez-Galán, C. Bossaer, G. Ernotte, A. M. Parks, R. E. F. Silva, D. M. Villeneuve, A. Staudte, T. Brabec, A. Luican-Mayer and G. Vampa, Orbital perspective on high-harmonic generation from solids, *Nature Communications* **14**, 1, p. 8421 (2023).

[141] G. G. Brown, Á. Jiménez-Galán, R. E. F. Silva and M. Ivanov, Ultrafast dephasing in solid state high harmonic generation: macroscopic origin revealed by real-space dynamics, *arXiv preprint arXiv:2310.17005* (2023).

[142] G. G. Brown, R. E. F. Silva, Á. Jiménez-Galán and M. Ivanov, Unambigous discrimination of high harmonic generation mechanisms in solids, *arXiv preprint arXiv:2310.17007* (2023).

[143] R. E. F. Silva, F. Martín and M. Ivanov, High harmonic generation in crystals using maximally localized Wannier functions, *Physical Review B* **100**, p. 195201 (2019).

https://doi.org/10.1142/9789811279560_0003

# Chapter 3

# Wannier Approach to the Semiconductor Bloch Equations

Eduardo B. Molinero[1], Álvaro Jiménez-Galán[1,2] and Rui E. F. Silva[1,2]

[1] *Instituto de Ciencia de Materiales de Madrid (ICMM), Consejo Superior de Investigaciones Científicas (CSIC), Sor Juana Inés de la Cruz 3, 28049, Madrid, Spain*
[2] *Max-Born-Institut, Max-Born Strasse 2A, D-12489, Berlin, Germany*

## 3.1 Optical Response of Solids

Light-matter interactions are one of the most exciting fields of physics because it allows us to unveil how photons and electrons interact with each other. This is especially true when we combine condensed matter systems with intense laser fields, as the synergy between these two systems not only uncovers new physics but also provides deeper insights into previously understood phenomena [1–3].

From a theoretical perspective, this topic is interesting in its own right. Developing theories that allows one to describe the interaction between photons and electrons is not an easy task, and it is full of complexities and subtleties. One of the first and most prominent methods to approach the light-matter interaction problem is time-dependent density functional theory [4, 5], often abbreviated as TDDFT. Indeed, TDDFT has been confirmed to successfully compute the optical response of solids [6] not only to weak fields but also to strong, intense laser fields [7,8]. However, TDDFT is not without drawbacks; for example, it is computationally exceptionally demanding.

Nevertheless, we will take a different pathway and base our approach on the Semiconductor Bloch Equations (SBEs) [9, 10]. The SBEs are a set of differential equations which model the microscopic response of materials (despite their name they can be applied to a various range of materials) to classical light. The equations are derived from the equation of motion of the one-particle reduced density matrix [9, 11]. Therefore, they constitute a closed set of differential equations for the non-interacting case [11, 12] and an integro-differential set for the interacting case [13]. While they have been proven useful and accurate in obtaining both the linear [9, 11] and the nonlinear optical response [12, 14] of materials, they do present a central problem. As we will see, this problem is a physical consequence of dealing with periodic crystals. The main idea of this chapter is to show that the best approach to circumvent this problem is by using something referred to as the *Wannier gauge.*

When we refer to this Wannier gauge, we are talking about a *structural gauge* and not about the electromagnetic gauge. The laser gauge does not present such a physical issue and one can choose between the velocity or the length gauge [14, 15]. Nevertheless, we will work through this whole chapter in the length gauge because it converges faster with the number of bands [15, 16] and it's easier to include dephasing terms onto the equations.

In order to really get a grasp of why the so-called Wannier gauge is a proper solution to the problem, one should understand why such a problem arises in the very first place.

## 3.2 Wannier Functions and Gauges

When we perform calculations on solids it is common to use periodic boundary conditions since it is the most natural choice. Then, the Bloch theorem applies [17, 18] and the eigenstates $|\psi_{n\boldsymbol{k}}\rangle$ of the single-particle Hamiltonian $H$ fulfills that

$$\langle \boldsymbol{r}|\psi_{n\boldsymbol{k}}\rangle = \psi_{n\boldsymbol{k}}(\boldsymbol{r}) = u_{n\boldsymbol{k}}(\boldsymbol{r})e^{i\boldsymbol{k}\cdot\boldsymbol{r}}, \tag{3.1}$$

where the $u_{n\boldsymbol{k}}(\boldsymbol{r})$ functions have the same periodicity as the lattice of the crystal. From the set of states $|\psi_{n\boldsymbol{k}}\rangle$, one can then obtain the electronic band structure of the system and from it, the majority of the properties (at a single-particle level) of the material can be extracted. Therefore, it may seem obvious that when we want to compute the response to the electromagnetic field, we should work with those states. However, there are some subtleties that become extremely relevant in the context of the optical response.

Recall that one consequence of the periodic boundary conditions in real space is a redundancy in its momentum counterpart. Let's explain this statement more clearly. The Bloch states transform under translations as

$$T_{\boldsymbol{R}} |\psi_{n\boldsymbol{k}}\rangle = e^{i\boldsymbol{k}\cdot\boldsymbol{R}} |\psi_{n\boldsymbol{k}}\rangle, \tag{3.2}$$

thus $|\psi_{n\boldsymbol{k}}\rangle$ are eigenstates of the translation operator. However, if we display the crystal momentum $\boldsymbol{k}$ by a vector of the reciprocal lattice $\boldsymbol{G}$, one can see that both $|\psi_{n\boldsymbol{k}}\rangle$ and $|\psi_{n\boldsymbol{k}+\boldsymbol{G}}\rangle$ have the same eigenvalue of the $T_{\boldsymbol{R}}$ operator (recall that the dual vectors satisfy $\boldsymbol{G}\cdot\boldsymbol{R} = 2\pi$ and thus $e^{i\boldsymbol{G}\cdot\boldsymbol{R}} = 1$). Furthermore, both states have the same energy since $[H, T_{\boldsymbol{R}}] = 0$.

Therefore, all possible *physical* states of the system are contained within a zone of the reciprocal space named the *Brillouin zone* [17]. We must admit this a twisted way of defining the Brillouin zone but it allows us to reinterpret the concept in a more general way. Namely, one can understand the Brillouin zone as a *manifold* (in its geometric sense) whose boundaries are connected to other Brillouin zones displaced by a lattice vector $\boldsymbol{G}$. As for any manifold with boundaries, it should satisfy some kind of *boundary conditions*. Let's try to see what those conditions are based on the physics of the system. As mentioned before, the states $|\psi_{n\boldsymbol{k}}\rangle$ and $|\psi_{n\boldsymbol{k}+\boldsymbol{G}}\rangle$ are physically redundant to each other, so the only thing a state can gain when it "crosses" the boundary is a phase, otherwise the states would yield different observables. More formally, this means that

$$|\psi_{n\boldsymbol{k}+\boldsymbol{G}}\rangle = e^{-i\theta(\boldsymbol{k})} |\psi_{n\boldsymbol{k}}\rangle, \tag{3.3}$$

where $\theta(\boldsymbol{k})$ is an arbitrary function whose specific form is physically unimportant. We will see that acquired phase can be regarded as a *gauge transformation.* Nevertheless, this phase will cause two "problems" of very different natures: a physical and a practical one.

Actually, the first is not a problem but rather is the cause of the emergence of one of the most recent and interesting fields in the condensed matter community: the existence of *topological phases in quantum materials* [19–21]. Those phases are characterized by the presence of a topological invariant that remains unchanged if we perform continuous transformation in the Hamiltonian of the system. Therefore, they cannot be characterized by a local order parameter using the well-established Landau's theory of phase transitions. The presence of such non-local order parameter gives rise to very unique materials, for example, the existence of topological insulators, materials that should be insulators but present gapless edge states [22]. For the sake of simplicity we will now focus on systems whose topological invariant is the Chern number. However, the discussion can be extended to other invariants (such classification is often referred to as the *periodic table of topological invariants* [23, 24]). The Chern number is defined (for a crystalline solid) as

$$C = \frac{1}{2\pi}\int_{\mathrm{BZ}} \boldsymbol{\Omega}_n(\boldsymbol{k}) \cdot d\boldsymbol{k}, \tag{3.4}$$

where $\boldsymbol{\Omega}(\boldsymbol{k})$ is the so-called Berry curvature [25] which is computed as

$$\boldsymbol{\Omega}_n(\boldsymbol{k}) = \nabla_{\boldsymbol{k}} \times \boldsymbol{\mathcal{A}}_n(\boldsymbol{k}). \tag{3.5}$$

The last term is referred to as the Berry connection (we will clarify why this quantity is referred to as a connection) and it is defined as

$$\boldsymbol{\mathcal{A}}_n(\boldsymbol{k}) = i\,\langle u_{n\boldsymbol{k}}|\nabla_{\boldsymbol{k}} u_{n\boldsymbol{k}}\rangle. \tag{3.6}$$

One can see a clear analogy with electromagnetism here. The role of the vector potential $\boldsymbol{A}$ is now replaced by the Berry connection while the magnetic field $\boldsymbol{B}$ is replaced with the Berry curvature. Indeed, if we perform the transformation 3.3, we can see that the Berry connection transforms as

$$\begin{aligned}\boldsymbol{\mathcal{A}}'_n(\boldsymbol{k}) &= i\langle u'_{n\boldsymbol{k}}|\nabla_{\boldsymbol{k}} u'_{n\boldsymbol{k}}\rangle = ie^{i\theta(\boldsymbol{k})}\langle u_{n\boldsymbol{k}}|\nabla_{\boldsymbol{k}}(e^{-i\theta(\boldsymbol{k})}u_{n\boldsymbol{k}})\rangle \\ &= i\,\langle u_{n\boldsymbol{k}}|\nabla_{\boldsymbol{k}} u_{n\boldsymbol{k}}\rangle + \nabla_{\boldsymbol{k}}\theta(\boldsymbol{k})\,\langle u_{n\boldsymbol{k}}|u_{n\boldsymbol{k}}\rangle = \boldsymbol{\mathcal{A}}_n(\boldsymbol{k}) + \nabla_{\boldsymbol{k}}\theta(\boldsymbol{k}),\end{aligned} \tag{3.7}$$

while the Berry curvature remains invariant

$$\begin{aligned}\boldsymbol{\Omega}'_n(\boldsymbol{k}) &= \nabla_{\boldsymbol{k}} \times \boldsymbol{\mathcal{A}}'_n(\boldsymbol{k}) = \nabla_{\boldsymbol{k}} \times (\boldsymbol{\mathcal{A}}_n(\boldsymbol{k}) + \nabla_{\boldsymbol{k}}\theta(\boldsymbol{k})) \\ &= \nabla_{\boldsymbol{k}} \times \boldsymbol{\mathcal{A}}_n(\boldsymbol{k}) = \boldsymbol{\Omega}_n(\boldsymbol{k}).\end{aligned} \tag{3.8}$$

Therefore, it is accurate to describe the transformation 3.3 as a *gauge transformation* [26]. Formally, the gauge theory is $\mathcal{U}(1)$, given that we are in the non-degenerate band structure case. We will comment afterwards what this means and its extension to the degenerate case. However, for the sake of simplicity, we will restrict it now to the non-degenerate case.

As for any gauge theory, one could potentially fix the gauge condition so that it takes the simplest form. In our case, such form would be

$$|\psi_{n\boldsymbol{k}+\boldsymbol{G}}\rangle = |\psi_{n\boldsymbol{k}}\rangle, \tag{3.9}$$

often referred to as *periodic boundary condition.* But can we always make this choice? Naively, one would answer yes due to the gauge freedom in our system. However, we will show that, sadly, this is not the case. Suppose that we divide the Brillouin zone, $\mathcal{B}$, into two sections namely $\mathcal{B}_1$ and $\mathcal{B}_2$

such that $\mathcal{B} = \mathcal{B}_1 + \mathcal{B}_2$. States in one zone can be mapped to other sections by (also) a gauge transformation

$$|\psi_{n\boldsymbol{k}}\rangle_{\mathcal{B}_1} = e^{-i\theta(\boldsymbol{k})}\,|\psi_{n\boldsymbol{k}}\rangle_{\mathcal{B}_2}. \tag{3.10}$$

To understand why we can do this, it is helpful to visualize the two "new" zones as their periodic counterparts. Think of it in a simpler way: one can always translate any momentum $\boldsymbol{k}$ into any other momentum $\boldsymbol{k}'$ (without crossing the boundary of the Brillouin zone) by a translation using a reciprocal basis vector $\boldsymbol{G}$. If one now computes the Chern number for the whole Brillouin zone

$$2\pi C = \int_{\mathcal{B}} \boldsymbol{\Omega}(\boldsymbol{k}) \cdot d\boldsymbol{k} = \int_{\mathcal{B}_1} \boldsymbol{\Omega}_{\mathcal{B}_1}(\boldsymbol{k}) \cdot d\boldsymbol{k} + \int_{\mathcal{B}_2} \boldsymbol{\Omega}_{\mathcal{B}_2}(\boldsymbol{k}) \cdot d\boldsymbol{k}, \tag{3.11}$$

and then applies Stokes' theorem, one obtains the following result

$$2\pi C = \oint_P \left(\boldsymbol{\mathcal{A}}_{\mathcal{B}_1}(\boldsymbol{k}) - \boldsymbol{\mathcal{A}}_{\mathcal{B}_2}(\boldsymbol{k})\right) \cdot d\boldsymbol{k}, \tag{3.12}$$

where $P$ is a closed path that traces the boundary between the two subzones. Note that the sign in front of each $\boldsymbol{\mathcal{A}}$ is arbitrary (since it depends on the particular election of the direction of the path); however, both quantities will always have opposite sign. Since we know how the Berry connection changes after a gauge transformation, we can simplify the previous equation into

$$C = \frac{1}{2\pi} \oint_P \nabla_{\boldsymbol{k}}\theta(\boldsymbol{k}) \cdot d\boldsymbol{k}. \tag{3.13}$$

Even though the previous equation seems rather obscure, it is the most insightful one when it comes to clarifying topological quantum matter. Imagine that we want to impose in the system the condition 3.9, that would mean that the phase $\theta(\boldsymbol{k})$ must be $2\pi$ times an integer. However, this condition directly implies that the Chern number vanishes. Think of it the other way around: if we were to have a system with non-zero Chern number, it would mean that we couldn't impose a periodic gauge on the system! This is the reason why systems with $C = 0$ are called *topologically trivial* because their Brillouin zone is periodic along all of its directions, i.e., it has the *topology* of a torus in $d$ dimensions. In the opposite case, when the system is topologically non-trivial ($C \neq 0$), *the Brillouin zone must have some discontinuities*; thus, it is physically impossible to construct a smooth and periodic gauge [27, 28].

It is common to draw an analogy between Berry's physics and that of electromagnetism, as mentioned earlier. However, an avid reader might pose the following question: Why don't we observe these physics in electromagnetism? The answer lies in the absence of magnetic monopoles [29]. The Berry connection behaves akin to a magnetic field that possesses point-like charges; more formally, it exhibits topological charges [26,30]. This is truly the root of topological matter: the electrons are behaving as if they were subjected to a magnetic field (Berry curvature) spanned by a magnetic charge (topological charge) which gives rise to their unique properties. Nevertheless, even though the field of topological matter is filled with rich physics, those are beyond the scope of this chapter. We refer the reader to the wonderful books of David B. Vanderbilt [28] and M. Araújo and P. Sacramento [31].

The practical problem is also related to the topology of the Brillouin zone, but is a bit more subtle (and more specific to our interests). Recall that in the dipole approximation and in the length gauge [32, 33], the electric field $\boldsymbol{E}(t)$ enters the equations of motion by coupling with the position operator $\boldsymbol{r}$. In atoms and molecules, such operator poses no problems but that is not the case for crystals, due (again) to their intrinsic *periodicity*. To better understand this issue, let's consider a one-dimensional lattice with periodic boundary conditions, represented as a ring. In this lattice, at any given position along the ring, there exist two distinct distances to its closest neighbors. The first distance corresponds to the shortest path since the neighboring lattice points are in close proximity. However, there is also a second distance that extends across the entire ring, reflecting the periodic nature of the lattice. Thus, it is rather problematic to define the position operator $\boldsymbol{r}$ in the Bloch basis [34]. However, if one processes carefully it can be expressed [11, 16, 35] as

$$\langle\psi_{n\boldsymbol{k}}|\, \boldsymbol{r}\, |\psi_{n\boldsymbol{k}}\rangle = i\left[\delta_{mn}\nabla_{\boldsymbol{k}}\delta(\boldsymbol{k}-\boldsymbol{k}') - i\boldsymbol{\mathcal{A}}_n(\boldsymbol{k})\right]. \tag{3.14}$$

It is natural to see then the position operator as a covariant derivative [26]. Indeed, if we apply the gauge transformation 3.3, the position operator remains gauge invariant (as any covariant derivative should be) and the Berry connection transforms as 3.7, i.e., it transforms as a *connection* [26] and hence the name Berry connection. In the context of differential geometry, the connection contains the information about how the manifold itself changes when we perform a parallel transport of some vector along a path. Indeed, the Berry connection describes how the states change along some path in the Hilbert space of the system.

The reader can ask (rightfully) now the following question: Where is the problem at all? We will show that in the SBEs, the dipole term goes like

$$\partial_t \rho_{nm}(\boldsymbol{k},t)|_{\text{dipole}} \sim \boldsymbol{E}(t)\cdot\nabla_{\boldsymbol{k}}\rho_{nm}(\boldsymbol{k},t) + \boldsymbol{E}(t)\cdot\boldsymbol{\mathcal{A}}_n(\boldsymbol{k}). \tag{3.15}$$

Numerically, this entails two key considerations. Firstly, it is essential to have a set of functions denoted as $|\psi_{n\boldsymbol{k}}\rangle$ (which will be utilized for constructing the density matrix) that are both smooth and periodic across the entire Brillouin zone. Failure to meet this criterion will not only render the gradient term in the SBEs ill-defined, but also the Berry connection as it can be deduced from Eq. (3.7). However, given the gauge freedom inherent in the eigenstates of the system, there may be instances where the states obtained from any kind of diagonalization techniques (such as DFT calculations) do not meet this requirement. Moreover, in scenarios involving topological systems, the creation of such a smooth gauge is physically prohibited, as previously mentioned. Secondly, there will be situations in which the Berry connection terms $\boldsymbol{\mathcal{A}}_n(\boldsymbol{k})$ are not smooth across the entire Brillouin zone, as discussed earlier.

How can one solve this? There are several approaches to this problem. The most obvious one is to impose the 3.9 condition, i.e., assuming perfect periodic boundary condition on the system, which as stated before is only valid for *trivial systems.* One can also use the twisted parallel transport gauge [14,15,28] which allows for the construction of a smooth gauge *inside* the Brillouin zone. It consists of assuming that neighboring states (i.e., $|u_{n\boldsymbol{k}}\rangle$ and $|u_{n\boldsymbol{k}+d\boldsymbol{k}}\rangle$) have zero overlap. Physically, this means that infinitesimally close states are perfectly aligned with each other. Then, the Berry phase is distributed smoothly along all the states of the Hilbert space. Another option is to construct a global smooth gauge by patching different gauges (which are continuous in its own) into one smooth gauge. This approach was taken to try to solve the problem of the gauge freedom in the Haldane model [36]. Nevertheless, all these approaches face the same issue: they are system-dependent. Each of them should be tailored specifically to the system we are treating in each moment. Therefore, one would desire for an approach that is system-independent. This is where Wannier functions enter the picture.

As for the majority of problems in quantum mechanics, the solution lies in choosing the *appropriate basis.* So far, we have been working in the eigenbasis of the system, i.e., the Bloch basis where all the gauge-freedom problems arose. However, there is no necessity to work in that base since the SBEs are basis-independent. This problem has long been known in the field of quantum molecular dynamics [37] as adiabatic versus diabatic representations. Let's illustrate this similarity with a simple example. Assume that we have a a diatomic molecule with two distinct electronic states. Its Hamiltonian is then simply

$$H = \begin{pmatrix} h_{11}(R) & h_{12}(R) \\ h_{21}(R) & h_{22}(R) \end{pmatrix}, \tag{3.16}$$

where $R$ is the internuclear distance. Diagonalizing $H$ gives us the energies of the system

$$E_n(R) = \frac{h_{11} + h_{22}}{2} \pm \frac{1}{2}\sqrt{(h_{11} - h_{22})^2 + 4|h_{12}|^2}, \tag{3.17}$$

where $n$ labels its eigenvalue index. Its energy splitting will be

$$\Delta E = \sqrt{(h_{11} - h_{22})^2 + 4|h_{12}|^2}. \tag{3.18}$$

These two levels will only cross when $h_{11}(R) = h_{22}(R)$ and $h_{12}(R) = 0$, i.e., at a degeneracy point where the two states possess the same energy. However, since we have two equations but only one independent parameter, those conditions cannot be fulfilled and hence, the two energy levels will not cross. This is the simplest example of an avoided crossing [38]. Does this imply that everything is smooth and without problems in the system? The answer is no. To illustrate why this is the case, we will parametrize the matrix elements of $H$ as $h_{11} - h_{22} = \alpha R$ and $dh_{12}(R)/dR = 0$. This parametrization was chosen so to avoid any loss of generality, following [38].

In Fig. 3.1, the energy levels of the system are depicted in terms of the parameter $R$, alongside the physical character of the eigenstates. Such character is obtained by projecting the eigenstates onto the $|0\rangle$ state, i.e., onto the lowest electronic state. One could have chosen to project the states onto the other electronic state, namely $|1\rangle$, but since we only have two possible physical states, the choice does not matter. From Fig. 3.1, it is evident that the states $|\psi_{1,2}\rangle$ start with opposite character. However, as we vary the $R$ parameter, we observe a change in their physical character after the point where the avoided crossing is, i.e., at $R = 0$. Now, let's consider the scenario where a time-dependent perturbation is applied to the system, for example a laser field. To perform the time propagation we will need to obtain terms [37] such as $\langle\psi_m|\nabla_R\psi_n\rangle$. However, if these states comes from a numerical diagonalization, such procedure may mix their physical character (in fact, it *will* mix them) leading to numerical instabilities at the avoided crossing, giving rise to ill-defined $\langle\psi_m|\nabla_R\psi_n\rangle$

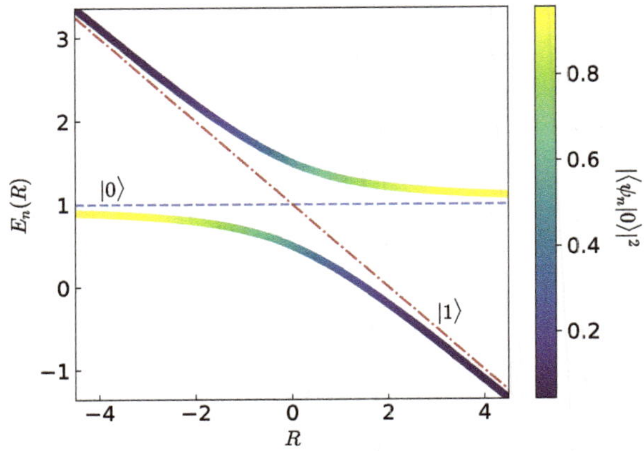


Fig. 3.1 Energy levels of the diatomic molecule in terms of the internuclear distance $R$. The color depicts the absolute value of the scalar product $\langle\psi_n|0\rangle$.

terms. This is because the adiabatic (in this case it is the eigenbasis $|\psi_{1,2}\rangle$) electronic states are defined to preserve the ordering of the energy levels, but in general not their physical character.

The solution found to this problem was to switch to the so-called *diabatic basis* [37,38]. In such basis, the states now do cross each other but *conserve their physical character*. In our example, the diabatic basis is simply the physical one $\{|0\rangle, |1\rangle\}$. Although one cannot refer to energy levels outside the eigenbasis, it is illustrative to observe the "energies" within this basis, represented by the blue and red lines in Fig. 3.1. In this scenario, the energies will be simply $h_{11}$ and $h_{22}$. They do indeed cross each other while maintaining their physical character alongside all the parameter space. As a rule of thumb, the diabatic basis, $|\psi'_n\rangle$, will be such that the expected values $\langle\psi'_m|\nabla_R\psi'_n\rangle$ are as small as possible. Coming back to our example it is immediately seen that if our diabatic basis is $\{|\psi'_n\rangle\} = \{|0\rangle, |1\rangle\}$, all the couplings $\langle\psi'_m|\nabla_R\psi'_n\rangle$ vanish exactly. Therefore, by not diagonalizing the Hamiltonian we have avoided the numerical instabilities caused by the avoided crossing. The attentive reader has likely already recognized the significant similarities between this subfield of quantum chemistry and solid-state physics. Both areas suffer from ill-defined gradients, which have the same origin. Furthermore, the change of physical character is also a relevant problem in solids. When two, or more, bands intersect it is extremely complicated [39] to keep track of their orbital character ($s$ vs. $p$, etc.). Therefore, we will apply to solids the same idea as that in the diatomic molecule example. We will look for a diabatic basis in the sense that the expected values $\langle u_{n\boldsymbol{k}}|\nabla_{\boldsymbol{k}} u_{n\boldsymbol{k}}\rangle$, i.e., the *Berry connections*, are as small as possible; that specific condition will be fulfilled only when the chosen basis respects the physical character of the electronic states. Such basis will be the Wannier basis.

Suppose for the sake of simplicity that our bands are isolated in the sense that there are no band crossings. Then, the Wannier functions are defined as

$$|n\boldsymbol{R}\rangle = \sum_{\boldsymbol{k}\in\mathrm{BZ}} e^{-i\boldsymbol{k}\cdot\boldsymbol{R}} |\psi_{n\boldsymbol{k}}\rangle, \tag{3.19}$$

where $\boldsymbol{R}$ is Bravais lattice vector, $V$ is the volume of the unit cell, $d$ is the dimensionality of our crystal and for the sake of clarity we have used the following compact notation

$$\sum_{\boldsymbol{k}\in\mathrm{BZ}} \equiv \frac{V}{(2\pi)^d} \int_{\mathrm{BZ}} d^d k.$$

These functions have a set of important properties: 1) they form an orthonormal set $\langle m\boldsymbol{R}'|n\boldsymbol{R}\rangle = \delta_{mn}\delta_{\boldsymbol{R}'\boldsymbol{R}}$, 2) they transform under translation as $T_{\boldsymbol{R}'}|n\boldsymbol{R}\rangle = |n\boldsymbol{R}-\boldsymbol{R}'\rangle$ and 3) they are localized in real space given that they are smooth enough in momentum space. Furthermore, the expected value of some operators has a particular simple form within this set of functions. Let's start with the Hamiltonian of the system, $H$. Its matrix elements are

$$\begin{aligned}\langle n\boldsymbol{0}| H |n\boldsymbol{R}\rangle &= \sum_{\boldsymbol{k}\in\mathrm{BZ}} \sum_{\boldsymbol{k}'\in\mathrm{BZ}} e^{-i(\boldsymbol{k}-\boldsymbol{k}')\cdot\boldsymbol{R}} \langle\psi_{n\boldsymbol{k}'}| H |\psi_{n\boldsymbol{k}}\rangle \\ &= \sum_{\boldsymbol{k}\in\mathrm{BZ}} e^{-i\boldsymbol{k}\cdot\boldsymbol{R}} \langle u_{n\boldsymbol{k}}| H |u_{n\boldsymbol{k}}\rangle, \end{aligned} \tag{3.20}$$

where we have used the following property: $\langle\psi_{n\boldsymbol{k}'}|\psi_{n\boldsymbol{k}}\rangle = \frac{(2\pi)^d}{V}\delta(\boldsymbol{k}-\boldsymbol{k}')\langle u_{n\boldsymbol{k}'}|u_{n\boldsymbol{k}}\rangle$. It is inmediate to realize that the matrix elements, $\langle u_{n\boldsymbol{k}}| H |u_{n\boldsymbol{k}}\rangle$, are the energies of the system. Therefore, the matrix elements $\langle n\boldsymbol{0}| H |n\boldsymbol{R}\rangle$ are just the *Fourier transform of the energies of the system.* A similar calculation [34] can be carried out for the position operator $\boldsymbol{r}$,

$$\begin{aligned}\langle n\boldsymbol{0}| \boldsymbol{r} |n\boldsymbol{R}\rangle &= \sum_{\boldsymbol{k}\in\mathrm{BZ}} \langle u_{n\boldsymbol{k}}| \boldsymbol{r} \left(e^{-i\boldsymbol{k}\cdot\boldsymbol{R}} |u_{n\boldsymbol{k}}\rangle\right) \sum_{\boldsymbol{k}\in\mathrm{BZ}} \langle u_{n\boldsymbol{k}}| i\nabla_{\boldsymbol{k}} \left(e^{-i\boldsymbol{k}\cdot\boldsymbol{R}} |u_{n\boldsymbol{k}}\rangle\right) \\ &= \sum_{\boldsymbol{k}\in\mathrm{BZ}} \langle u_{n\boldsymbol{k}}| e^{-i\boldsymbol{k}\cdot\boldsymbol{R}} \left(\boldsymbol{R}|u_{n\boldsymbol{k}}\rangle + i\,|\nabla_{\boldsymbol{k}} u_{n\boldsymbol{k}}\rangle\right)\end{aligned}$$

$$= \sum_{\boldsymbol{k}\in \mathrm{BZ}} e^{-i\boldsymbol{k}\cdot\boldsymbol{R}} \left(\boldsymbol{R} + i \left\langle u_{n\boldsymbol{k}} \middle| \nabla_{\boldsymbol{k}} u_{n\boldsymbol{k}} \right\rangle\right)$$

$$= \sum_{\boldsymbol{k}\in \mathrm{BZ}} e^{-i\boldsymbol{k}\cdot\boldsymbol{R}} i \left\langle u_{n\boldsymbol{k}} \middle| \nabla_{\boldsymbol{k}} u_{n\boldsymbol{k}} \right\rangle = \sum_{\boldsymbol{k}\in \mathrm{BZ}} e^{-i\boldsymbol{k}\cdot\boldsymbol{R}} \boldsymbol{\mathcal{A}}_n(\boldsymbol{k}). \tag{3.21}$$

Again, we see the same feature: the matrix elements of the position operator are nothing more than the *Fourier transform of the Berry connection*, i.e., $\langle n\boldsymbol{0}| \boldsymbol{r} |n\boldsymbol{R}\rangle = \boldsymbol{\mathcal{A}}_n(\boldsymbol{R})$. An important question now arises: how do these expected values behave under the gauge transformation of Eq. (3.3)? Let's start with the Hamiltonian matrix elements. They transform as

$$\langle n\boldsymbol{0}'| H |n\boldsymbol{R}'\rangle = \sum_{\boldsymbol{k}\in \mathrm{BZ}} e^{-i\boldsymbol{k}\cdot\boldsymbol{R}} \langle u_{n\boldsymbol{k}}| H |u_{n\boldsymbol{k}}\rangle$$

$$= \sum_{\boldsymbol{k}\in \mathrm{BZ}} e^{-i\boldsymbol{k}\cdot\boldsymbol{R}} \Big( \langle u'_{n\boldsymbol{k}}| e^{-i\theta_n(\boldsymbol{k})}\Big) H \Big(e^{i\theta_n(\boldsymbol{k})} |u'_{n\boldsymbol{k}}\rangle\Big)$$

$$= \sum_{\boldsymbol{k}\in \mathrm{BZ}} e^{-i\boldsymbol{k}\cdot\boldsymbol{R}} \langle u'_{n\boldsymbol{k}}| H |u'_{n\boldsymbol{k}}\rangle = \langle n\boldsymbol{0}| H |n\boldsymbol{R}\rangle \tag{3.22}$$

and hence, the $E_n(\boldsymbol{R})$ are invariant under gauge transformation. In the case of the position operator, one can proceed in the same fashion to obtain

$$\langle n\boldsymbol{0}'| \boldsymbol{r} |n\boldsymbol{R}'\rangle = \sum_{\boldsymbol{k}\in \mathrm{BZ}} e^{-i\boldsymbol{k}\cdot\boldsymbol{R}} \boldsymbol{\mathcal{A}}_n(\boldsymbol{k}) = \sum_{\boldsymbol{k}\in \mathrm{BZ}} e^{-i\boldsymbol{k}\cdot\boldsymbol{R}} \left(\boldsymbol{\mathcal{A}}'_n(\boldsymbol{k}) - \nabla_{\boldsymbol{k}}\theta(\boldsymbol{k})\right)$$

$$= \langle n\boldsymbol{0}| \boldsymbol{r} |n\boldsymbol{R}\rangle - \sum_{\boldsymbol{k}\in \mathrm{BZ}} e^{-i\boldsymbol{k}\cdot\boldsymbol{R}} \nabla_{\boldsymbol{k}}\theta_n(\boldsymbol{k})$$

$$= \langle n\boldsymbol{0}| \boldsymbol{r} |n\boldsymbol{R}\rangle - \sum_{\boldsymbol{k}\in \mathrm{BZ}} \nabla_{\boldsymbol{k}} \left(e^{-i\boldsymbol{k}\cdot\boldsymbol{R}}\theta_n(\boldsymbol{k})\right) + i\boldsymbol{R}\theta_n(\boldsymbol{k}) \tag{3.23}$$

$$= \langle n\boldsymbol{0}| \boldsymbol{r} |n\boldsymbol{R}\rangle - i\boldsymbol{R} \sum_{\boldsymbol{k}\in \mathrm{BZ}} \theta_n(\boldsymbol{k}), \tag{3.24}$$

where the second term in Eq. (3.23) vanished identically since it is a pure surface term. As one can see it is not gauge invariant; nonetheless, for the diagonal term ($\boldsymbol{R} = 0$) the second term vanishes and we obtain

$$\langle n\boldsymbol{0}'| \boldsymbol{r} |n\boldsymbol{0}'\rangle = \langle n\boldsymbol{0}| \boldsymbol{r} |n\boldsymbol{0}\rangle = \boldsymbol{\tau}_n. \tag{3.25}$$

Therefore, the diagonal term is gauge invariant. The quantity $\boldsymbol{\tau}_n$ is often referred to as the *Wannier center*. Physically, it can be understood as where the Wannier function is centered around in the real-space lattice of the crystal. Furthermore, it can be understood as an analog of the center of mass.

It seems that we have gained nothing by switching to the Wannier functions but let's work in the opposite direction. Suppose that we already have the matrix elements $\langle n\mathbf{0}| H |n\boldsymbol{R}\rangle$ and $\langle n\mathbf{0}| \boldsymbol{r} |n\boldsymbol{R}\rangle$. Indeed, those elements can be obtained from either an *ab initio* calculation followed by a Wannierization procedure [39, 40] or derived from analytical tight-binding models. Furthermore, since we are interested in the optical response of the materials we only need those two sets of matrix elements. We will focus on the position operator since it is where the majority of the problems are. In momentum space (as we have showed earlier) it can be simply obtained as

$$\boldsymbol{\mathcal{A}}_n^{(W)}(\boldsymbol{k}) = \sum_{\boldsymbol{R}} e^{i\boldsymbol{k}\cdot\boldsymbol{R}} \langle n\mathbf{0}| \boldsymbol{r} |n\boldsymbol{R}\rangle, \tag{3.26}$$

where the $(W)$ superscript stands for Wannier. To see the usefulness of this representation let us make the following approximation

$$\langle n\mathbf{0}| \boldsymbol{r} |n\boldsymbol{R}\rangle = \delta_{\mathbf{0},\boldsymbol{R}}\boldsymbol{\tau}_n. \tag{3.27}$$

Such approximation is known as the *tight-binding approximation*. The Berry connection is then simply $\boldsymbol{\mathcal{A}}_n^{(W)}(\boldsymbol{k}) = \boldsymbol{\tau}_n$ and thus gauge-invariant. While it is may seem rather crude, it does capture the qualitative physics of the optical response of solids [12, 41–43]. However, for very specific materials one should go beyond this approximation [44]. Therefore, with a minimal $\boldsymbol{r}$ operator (recall that we are only keeping the diagonal terms), we obtain a faithful optical response.

Nevertheless, even in the general case Eq. (3.26) is well-defined and free of singularities. In the case of a trivial system, the Wannier funciton will be then exponentially localized [34]. Therefore, the Berry connection, $\boldsymbol{\mathcal{A}}_n^{(W)}(\boldsymbol{k})$, will not have any kind of ill-defined phase, even though there are topological obstructions to construct Wannier functions in a non-trivial system [45–47]. Such obstruction is intrinsically linked to the fact that we cannot construct a smooth gauge in momentum space. Does this mean that we cannot use this approach for topological systems? The answer is that yes, we can but one should proceed with care. The main idea is that instead of trying to Wannierize the whole band structure, look for subsets of bands that independently have vanishing Chern number. Then, disentangle them using the maximally localized Wannier function procedure [39].

However, this approach is not empty of drawbacks. As we are working in the diabatic basis (the Wannier one), the Hamiltonian will not be diagonal, as for the diatomic molecule example. Nevertheless, that is not a major issue; while a diagonal matrix is typically desired for numerical propagation, having a denser matrix without any kind of numerical instabilities is often preferred.

Another important concern is where all the singularities have gone. However, let us first formalize our discussion. So far, we have only dealt with the case of non-degenerate bands, but it is time to move to the general case. In the multiband case, the Berry connection is defined as

$$\boldsymbol{\mathcal{A}}_{mn}(\boldsymbol{k}) = i \langle u_{m\boldsymbol{k}} | \nabla_{\boldsymbol{k}} u_{n\boldsymbol{k}} \rangle, \tag{3.28}$$

and hence the gauge symmetry is lifted from $\mathcal{U}(1)$ to $\mathcal{U}(N)$ (for a discussion on why this is the case we refer the reader to the appendix of [35]) where $N$ is the number of bands. Thus, we have moved from an abelian connection to a non-abelian one, increasing the complexity of the gauge freedom. Nevertheless, all our previous discussions hold true, with the only exception that now, one needs to consider couplings between different bands. More formally, this means that we will need $\langle m\mathbf{0}| H |n\boldsymbol{R}\rangle$ and $\langle m\mathbf{0}| \boldsymbol{r} |n\boldsymbol{R}\rangle$ to describe the system accurately. If we now wish to diagonalize the Hamiltonian,

$$H_{mn}^{(W)}(\boldsymbol{k}) = \sum_{\boldsymbol{R}} e^{i\boldsymbol{k}\cdot\boldsymbol{R}} \langle m\mathbf{0}| H |n\boldsymbol{R}\rangle, \tag{3.29}$$

one can see that is nothing more than moving to the Bloch basis [12, 39, 48]. Therefore, the Hamiltonian in the Bloch basis, denoted by the $(H)$ superscript, is simply

$$H^{(H)}(\boldsymbol{k}) = U^{\dagger}(\boldsymbol{k}) H^{(W)}(\boldsymbol{k}) U(\boldsymbol{k}), \tag{3.30}$$

where $U(\boldsymbol{k})$ is the unitary matrix that diagonalizes the Hamiltonian. Therefore, we will refer to this gauge as the Hamiltonian gauge. Both bases are related to each other as

$$\left|\psi_{n\boldsymbol{k}}^{(H)}\right\rangle = \sum_{m} U_{mn}(\boldsymbol{k}) \left|\psi_{m\boldsymbol{k}}^{(W)}\right\rangle. \tag{3.31}$$

The definition of Eq. (3.19) is still valid but the equivalence between the two gauges is simply given by

$$|n\boldsymbol{R}\rangle = \sum_{\boldsymbol{k}\in\mathrm{BZ}} e^{-i\boldsymbol{k}\cdot\boldsymbol{R}} \left|\psi_{n\boldsymbol{k}}^{(W)}\right\rangle = \sum_{\boldsymbol{k}\in\mathrm{BZ}} \sum_{m} U_{mn}^{\dagger}(\boldsymbol{k}) \left|\psi_{m\boldsymbol{k}}^{(H)}\right\rangle. \tag{3.32}$$

With regard to the Berry connection, it can be shown that it transforms as [48]

$$\boldsymbol{\mathcal{A}}^{(H)}(\boldsymbol{k}) = U^{\dagger}(\boldsymbol{k})\boldsymbol{\mathcal{A}}^{(W)}(\boldsymbol{k})U(\boldsymbol{k}) + iU^{\dagger}(\boldsymbol{k})\nabla_{\boldsymbol{k}}U(\boldsymbol{k}). \tag{3.33}$$

Since we have a gauge freedom on the Bloch states (Eq. (3.3)), it follows that the matrices $U(\boldsymbol{k})$ are not unique. This is where the phase problem encoded: while $\boldsymbol{\mathcal{A}}^{(W)}(\boldsymbol{k})$ is perfectly well-defined and smooth, when we rotate into the eigenbasis of the system we introduce the non-smoothness through the $iU^{\dagger}(\boldsymbol{k})\nabla_{\boldsymbol{k}}U(\boldsymbol{k})$ term due to its non-uniqueness. Physically what is happening is the following: in the Wannier gauge, $(W)$, we are fixing the position and the character of the electronic orbital (since it is the diabatic basis); however, when we move the Hamiltonian gauge, $(H)$, what we are fixing is the *ordering* of the energies (since it is the adiabatic basis) and hence, the physical character of the electronic states is left free; the character of the states are not completely fixed and the gauge freedom arises.

Using Wannier functions not only resolves this problem but also provides new physical insights. By employing localized atomic sites for valence states in conjunction with Bloch states for conduction states [49], a clear analogy with high-order harmonic generation from atomic gases emerges. Specifically, it is demonstrated that the primary contribution to the emission spectra comes from a recombination process where electrons return to the site from which they were initially ejected. This parallels the scenario observed in atomic gases. Initially, the electron is localized (whether in a Wannier function or an atomic orbital), then it tunnels into a delocalized state, and finally, it recollides with the parent ion.

## 3.3 Semiconductor Bloch Equations in the Wannier Gauge

To obtain the optical response of a system, one needs to compute the mean value of the current operator, $\boldsymbol{J}$, defined as

$$\boldsymbol{J} = -\frac{|e|}{i}\left[\boldsymbol{r}, H(t)\right] \tag{3.34}$$

which is a one-body operator, where $e$ is the electron charge. Therefore, all the information about one-body operators is contained within the reduced density matrix (RDM) [50, 51]

$$\rho_{nm}(\boldsymbol{k}, t) = \left\langle c^{\dagger}_{m\boldsymbol{k}} c_{n\boldsymbol{k}} \right\rangle, \tag{3.35}$$

where $c^\dagger_{m\boldsymbol{k}}/c_{n\boldsymbol{k}}$ is the electronic creation/annihilation operator of a Bloch state on the $n$-th band. The most general non-interacting Hamiltonian in the presence of an laser field, $\boldsymbol{E}(t)$, is

$$H(t) = \sum_{\boldsymbol{k}\in\mathrm{BZ}} \sum_{nm} c^\dagger_{m\boldsymbol{k}} H_{nm}(\boldsymbol{k}) c_{n\boldsymbol{k}} + |e|\boldsymbol{E}(t) \cdot c^\dagger_{m\boldsymbol{k}} \boldsymbol{r}_{mn}(\boldsymbol{k}) c_{n\boldsymbol{k}}, \tag{3.36}$$

where $H_{nm}(\boldsymbol{k})$ and $\boldsymbol{r}_{mn}(\boldsymbol{k})$ are, respectively, the matrix elements of the single-particle Hamiltonian and the position operator (Eq. (3.14)) in a momentum basis; it can be either the Wannier or the Bloch basis.

While the derivation of the equations of motion for the RDM is not complicated, it is rather lengthy. For a step-by-step derivation we refer the reader to [16]. Those equations are

$$\begin{aligned} i\partial_t \rho^{(G)}_{nm}(\boldsymbol{k},t) &= \left\langle [c^\dagger_{m\boldsymbol{k}} c_{n\boldsymbol{k}}, H(t)] \right\rangle = [H^{(G)}(\boldsymbol{k}), \rho^{(G)}(\boldsymbol{k},t)]_{mn} \\ &\quad + |e|\boldsymbol{E}(t) \cdot \left(i\nabla_{\boldsymbol{k}} \rho^{(G)}_{nm}(\boldsymbol{k},t) + [\boldsymbol{\mathcal{A}}^{(G)}(\boldsymbol{k}), \rho^{(G)}(\boldsymbol{k},t)]_{mn}\right), \end{aligned} \tag{3.37}$$

where $(G)$ denotes which gauge are we in, either the Wannier gauge $(W)$ or the Hamiltonian one $(H)$. This set of differential equations are also known in the literature as the SBEs [9,10]. In their original form, the SBEs treated the electronic density (diagonal terms of $\rho_{nm}$) and the electronic polarization (off-diagonal terms of $\rho_{nm}$) as different, yet coupled, quantities. However, within this formalism both quantities are treated in equal footing.

Nevertheless, the previous equations are not a complete description of the material optical response. This is because Eqs. (3.37) do not take into account many-body couplings such as electron-electron interactions and electron-phonon interaction or macroscopic effects. Such interactions lead to decoherence between electronic states. Hence, if one wants a realistic description of fast optical processes, decoherence must be taken into account. Indeed, it has been shown that if one wants to match the experimental findings, a *phenomenological* dephasing must be included [14,52,53]. Such term is included in the following way

$$\begin{aligned} i\partial_t \rho^{(H)}_{nm}(\boldsymbol{k},t) &= [H^{(H)}(\boldsymbol{k}), \rho^{(H)}(\boldsymbol{k},t)]_{mn} + i|e|\boldsymbol{E}(t) \cdot \nabla_{\boldsymbol{k}} \rho^{(H)}_{nm}(\boldsymbol{k},t) \\ &\quad + e|\boldsymbol{E}(t) \cdot [\boldsymbol{\mathcal{A}}^{(H)}(\boldsymbol{k}), \rho^{(H)}(\boldsymbol{k},t)]_{mn} - i\left(\frac{1-\delta_{mn}}{T_2}\right) \rho^{(H)}_{nm}(\boldsymbol{k},t), \end{aligned} \tag{3.38}$$

where $T_2$ is the dephasing time. It is important to stress that Eqs. (3.38) can only be included in the Hamiltonian gauge. Physically, this is rooted in the fact that in order to kill coherence terms (off-diagonal terms of $\rho_{nm}$), one needs to know which are the states that the system can really be in.

As for any differential equation, one must also specify one initial condition. It is completely safe to assume that the electrons in a solid are in thermal equilibrium before the application of the time-dependent perturbation. Hence our initial state is a thermal (and thus mixed) state of the form

$$\rho_{nm}^{(H)}(\boldsymbol{k}, t_0) = \delta_{nm} s_n p_{n\boldsymbol{k}}, \tag{3.39}$$

where $s_n = 1, 2$ takes into account spin degenerancy, $p_{n\boldsymbol{k}} = 1/(e^{[(\varepsilon_n(\boldsymbol{k})-\mu)/k_B T]})$ is the Fermi-Dirac distribution and $u$ is the chemical potential.

We are currently confronted with a dilemma. Both the dephasing term and the initial condition must be specified within the Hamiltonian gauge. However, it is desirable to work in the Wannier gauge. So, how can we reconcile these two sides? As for any other observable, we can always transform the RDM from one gauge to the other [11, 12]. Such relationship is given by

$$\rho_{nm}^{(H)}(\boldsymbol{k}, t) = \sum_{ab} U_{nb}^{\dagger}(\boldsymbol{k}) \rho_{ba}^{(W)}(\boldsymbol{k}, t) U_{am}(\boldsymbol{k}), \tag{3.40}$$

$$\rho_{nm}^{(W)}(\boldsymbol{k}, t) = \sum_{ab} U_{nb}(\boldsymbol{k}) \rho_{ba}^{(H)}(\boldsymbol{k}, t) U_{am}^{\dagger}(\boldsymbol{k}). \tag{3.41}$$

Several consequences arise from these transformations. First of all, they are *time-independent*; once we have constructed the matrices $U_{mn}(\boldsymbol{k})$, they can be used at any time without any increased cost. Therefore, whenever we need to switch to the Hamiltonian gauge (for example to obtain the populations of a desired band), we simply perform the transformation. Secondly, even though there is a gauge freedom in the choice of $U$, the initial state in the Wannier gauge is uniquely defined. This is because of the absence of coherences in the initial state, namely that $\left\langle c_{m\boldsymbol{k}}^{(H)\dagger} c_{n\boldsymbol{k}}^{(H)} \right\rangle = 0$ when $n \neq m$. However, this is not unique to the initial state, as for any time $t$, the matrix $\rho_{nm}^{(W)}(\boldsymbol{k}, t)$ is unequivocally defined. This relies on the fact that any phase constrained in $U_{mn}(\boldsymbol{k})$ will be canceled out by its Hermitian conjugate $U_{mn}^{\dagger}(\boldsymbol{k})$, therefore making the RDM in the Wannier gauge *robust under phase transformations*.

Another advantage is that for any observable of the form

$$O = \mathrm{Tr}\left(\hat{O}^{(G)} \rho^{(G)}\right), \tag{3.42}$$

it remains invariant under a gauge transformation due to the basis independence of the trace. Nevertheless, some observables have more physical meaning in one gauge than in the other. The best example for this behaviour is the current operator Eq. (3.34). Its matrix elements in a momentum basis are

$$\boldsymbol{J}_{mn}^{(G)}(\boldsymbol{k}) = -|e|\big(\nabla_{\boldsymbol{k}} H_{mn}^{(G)}(\boldsymbol{k}) - i[\boldsymbol{\mathcal{A}}^{(G)}(\boldsymbol{k}), H^{(G)}(\boldsymbol{k})]_{mn}\big). \tag{3.43}$$

However, in the Hamiltonian gauge (where $H_{mn}^{(G)}(\boldsymbol{k})$ really represents the energies of the system) the two contributions can be asigned distinct physical meaning. These two terms will be the intraband and interband terms:

$$\boldsymbol{J}_{\text{intra},mn}^{(H)}(\boldsymbol{k}) = -|e|\nabla_{\boldsymbol{k}} H_{mn}^{(H)}(\boldsymbol{k}), \tag{3.44}$$

$$\boldsymbol{J}_{\text{inter},mn}^{(H)}(\boldsymbol{k}) = i|e|[\boldsymbol{\mathcal{A}}^{(H)}(\boldsymbol{k}), H^{(H)}(\boldsymbol{k})]_{mn}. \tag{3.45}$$

We want to stress that while none of these terms are physical observables [54, 55], they do provide physical insight.

## 3.4 Final Remarks

So far, we have shown the advantages of moving to the Wannier gauge to obtain the optical response of solids. We started by signaling the roots of the gauge freedom problem in solids as both a practical and a physical problem. Then, we showed that this was caused because we were working on the basis that the physical character of the states was not respected: the adiabatic basis. By then moving to a basis where such physical character was preserved (the Wannier basis), we showed that we could get rid of the phase problem in a general manner. Lastly, we formulated the SBEs in such gauge, and demonstrated that it is not only a useful approach but also an efficient one, numerically speaking.

While the Wannier gauge was originally developed for computing time-independent observables, such as the anomalous quantum Hall current in [48], its application to the SBEs was established in the paper by [12]. It was then shown to be resolute when it comes to unveiling new physics. Relevant examples include: the detection of topological phases using high-order harmonic generation [56], lightwave engineering of topological properties [57], observation of attosecond dynamics in solids [58], tracking photo-induced dynamics in complex materials [59–61], observation of light-driven band structure by high harmonic spectroscopy [62], and lastly, the possibility of an all-optical characterization of magic-angle twisted bilayer graphene [63].

Several extensions has been applied to this method including: its extension to the interacting case [13], which has led to the discovery of sub-cycle dynamics of excitons under strong laser fields [64]; its formulation in the velocity gauge [65] and also its formulation in a semiclassical picture [66].

## Acknowledgments

E. B. M. and R. E. F. S. acknowledge support from the fellowship LCF/BQ/PR21/11840008 from "La Caixa" Foundation (ID 100010434). This research was supported by Grant PID2021-122769NB-I00 funded by MCIN/AEI/10.13039/501100011033. A. J. G. acknowledges financial support from the Comunidad de Madrid through TALENTO Grant 2022-T1/IND-24102.

## References

[1] S. Y. Kruchinin, F. Krausz and V. S. Yakovlev, Colloquium: Strong-field phenomena in periodic systems, *Reviews of Modern Physics* **90**, 2, p. 021002 (2018).

[2] S. Ghimire and D. A. Reis, High-harmonic generation from solids, *Nature Physics* **15**, 1, pp. 10–16 (2018).

[3] E. Goulielmakis and T. Brabec, High harmonic generation in condensed matter, *Nature Photonics* **16**, 6, pp. 411–421 (2022).

[4] M. A. Marques, C. A. Ullrich, F. Nogueira, A. Rubio, K. Burke and E. K. U. Gross, *Time-Dependent Density Functional Theory*. Springer (2006).

[5] M. Marques, N. Maitra, F. Nogueira, E. Gross and A. Rubio, *Fundamentals of Time-Dependent Density Functional Theory*. Springer (2012).

[6] G. Onida, L. Reining and A. Rubio, Electronic excitations: Density-functional versus many-body Green's-function approaches, *Reviews of Modern Physics* **74**, pp. 601–659 (2002).

[7] N. Tancogne-Dejean, M. A. Sentef and A. Rubio, Ultrafast modification of Hubbard $U$ in a strongly correlated material: Ab initio high-harmonic generation in NiO, *Physical Review Letters* **121**, p. 097402 (2018).

[8] N. Tancogne-Dejean and A. Rubio, Atomic-like high-harmonic generation from two-dimensional materials, *Science Advances* **4**, 2 (2018).

[9] M. Lindberg and S. W. Koch, Effective Bloch equations for semiconductors, *Physical Review B* **38**, pp. 3342–3350 (1988).

[10] H. Haug and S. Koch, *Quantum Theory Of The Optical And Electronic Properties Of Semiconductors (5th Edition)*. World Scientific (2009).

[11] G. B. Ventura, D. J. Passos, J. M. B. L. dos Santos, J. M. V. P. Lopes and N. M. R. Peres, Gauge covariances and nonlinear optical responses, *Physical Review B* **96**, 3 (2017).
[12] R. E. F. Silva, F. Martín and M. Ivanov, High harmonic generation in crystals using maximally localized Wannier functions, *Physical Review B* **100**, 19 (2019).
[13] G. Cistaro, M. Malakhov, J. J. Esteve-Paredes, A. J. Uría-Álvarez, R. E. F. Silva, F. Martín, J. J. Palacios and A. Picón, Theoretical approach for electron dynamics and ultrafast spectroscopy (EDUS), *Journal of Chemical Theory and Computation* **19**, 1, pp. 333–348 (2022).
[14] L. Yue and M. B. Gaarde, Introduction to theory of high-harmonic generation in solids: tutorial, *Journal of the Optical Society of America B* **39**, 2, p. 535 (2022).
[15] L. Yue and M. B. Gaarde, Structure gauges and laser gauges for the semiconductor Bloch equations in high-order harmonic generation in solids, *Physical Review A* **101**, 5 (2020).
[16] G. B. Ventura, *Nonlinear optical response of two-dimensional crystals*, Ph.D. thesis, Universidade do Porto (2021), https://repositorio-aberto.up.pt/bitstream/10216/139780/2/531840.pdf.
[17] P. W. Anderson, *Concepts in Solids: Lectures on the Theory of Solids*, Vol. 58. World Scientific (1997).
[18] M. P. Marder, *Condensed Matter Physics*. John Wiley & Sons (2010).
[19] M. Z. Hasan and C. L. Kane, Colloquium: Topological insulators, *Reviews of Modern Physics* **82**, pp. 3045–3067 (2010).
[20] X.-L. Qi and S.-C. Zhang, Topological insulators and superconductors, *Reviews of Modern Physics* **83**, pp. 1057–1110 (2011).
[21] A. Bansil, H. Lin and T. Das, Colloquium: Topological band theory, *Reviews of Modern Physics* **88**, p. 021004 (2016).
[22] B. A. Bernevig, T. L. Hughes and S.-C. Zhang, Quantum spin Hall effect and topological phase transition in HgTe quantum wells, *Science* **314**, 5806, pp. 1757–1761 (2006).
[23] A. Altland and M. R. Zirnbauer, Nonstandard symmetry classes in mesoscopic normal-superconducting hybrid structures, *Physical Review B* **55**, pp. 1142–1161 (1997).
[24] C.-K. Chiu, J. C. Y. Teo, A. P. Schnyder and S. Ryu, Classification of topological quantum matter with symmetries, *Reviews of Modern Physics* **88**, p. 035005 (2016).
[25] M. V. Berry, Quantal phase factors accompanying adiabatic changes, *Proceedings of the Royal Society of London Series A* **392**, 1802, pp. 45–57 (1984).
[26] M. Nakahara, *Geometry, Topology and Physics*. CRC Press (2018).
[27] M. Kohmoto, Topological invariant and the quantization of the Hall conductance, *Annals of Physics* **160**, 2, pp. 343–354 (1985).
[28] D. Vanderbilt, *Berry Phases in Electronic Structure Theory: Electric Polarization, Orbital Magnetization and Topological Insulators*. Cambridge University Press (2018).

[29] P. A. M. Dirac, Quantised singularities in the electromagnetic field, *Proceedings of the Royal Society of London A* **133**, 821, pp. 60–72 (1931).

[30] T. T. Wu and C. N. Yang, Concept of nonintegrable phase factors and global formulation of gauge fields, *Physical Review D* **12**, pp. 3845–3857 (1975).

[31] M. Araújo and P. Sacramento, *Topology in Condensed Matter.* World Scientific (2021).

[32] C. Joachain, N. Kylstra and R. Potvliege, *Atoms in Intense Laser Fields.* Cambridge University Press (2012).

[33] M. Ivanov, *Ionization in Strong Low-Frequency Fields*, chap. 6. John Wiley & Sons, pp. 177–200 (2014).

[34] E. Blount, Formalisms of band theory, in *Solid State Physics.* Elsevier, pp. 305–373 (1962).

[35] D. E. Parker, T. Morimoto, J. Orenstein and J. E. Moore, Diagrammatic approach to nonlinear optical response with application to Weyl semimetals, *Physical Review B* **99**, 4, p. 045121 (2019).

[36] A. Chacón, D. Kim, W. Zhu, S. P. Kelly, A. Dauphin, E. Pisanty, A. S. Maxwell, A. Picón, M. F. Ciappina, D. E. Kim, C. Ticknor, A. Saxena and M. Lewenstein, Circular dichroism in higher-order harmonic generation: Heralding topological phases and transitions in Chern insulators, *Physical Review B* **102**, p. 134115 (2020).

[37] D. Tannor, *Introduction to Quantum Mechanics.* University Science Books (2007).

[38] C. Zener, Non-adiabatic crossing of energy levels, *Proceedings of the Royal Society of London A* **137**, 833, pp. 696–702 (1932).

[39] N. Marzari, A. A. Mostofi, J. R. Yates, I. Souza and D. Vanderbilt, Maximally localized Wannier functions: Theory and applications, *Reviews of Modern Physics* **84**, pp. 1419–1475 (2012).

[40] A. A. Mostofi, J. R. Yates, G. Pizzi, Y.-S. Lee, I. Souza, D. Vanderbilt and N. Marzari, An updated version of wannier90: A tool for obtaining maximally-localised Wannier functions, *Computer Physics Communications* **185**, 8, pp. 2309–2310 (2014).

[41] M. Graf and P. Vogl, Electromagnetic fields and dielectric response in empirical tight-binding theory, *Physical Review B* **51**, pp. 4940–4949 (1995).

[42] T. G. Pedersen, K. Pedersen and T. B. Kriestensen, Optical matrix elements in tight-binding calculations, *Physical Review B* **63**, 20 (2001).

[43] B. A. Foreman, Consequences of local gauge symmetry in empirical tight-binding theory, *Physical Review B* **66**, p. 165212 (2002).

[44] J. Ibañez-Azpiroz, F. de Juan and I. Souza, Assessing the role of interatomic position matrix elements in tight-binding calculations of optical properties, *SciPost Physics* **12**, 2 (2022).

[45] D. J. Thouless, Wannier functions for magnetic sub-bands, *Journal of Physics C: Solid State Physics* **17**, 12, pp. L325–L327 (1984).

[46] T. Thonhauser and D. Vanderbilt, Insulator/Chern-insulator transition in the haldane model, *Physical Review B* **74**, p. 235111 (2006).

[47] A. A. Soluyanov and D. Vanderbilt, Wannier representation of $Z_2$ topological insulators, *Physical Review B* **83**, p. 035108 (2011).

[48] X. Wang, J. R. Yates, I. Souza and D. Vanderbilt, Ab-initio calculation of the anomalous Hall conductivity by Wannier interpolation, *Physical Review B* **74**, 19 (2006).

[49] E. N. Osika, A. Chacón, L. Ortmann, N. Suárez, J. A. Pérez-Hernández, B. Szafran, M. F. Ciappina, F. Sols, A. S. Landsman and M. Lewenstein, Wannier-Bloch approach to localization in high-harmonics generation in solids, *Physical Review X* **7**, p. 021017 (2017).

[50] H. Bruus and K. Flensberg, *Many-Body Quantum Theory in Condensed Matter Physics: An Introduction*. Oxford University Press (2004).

[51] P. Coleman, *Introduction to Many-Body Physics*. Cambridge University Press (2015).

[52] G. Vampa, C. R. McDonald, G. Orlando, D. D. Klug, P. B. Corkum and T. Brabec, Theoretical analysis of high-harmonic generation in solids, *Physical Review Letters* **113**, p. 073901 (2014).

[53] I. Floss, C. Lemell, G. Wachter, V. Smejkal, S. A. Sato, X.-M. Tong, K. Yabana and J. Burgdorfer, Ab initio multiscale simulation of high-order harmonic generation in solids, *Physical Review A* **97**, p. 011401 (2018).

[54] P. Foldi, Gauge invariance and interpretation of interband and intraband processes in high-order harmonic generation from bulk solids, *Physical Review B* **96**, p. 035112 (2017).

[55] G. Ernotte, T. J. Hammond and M. Taucer, A gauge-invariant formulation of interband and intraband currents in solids, *Physical Review B* **98**, p. 235202 (2018).

[56] R. E. F. Silva, Á. Jiménez-Galán, B. Amorim, O. Smirnova and M. Ivanov, Topological strong-field physics on sub-laser-cycle timescale, *Nature Photonics* **13**, 12, pp. 849–854 (2019).

[57] Á. Jiménez-Galán, R. E. F. Silva, O. Smirnova and M. Ivanov, Lightwave control of topological properties in 2D materials for sub-cycle and non-resonant valley manipulation, *Nature Photonics* **14**, 12, pp. 728–732 (2020).

[58] A. J. Uzan, G. Orenstein, A. Jimenez-Galan, C. McDonald, R. E. F. Silva, B. D. Bruner, N. D. Klimkin, V. Blanchet, T. Arusi-Parpar, M. Kruger, A. N. Rubtsov, O. Smirnova, M. Ivanov, B. Yan, T. Brabec and N. Dudovich, Attosecond spectral singularities in solid-state high-harmonic generation, *Nature Photonics* **14**, 3, pp. 183–187 (2020).

[59] M. R. Bionta, E. Haddad, A. Leblanc, V. Gruson, P. Lassonde, H. Ibrahim, J. Chaillou, N. Émond, M. R. Otto, A. Jiménez-Galán, R. E. F. Silva, M. Ivanov, B. J. Siwick, M. Chaker and F. M. C. Légaré, Tracking ultrafast solid-state dynamics using high harmonic spectroscopy, *Physical Review Research* **3**, p. 023250 (2021).

[60] E. B. Molinero and R. E. F. Silva, Coherent control of the photoinduced transition in a strongly correlated material, *Physical Review Research* **4**, p. 043198 (2022).

[61] Z. Nie, L. Guery, E. B. Molinero, P. Juergens, T. J. van den Hooven, Y. Wang, A. Jimenez Galan, P. C. M. Planken, R. E. F. Silva and P. M. Kraus, Following the nonthermal phase transition in niobium dioxide

by time-resolved harmonic spectroscopy, *Physical Review Letters* **131**, p. 243201 (2023).

[62] A. J. Uzan-Narovlansky, Á. Jiménez-Galán, G. Orenstein, R. E. F. Silva, T. Arusi-Parpar, S. Shames, B. D. Bruner, B. Yan, O. Smirnova, M. Ivanov and N. Dudovich, Observation of light-driven band structure via multiband high-harmonic spectroscopy, *Nature Photonics* **16**, 6, pp. 428–432 (2022).

[63] E. B. Molinero, A. Datta, M. J. Calderón, E. Bascones and R. E. F. Silva, High-harmonic generation with a twist: All-optical characterization of magic-angle twisted bilayer graphene, *Optica* **11**, 2, pp. 171–175 (2024).

[64] E. B. Molinero, B. Amorim, M. Malakhov, G. Cistaro, Á. Jiménez-Galán, M. Ivanov, A. Picón, P. San-José and R. E. F. Silva, Formation, stability, and highly nonlinear optical response of excitons to intense light fields interacting with two-dimensional materials, *arXiv preprint arXiv:2307.16647* (2023).

[65] Y. Zhang, M. Zhang, W. Yang, H. Yu, M. S. Si, S. Xue and H. Du, Defects of the nearest-neighbor tight-binding model in the study of solid harmonics, *Physical Review A* **108**, p. 043508 (2023).

[66] A. M. Parks, G. Ernotte, A. Thorpe, C. R. McDonald, P. B. Corkum, M. Taucer and T. Brabec, Wannier quasi-classical approach to high harmonic generation in semiconductors, *Optica* **7**, 12, pp. 1764–1772 (2020).

https://doi.org/10.1142/9789811279560_0004

# Chapter 4

# Theory of Topological Effects on High-Order Harmonic Generation in Solids

Hannah Jürß, Helena Drüeke, Dieter Bauer and Francisco Navarrete

*Institute of Physics, University of Rostock, 18051 Rostock, Germany*

## 4.1 Introduction

Studies on high-order harmonic generation (HHG) in solids have grown substantially since its experimental determination [1], which has motivated numerous contributions both theoretically and experimentally. Independently there has been a growing interest in modern condensed matter physics concerning geometrical phases and topology [2–4]. This scenario naturally raised the question of whether topological effects can be probed in solids using strong-field techniques such as HHG spectroscopy.

In a pioneer theoretical analysis of topological effects on HHG in solids [5], finite one-dimensional systems were analyzed (instead of infinite periodic solids, i.e., bulk properties), which led to the finding that two materials with the same bulk structure but different configuration of their edges[1] might lead to one phase with and one without edge states,[2] which in turn implies dramatic differences in the HHG yield for each phase which are very robust [6], as we will elucidate later on in this chapter. The existence of differences in the physical effects observed in each phase is deemed topological because, apart from the rearrangement of the edges, both have the same bulk structure. These two phases are qualitatively different, i.e., they are characterized by a different topological invariant. Here it is worth

[1] This qualitative difference defines two different "phases" of the same material.

[2] The fact that the non-trivial phase has edge states is a band-structure property known as bulk-boundary correspondence.

highlighting that it has been found [7] that in one-dimensional systems the presence or absence of edge states in the band structure is more relevant for the observation of strong-field topological effects than the topological invariant (i.e., the specific phase). Nevertheless, for higher dimensional systems (e.g., Haldanite, which is bidimensional), topological invariants start playing a major role in the characteristics of the HHG spectrum, even when edge properties are neglected [8]. However, the influence of a topological phase transition on the harmonic spectrum strongly depends on the material itself. So far, no universal feature from a topological phase transition in HHG was found [9].

Before we start our theoretical analysis of the topological effects on HHG in solids, it is timely to describe their experimental realizations as well as the potential technological applications they might enable. Regarding their experimental determination, it has been found in the laboratory that topological phases can be measured through circular dichroism [10]. Also, studies on $Bi_2T_3$ found that its HHG spectrum, which depends on the carrier-envelope phase and is composed of non-integer frequencies, is generated by topological surface states [11]. Besides, experiments on $Bi_2Se_3$ [12, 13] and $BiSbTeSe_2$ [14] have explored the dependence of surface and bulk harmonic spectra on the orientation and polarization of the driving laser. Concerning applications, among the promising approaches exploring the mergers of modern condensed matter physics and strong-field attosecond science so far, we can mention light-driven ultrafast currents [15–18], in particular those topologically protected [19], ultrafast valleytronics [20] and controlling topological properties via laser dressing [21, 22]. Finally, specifically related to the HHG process, it has been found that hexagonal nanoribbons in two different edge configurations might produce harmonics with different polarization patterns [23] and that the helicity of the emitted photons is determined by the topological phase of the system [8, 24–26], which might be useful for the production of complex-shaped harmonics. Furthermore, the surface of three-dimensional topological insulators emits harmonics with a different dependence on the ellipticity than that of the bulk [12], which might allow studying helical surface currents on topological materials with implications for light-wave-driven nanoelectronics. The above-mentioned experimental evidence and technological potential represent a further motivation to continue the exploration of this exciting subject.

## 4.2 HHG in One-Dimensional Topological Systems

In our description of topological effects on HHG in solids, we will start by describing simple systems before going to targets of increasing complexity, both in dimensionality and constituents. This approach allows us to individualize purely topological effects from other intervening mechanisms.

### 4.2.1 *Su-Schrieffer-Heeger Chain without Electromagnetic Fields*

The simplest system to investigate is a one-dimensional one, such as the Su-Schrieffer-Heeger (SSH) chain [27], sketched in Fig. 4.1.

First conceived of as a model for polyacetylene, the SSH chain's unit cell is a dimer with average lattice parameter $a$. We label one site with odd index $i$ and the other with even index $i$. By shifting alternatingly to the right (sites with odd $i$) and to the left (sites with even $i$) by $\delta$ the system undergoes a Peierls transition.

If $\delta > 0$, the intracell spacing $a - 2\delta$ is smaller than the intercell spacing $a + 2\delta$, all sites are bound in pairs, and the chain is in its trivial insulator phase. The topological insulator phase occurs for $\delta < 0$ and exhibits unpaired sites at both edges. At $\delta = 0$, the system is metallic, but this phase is not stable. We use density functional theory to find the energetically most favorable $\delta > 0$ for the trivial phase and $\delta < 0$ for the topological phase [6]. In the following, we investigate the trivial and topological phases at these energetic minima.

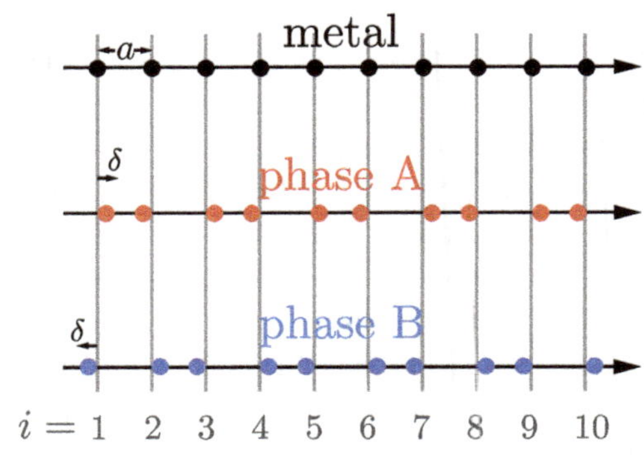


Fig. 4.1 Su-Schrieffer-Heeger chain in the metallic, trivial (phase A), and topological (phase B) configuration. Reproduced with permission from [6].

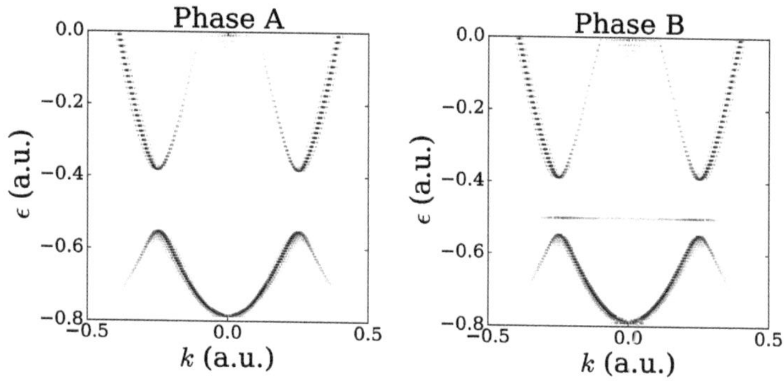


Fig. 4.2 Eigenstates of the SSH chain in the trivial (A) and topological (B) configuration. Reproduced with permission from [6].

The eigenstates are shown in reciprocal ($k$-)space in Fig. 4.2. Both the trivial and the topological insulator phases exhibit a fully occupied valence band and a fully unoccupied conduction band, with a band gap between them. The defining feature of the topological phase is the existence of two energetically degenerate edge states within the band gap. One of these edge states is occupied, and the other one is not. The edge states are located on the unpaired sites at the ends of the chain in real space.

### 4.2.2 *HHG in the Su-Schrieffer-Heeger Chain*

Here we show calculated HHG spectra from the interaction of an intense infrared laser with a one-dimensional SSH chain. Due to the dimensionality of the system for which the calculations are made, in higher dimensions, these results would correspond either to a linearly polarized laser along the chain, as well as bidimensional targets when both the target and the field share the same symmetries [28] (e.g., a circularly symmetric chain interacting with a circularly polarized field).

Looking at the harmonic spectra of the trivial and topological phase in Fig. 4.3, we see a dip in the trivial spectrum, which extends up to the 22nd harmonic. This is the expected behavior because the minimum band gap in the trivial phase corresponds to $\simeq$22 times the energy of an incoming photon. Harmonics below this threshold are due to time-dependent populations and anharmonic intraband motion. In the topological phase, the presence of the two degenerate edge states strongly affects its HHG spectra because (i) there is one electron in the edge states, which sees a reduced band gap

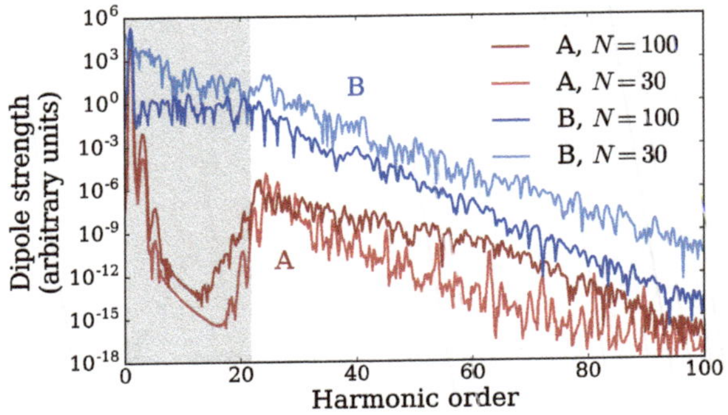


Fig. 4.3 High-order harmonic spectra of the trivial (A) and topological (B) phase of the SSH chain for $N = 100$ and $N = 30$ ions. The gray shade indicates the sub-band-gap regime of the trivial phase. Reproduced with permission from [6].

to the conduction band, and (ii) the edge states can serve as a stepping stone for other electrons in the valence band. Fact (i) explains why, in the topological phase, strong harmonics are already seen well below harmonic order 22, and fact (ii) why the harmonic yield is higher for all harmonic orders.

As shown in Fig. 4.3, these topological effects hold even for short chains of $N = 30$ lattice sites. They are also robust against disorder in the sites' positions and potentials [6].

### 4.2.3 *Tight-Binding Approach*

Calculating strong-field phenomena for finite systems using time-dependent density functional theory can become quite computationally expensive for larger systems and higher dimensions. It is computationally much cheaper to use a tight-binding approach as we no longer have to resolve the wave functions on a fine spatial grid in between and around the sites. For the SSH chain above, the different distances $a \pm 2\delta$ have the effect that the tunneling amplitude between neighboring sites is larger for a distance $a - 2\delta$ than for a distance $a + 2\delta$. In the tight-binding approach, this different "hopping" between sites is put by hand, using hopping amplitudes $v$ and $w$. We note in passing that in the original SSH work [27] the tight-binding description was used for the electronic degrees of freedom. We applied the tight-binding approach to HHG in SSH chains [7] and obtained results that are very similar to those using time-dependent density function theory.

## 4.3 HHG in Two-Dimensional Topological Insulators

What makes topological insulators unique is the coexistence of an insulating bulk and conducting edges. However, one-dimensional materials only have zero-dimensional edges while an electronic edge current requires at least a one-dimensional edge, which can be present in materials with two dimensions or more.

One conceptually important two-dimensional topological insulator derives from the Haldane model [29], which we refer to as Haldanite. It was originally proposed as a toy model but was later realized using cold atoms [30]. Photonic waveguides also allow realizations of topological insulators [31–33]. Extensions of Haldanite that take spin and spin-orbit coupling into account bring the model closer to naturally existing condensed matter systems [34, 35].

### 4.3.1 *Static Haldane System*

Haldanite has a hexagonal structure and consists of two different atomic elements, similar to hexagonal boron nitride. Figure 4.4 shows three different finite Haldanite systems with different shapes. The different

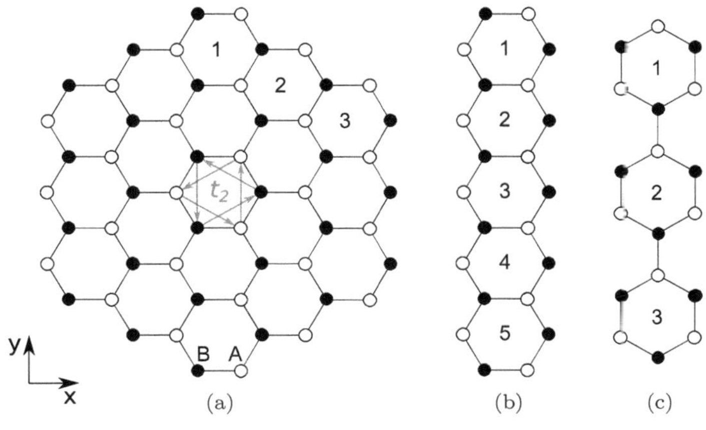


Fig. 4.4 (a) Sketch of a finite Haldanite flake with highlighted sublattice sites A (unfilled circles) and B (filled circles), nearest neighbor hopping by a solid line and directed next-nearest neighbor hopping by the arrows. The arrows are only shown within one hexagon for visibility. The size of the flake is defined by the number of hexagons along one edge, here $N = 3$, indicated by the numbers along the edge. (b) Finite nanoribbon with zig-zag edges of size $N_{\text{hex}} = 5$. (c) Finite nanoribbon with armchair edges of size $N_{\text{hex}} = 3$. The laser field is linearly polarized in $y$-direction for all systems (a–c).

elements are indicated by filled (sublattice site A) and unfilled (sublattice site B) circles and result in different on-site potentials, $M$ for site A and $-M$ for site B. Those different atoms open a band gap between the valence and the conduction band, which makes the solid insulating. In Ref. [36] it was shown that the tight-binding approximation is sufficient to calculate HHG in such materials by comparing to results obtained with real-space time-dependent Schrödinger calculations [23].

To make the system topologically non-trivial, a complex next-nearest neighbor hopping is introduced. In Fig. 4.4(a), the arrows indicate the complex hoppings: along the arrows, the hopping is given by $t_2$, in the opposite direction by $t_2^*$. The complex hopping breaks time-reversal symmetry, which is essential for a topological insulator [29]. External magnetic fields break time-reversal symmetry, which is key to understanding the robustness of the resistivity plateaus in the quantum Hall effect. Spin-orbit coupling can also break time-reversal symmetry [35], leading to the quantum spin Hall effect.

Edge states are present only when considering finite systems. Those finite systems can have different edge configurations. The two most fundamental ones are zig-zag and armchair edges. Nanoribbons with zig-zag and armchair edges are shown in Fig. 4.4(b) and (c), respectively. Graphene nanoribbons (i.e., $M = t_2 = 0$) that are wider (i.e., more extended in $x$-direction) can already show topological effects [37,38]. However, this chapter will focus on topological effects from the complex next-nearest neighbor hopping, i.e., from Haldanite.

### 4.3.2 *Generation of High-Order Harmonics*

#### 4.3.2.1 *Influence to the Emitted Photon Helicity*

In a semiclassical picture, the electron (and hole) need to move according to the band structure given by $E_n(\boldsymbol{k})$ ($n$ denotes the band). The electron (and hole) are shifted from their original momentum $\boldsymbol{k}_0$ in the presence of an external field to $\boldsymbol{k}(t) = \boldsymbol{k}_0 + \boldsymbol{A}(t)$, where $\boldsymbol{A}(t)$ is the vector potential. For the presented results, the vector potential is linearly polarized in the $y$-direction, see Fig. 4.4. Additionally, the dynamical processes of electrons and holes are determined by the Berry curvature $\boldsymbol{\Omega}_n(\boldsymbol{k}(t))$, which can be influenced by the topology. The semiclassical expression for the velocity in solids is given by [39, 40]

$$\dot{\boldsymbol{r}}(t) = \boldsymbol{\nabla}_{\boldsymbol{k}} E(\boldsymbol{k}(t)) + \boldsymbol{E}(t) \times \boldsymbol{\Omega}_n(\boldsymbol{k}(t)), \tag{4.1}$$

where

$$\tilde{\boldsymbol{d}} = \boldsymbol{d}(\boldsymbol{k} + \boldsymbol{A}(t)),$$

and

$$E_\pm(\tilde{\boldsymbol{d}})\mathbf{C}_\pm(\tilde{\boldsymbol{d}}) = \mathbf{H}(\tilde{\boldsymbol{d}})\mathbf{C}_\pm(\tilde{\boldsymbol{d}}).$$

With one additional substitution that removes the diagonal elements in the Hamiltonian,

$$\eta_{\boldsymbol{k}\pm}(t) = \alpha_{\boldsymbol{k}\pm}(t)\, e^{i\int^t (E_\pm(\tilde{\boldsymbol{d}}_{t'}) - i\mathbf{C}_\pm^\dagger(\tilde{\boldsymbol{d}}_{t'})\cdot\dot{\mathbf{C}}_\pm(\tilde{\boldsymbol{d}}_{t'})\,dt'},$$

we arrive at the equation of motion

$$i\begin{pmatrix}\dot{\eta}_{+\boldsymbol{k}}(t)\\ \dot{\eta}_{-\boldsymbol{k}}(t)\end{pmatrix} = \begin{pmatrix}0 & \Xi^*(\tilde{\boldsymbol{d}})\\ \Xi(\tilde{\boldsymbol{d}}) & 0\end{pmatrix}\begin{pmatrix}\eta_{+\boldsymbol{k}}(t)\\ \eta_{-\boldsymbol{k}}(t)\end{pmatrix} \tag{4.3}$$

where

$$\begin{aligned}\Xi(\tilde{\boldsymbol{d}}) &= -\mathcal{A}_{-+}(\tilde{\boldsymbol{d}})\, e^{-i\int^t (\Delta E(\boldsymbol{d}_{t'}) - \Delta\mathcal{A}(\boldsymbol{d}_{t'}))\,dt'}\\ \Delta E(\tilde{\boldsymbol{d}}) &= E_+(\tilde{\boldsymbol{d}}) - E_-(\tilde{\boldsymbol{d}}) = 2|\tilde{\boldsymbol{d}}|, \quad \Delta\mathcal{A}(\tilde{\boldsymbol{d}}) = \mathcal{A}_{++}(\tilde{\boldsymbol{d}}) - \mathcal{A}_{--}(\tilde{\boldsymbol{d}})\\ \mathcal{A}_{\pm\pm}(\tilde{\boldsymbol{d}}) &= i\mathbf{C}_\pm^\dagger(\tilde{\boldsymbol{d}})\cdot\dot{\mathbf{C}}_\pm(\tilde{\boldsymbol{d}}), \quad \mathcal{A}_{\pm\mp}(\tilde{\boldsymbol{d}}) = i\mathbf{C}_\pm^\dagger(\tilde{\boldsymbol{d}})\cdot\dot{\mathbf{C}}_\mp(\tilde{\boldsymbol{d}}).\end{aligned}$$

While the appearance of the energy difference $\Delta E(\tilde{\boldsymbol{d}})$ is expected (whenever one transforms away the diagonal elements in a Hamiltonian), one sees how the ansatz using adiabatically following eigenstates bring the Berry connections $\mathcal{A}_{\pm\pm}$ and $\mathcal{A}_{\pm\mp}$ into play.

The expectation value $v_i$ for the velocity in $k_i$-direction reads

$$\begin{aligned}v_i(\boldsymbol{k}, t) &= \mathbf{C}^\dagger(\boldsymbol{v}, t)\,\mathbf{v}_i(\tilde{\boldsymbol{d}})\mathbf{C}(\boldsymbol{v}, t)\\ &= v_i^{--}(\boldsymbol{k}, t) + v_i^{++}(\boldsymbol{k}, t) + v_i^{-+}(\boldsymbol{k}, t) + v_i^{+-}(\boldsymbol{k}, t)\end{aligned} \tag{4.4}$$

with the velocity operator

$$\mathbf{v}_i(\tilde{\boldsymbol{d}}) = \partial_{k_i}\mathbf{H}(\boldsymbol{k}) = \partial_{k_i}\tilde{\boldsymbol{d}}\cdot\boldsymbol{\sigma}, \tag{4.5}$$

the intraband contribution

$$\begin{aligned}v_i^{\pm\pm}(\boldsymbol{k}, t) &= |\eta_{\boldsymbol{k}\pm}(t)|^2(\partial_{k_i}\tilde{\boldsymbol{d}})\cdot\mathbf{C}_\pm^\dagger(\tilde{\boldsymbol{d}})\boldsymbol{\sigma}\mathbf{C}_\pm(\tilde{\boldsymbol{d}})\\ &= \pm|\eta_{\boldsymbol{k}\pm}(t)|^2\partial_{k_i}\tilde{d} = |\eta_{\boldsymbol{k}\pm}(t)|^2\partial_{k_i}E_\pm(\boldsymbol{k}),\end{aligned} \tag{4.6}$$

and an interband contribution

$$\begin{aligned} v_i^{-+}(\boldsymbol{k},t) &= \left(v_i^{+-}(\boldsymbol{k},t)\right)^* \\ &= \eta_{\boldsymbol{k}-}^*(t)\eta_{\boldsymbol{k}+}(t)\, e^{-i\int^t (\Delta E(\tilde{\boldsymbol{d}})-\Delta\mathcal{A}(\tilde{\boldsymbol{d}}))\,dt'} \\ &\quad\times \left[(\partial_{k_i}\tilde{\boldsymbol{d}})\cdot \mathbf{C}_-^\dagger(\tilde{\boldsymbol{d}})\boldsymbol{\sigma}\mathbf{C}_+(\tilde{\boldsymbol{d}})\right] \\ &= -\eta_{\boldsymbol{k}-}^*(t)\eta_{\boldsymbol{k}+}(t)\, e^{-i\int^t [2\tilde{d}-\Delta\mathcal{A}(\tilde{\boldsymbol{d}})]\,dt'} \\ &\quad\times \left[\frac{1}{\tilde{d}_\perp}\left[\mathcal{D}_i(\tilde{d}_z,\tilde{d}) + i\mathcal{D}_i(\tilde{d}_x,\tilde{d}_y)\right]\right], \end{aligned} \tag{4.7}$$

with $\mathcal{D}_i(a,b) = a\partial_{k_i}b - b\partial_{k_i}a$ and $\Delta\mathcal{A}(\tilde{\boldsymbol{d}}) = \frac{\tilde{d}_z}{\tilde{d}\tilde{d}_\perp^2}\sum_j \dot{k}_j\mathcal{D}_j(\tilde{d}_x,\tilde{d}_y)$. The intraband contributions (4.6) have the expected form: band populations $|\eta_{\boldsymbol{k}\pm}(t)|^2$, multiplied by the respective group velocities $\partial_{k_i}E_\pm(\boldsymbol{k})$. It should be noted though that the intraband terms (4.6) not only generate low-order harmonics below the band gap but also in the plateau.

The interband contribution (4.7) is less intuitive. It contains the time-dependent amplitudes of both valence and conduction bands, which produce not only harmonics above the band gap (i.e., in the plateau) but also low orders. The topology will affect the interband velocities because of the presence of $\Delta\mathcal{A}$ in (4.7). However, the time evolution of the amplitudes $\eta_{\boldsymbol{k}\pm}(t)$ also depends on the Berry connections (see (4.3)) so that topological effects can also alter the intraband contributions to HHG through the band populations $|\eta_{\boldsymbol{k}\pm}(t)|^2$.

When applied to Haldanite, this model helped to clarify that bulk topological effects, i.e., only associated with the topological phase (or its associated invariant) can be responsible for qualitative changes (helicity flips) in the HHG spectrum [8]. This is in contrast to the SSH chain where edge states have to be explicitly present to observe differences between the spectrum for the topologically trivial and non-trivial phases [6, 7]. If only the SSH bulk is considered, there are no differences (unless one makes the mistake of choosing the expression for the current or velocity from which the HHG spectra are calculated that is inconsistent with the choice of the Bloch ansatz [8]).

### 4.4.3 *Lewenstein-Like HHG in a Two-Band System*

Notice that up to here we have not made any approximation (besides the two-band assumption in the first place) but only performed mathematical operations to obtain (4.4). Now we will make approximations that are well justified and analog to the approximations in the Lewenstein model for gas-phase HHG [44].

First, we assume a fully occupied valence band and an empty conduction band before the laser arrives, i.e., $\eta_{\boldsymbol{k}+}(0) = 0$ and $\eta_{\boldsymbol{k}-}(0) = 1$. Furthermore, neglecting depletion of the valence band (i.e., keeping the solid target intact), $\eta_{\boldsymbol{k}-}(t) \simeq 1$, we obtain

$$\eta_{\boldsymbol{k}+}(t) = i \int^{t} \left[\mathcal{A}_{+-}(\tilde{\boldsymbol{d}})\right] e^{i \int^{t'} [2\tilde{d} - \Delta\mathcal{A}(\tilde{\boldsymbol{d}})]\, dt''}\, dt'$$

and thus a closed expression for the interband velocity

$$v_i^{-+}(\boldsymbol{k}, t) \simeq \left[\frac{1}{\tilde{d}_\perp}\left[\mathcal{D}_i(\tilde{d}_x, \tilde{d}_y) - i\mathcal{D}_i(\tilde{d}_z, \tilde{d})\right]\right] \times \int^{t} \left[\frac{\sum_j \dot{k}_j \left(\mathcal{D}_j(\tilde{d}_x, \tilde{d}_y) + i\mathcal{D}_j(\tilde{d}_z, \tilde{d})\right)}{2\tilde{d}\tilde{d}_\perp}\right] e^{-i \int_{t'}^{t} [2\tilde{d} - \Delta\mathcal{A}(\tilde{\boldsymbol{d}})]\, dt''}\, dt'.$$

This equation has the same structure as the dipole derived in the Lewenstein paper [44]: integration over the "ionization time" $t'$ up to the recombination time $t$. The time integral in the phase under the integral is over the travel period $[t', t]$. The Fourier transform of the velocity performed to calculate the HHG spectrum brings in the integration over the recombination time. The integral over all possible initial $\boldsymbol{k}$ (i.e., over the Brillouin zone) that is needed to add up all the contributions from the electrons in the material coherently is analogous to the momentum integral present in the expression for the dipole in [44]. The latter point is not that trivial because the Lewenstein model for gas-phase HHG is a single-active electron theory, and the momentum over which is integrated is the momentum the electron has while it is in the continuum. The saddle-point approximation to that integral leads to an equation that ensures the return of the electron to the parent ion for recombination. Here, in our condensed-matter theory, the integral arises because of the many electrons and their coherent emission we need to consider. In particular, this integral is responsible for interfering away almost all the emissions due to the intraband motion of individual, non-interacting electrons in fully occupied bands.

It is worth to highlight that our result is neither restricted to a particular target dimension nor did we apply semiclassical wave-packet dynamics or use a single-band approximation. Therefore our result (4.4) is more general than the often employed (4.1). Further, we may use the well-known semi-classical approximations, i.e., the saddle-point approximation to integrals over time and momentum [45], to formulate a quasi-classical quantum orbit theory for HHG in topological condensed matter, which gives maximum intuitive insight into the electron dynamics.

## 4.5 Summary and Outlook

In this chapter, we discussed examples of topological aspects of HHG in finite systems, namely SSH chains and Haldanite flakes. In these systems, the presence of topologically protected edge states strongly affects HHG spectra. As a laser pulse that impinges on topological material necessarily interacts with the surface, the topologically protected edge states should be self-consistently taken into account in calculations, rather than added separately by hand into the equations. Besides searching for signatures of such edge states in HHG spectra, it will be interesting to control the edge-state electronics. This is appealing, as the edge currents are ballistic, robust, and dissipationless. This might pave the way for ultrafast light wave electronics with clock rates that are orders of magnitude faster than today. One important challenge here will be miniaturization, as the laser wavelengths are several micrometers. Employing near fields might be a solution to overcome this scaling problem in light-matter coupling.

Concerning bulk, we introduced a general two-band theory that is capable of describing (almost analytically) HHG, including all topological effects. It will be interesting to exploit this approach further with respect to multi-color pulses of various polarizations. The crux with "conventional" condensed matter is that it can hardly be tuned through a topological phase transition. Varying the degree of doping has proved to be an option here, though a challenging one. It would be advantageous to tune the same sample continuously in a purely optical, otherwise non-invasive way. That should be possible with Floquet topological insulators where one laser field controls the topology and another probes the charge-carrier dynamics by HHG or other means.

Finally, we are currently working on the inclusion of electron-electron interaction and coupling to phonons into the light-matter models for HHG developed by us. This is not only of interest to our strong-field community but of much wider and fundamental relevance because the standard theory of topological invariants requires a band structure, which ceases to exist for interacting systems.

## Acknowledgments

Funding by the German Research Foundation, SFB 1477 "Light-Matter Interactions at Interfaces," Project No. 441234705, is gratefully acknowledged. H. J. acknowledges financial support from the doctoral fellowship program of the University of Rostock. H. D. acknowledges financial support by the Studienstiftung des deutschen Volkes (German Academic Scholarship Foundation).

## References

[1] S. Ghimire, A. Dichiara, E. Sistrunk, P. Agostini, L. Dimauro and D. Reis, Observation of high-order harmonic generation in a bulk crystal, *Nature Physics* **7**, 2, pp. 138–141 (2011).

[2] M. Franz and L. Molenkamp (Eds.), *Topological Insulators*. Elsevier (2013).

[3] B. A. Bernevig and T. L. Hughes, *Topological Insulators and Topological Superconductors*. Princeton University Press (2013).

[4] D. Vanderbilt, *Berry Phases in Electronic Structure Theory: Electric Polarization, Orbital Magnetization and Topological Insulators*. Cambridge University Press (2018).

[5] D. Bauer and K. K. Hansen, High-harmonic generation in solids with and without topological edge states, *Physical Review Letters* **120**, 17, p. 177401 (2018).

[6] H. Drüeke and D. Bauer, Robustness of topologically sensitive harmonic generation in laser-driven linear chains, *Physical Review A* **99**, 5, p. 053402 (2019).

[7] H. Jürß and D. Bauer, High-harmonic generation in Su-Schrieffer-Heeger chains, *Physical Review B* **99**, 19, p. 195428 (2019).

[8] D. Moos, H. Jürß and D. Bauer, Intense-laser-driven electron dynamics and high-order harmonic generation in solids including topological effects, *Physical Review A* **102**, 5, p. 053112 (2020).

[9] O. Neufeld, N. Tancogne-Dejean, H. Hübener, U. De Giovannini and A. Rubio, Are there universal signatures of topological phases in high-harmonic generation? Probably not. *Physical Review X* **13**, p. 031011 (2023).

[10] A. Chacón, W. Zhu, S. P. Kelly, A. Dauphin, E. Pisanty, A. Picón, C. Ticknor, M. F. Ciappina, A. Saxena and M. Lewenstein, Observing topological phase transitions with high harmonic generation, *arXiv:1807.01616v3 [cond-mat, physics:quant-ph]* (2018).

[11] C. P. Schmid, L. Weigl, P. Grössing, V. Junk, C. Gorini, S. Schlauderer, S. Ito, M. Meierhofer, N. Hofmann, D. Afanasiev, J. Crewse, K. A. Kokh, O. E. Tereshchenko, J. Güdde, F. Evers, J. Wilhelm, K. Richter, U. Höfer and R. Huber, Tunable non-integer high-harmonic generation in a topological insulator, *Nature* **593**, 7859, pp. 385–390 (2021).

[12] D. Baykusheva, A. Chacón, J. Lu, T. P. Bailey, J. A. Sobota, H. Soifer, P. S. Kirchmann, C. Rotundu, C. Uher, T. F. Heinz, D. A. Reis and S. Ghimire, All-optical probe of three-dimensional topological insulators based on high-harmonic generation by circularly polarized laser fields, *Nano Letters* **21**, 21, pp. 8970–8978 (2021).

[13] D. Baykusheva, A. Chacón, D. Kim, D. E. Kim, D. A. Reis and S. Ghimire, Strong-field physics in three-dimensional topological insulators, *Physical Review A* **103**, 2, p. 023101 (2021).

[14] Y. Bai, F. Fei, S. Wang, N. Li, X. Li, F. Song, R. Li, Z. Xu and P. Liu, High-harmonic generation from topological surface states, *Nature Physics* **17**, 3, pp. 311–315 (2021).

[15] M. Garg, M. Zhan, T. Luu, H. Lakhotia, T. Klostermann, A. Guggenmos and E. Goulielmakis, Multi-petahertz electronic metrology, *Nature* **538**, 7625, pp. 359–363 (2016).

[16] T. Higuchi, C. Heide, K. Ullmann, H. Weber and P. Hommelhoff, Light-field-driven currents in graphene, *Nature* **550**, 7675, pp. 224–228 (2017).

[17] M. Schultze, E. M. Bothschafter, A. Sommer, S. Holzner, W. Schweinberger, M. Fiess, M. Hofstetter, R. Kienberger, V. Apalkov, V. S. Yakovlev, M. I. Stockman and F. Krausz, Controlling dielectrics with the electric field of light, *Nature* **493**, 7430, pp. 75–78 (2013).

[18] M. Baudisch, A. Marini, J. D. Cox, T. Zhu, F. Silva, S. Teichmann, M. Massicotte, F. Koppens, L. S. Levitov, F. J. García de Abajo and J. Biegert, Ultrafast nonlinear optical response of Dirac fermions in graphene, *Nature Communications* **9**, 1, p. 1018 (2018).

[19] J. Reimann, S. Schlauderer, C. P. Schmid, F. Langer, S. Baierl, K. A. Kokh, O. E. Tereshchenko, A. Kimura, C. Lange, J. Güdde, U. Höfer and R. Huber, Subcycle observation of lightwave-driven Dirac currents in a topological surface band, *Nature* **562**, 7727, pp. 396–400 (2018).

[20] A. Jiménez-Galán, R. E. F. Silva, O. Smirnova and M. Ivanov, Lightwave control of topological properties in 2D materials for sub-cycle and non-resonant valley manipulation, *Nature Photonics* **14**, 12, pp. 728–732 (2020).

[21] H. Hübener, M. A. Sentef, U. De Giovannini, A. F. Kemper and A. Rubio, Creating stable Floquet–Weyl semimetals by laser-driving of 3D Dirac materials, *Nature Communications* **8**, 1, p. 13940 (2017).

[22] U. D. Giovannini and H. Hübener, Floquet analysis of excitations in materials, *Journal of Physics: Materials* **3**, 1, p. 012001 (2019).

[23] H. Drüeke and D. Bauer, High-harmonic spectra of hexagonal nanoribbons from real-space time-dependent Schrödinger calculations, *The European Physical Journal Special Topics* **230**, 23, pp. 4065–4070 (2021).

[24] R. E. F. Silva, A. Jiménez-Galán, B. Amorim, O. Smirnova and M. Ivanov, Topological strong-field physics on sub-laser-cycle timescale, *Nature Photonics* **13**, 12, pp. 849–854 (2019).

[25] H. Jürß and D. Bauer, Helicity flip of high-order harmonic photons in Haldane nanoribbons, *Physical Review A* **102**, 4, p. 043105 (2020).

[26] H. Jürß and D. Bauer, Edge-state influence on high-order harmonic generation in topological nanoribbons, *The European Physical Journal D* **75**, 6, p. 190 (2021).

[27] W. P. Su, J. R. Schrieffer and A. J. Heeger, Solitons in polyacetylene, *Physical Review Letters* **42**, 25, pp. 1698–1701 (1979).

[28] D. Bauer and F. Ceccherini, A numerical ab initio study of harmonic generation from a ring-shaped model molecule in laser fields, *Laser and Particle Beams* **19**, 1, p. 85–90 (2001).

[29] F. D. M. Haldane, Model for a quantum Hall effect without Landau levels: condensed-matter realization of the "parity anomaly", *Physical Review Letters* **61**, 18, pp. 2015–2018 (1988).

[30] G. Jotzu, M. Messer, R. Desbuquois, M. Lebrat, T. Uehlinger, D. Greif and T. Esslinger, Experimental realization of the topological Haldane model with ultracold fermions, *Nature* **515**, 7526, pp. 237–240 (2014).

[31] F. D. M. Haldane and S. Raghu, Possible realization of directional optical waveguides in photonic crystals with broken time-reversal symmetry, *Physical Review Letters* **100**, p. 013904 (2008).

[32] A. B. Khanikaev, S. Hossein Mousavi, W.-K. Tse, M. Kargarian, A. H. MacDonald and G. Shvets, Photonic topological insulators, *Nature Materials* **12**, 3, pp. 233–239 (2013).

[33] M. C. Rechtsman, J. M. Zeuner, Y. Plotnik, Y. Lumer, D. Podolsky, F. Dreisow, S. Nolte, M. Segev and A. Szameit, Photonic Floquet topological insulators, *Nature* **496**, 7444, pp. 196–200 (2013).

[34] M. Hasan and C. Kane, Colloquium: Topological insulators, *Reviews of Modern Physics* **82**, 4, pp. 3045–3067 (2010).

[35] C. L. Kane and E. J. Mele, Quantum spin Hall effect in graphene, *Physical Review Letters* **95**, 22, p. 226801 (2005).

[36] H. Jürß and D. Bauer, High-order harmonic generation in hexagonal nanoribbons, *The European Physical Journal Special Topics* **230**, 23, pp. 4081–4089 (2021).

[37] T. Cao, F. Zhao and S. G. Louie, Topological phases in graphene nanoribbons: Junction states, spin centers, and quantum spin chains, *Physical Review Letters* **119**, 7, p. 076401 (2017).

[38] D. J. Rizzo, G. Veber, T. Cao, C. Bronner, T. Chen, F. Zhao, H. Rodriguez, S. G. Louie, M. F. Crommie and F. R. Fischer, Topological band engineering of graphene nanoribbons, *Nature* **560**, 7717, pp. 204–208 (2018).

[39] D. Xiao, M.-C. Chang and Q. Niu, Berry phase effects on electronic properties, *Reviews of Modern Physics* **82**, pp. 1959–2007 (2010).

[40] M.-C. Chang and Q. Niu, Berry phase, hyperorbits, and the Hofstadter spectrum, *Physical Review Letters* **75**, pp. 1348–1351 (1995).

[41] P. Gosselin, F. Ménas, A. Bérard and H. Mohrbach, Semiclassical dynamics of electrons in magnetic Bloch bands: A Hamiltonian approach, *Europhysics Letters* **76**, 4, p. 651 (2006).

[42] G. Sundaram and Q. Niu, Wave-packet dynamics in slowly perturbed crystals: Gradient corrections and Berry-phase effects, *Physical Review B* **59**, pp. 14915–14925 (1999).

[43] H. Jürß and D. Bauer, Topological edge-state contributions to high-order harmonic generation in finite flakes, *Physical Review B* **106**, 5, p. 054303 (2022).

[44] M. Lewenstein, P. Balcou, M. Y. Ivanov, A. L'Huillier and P. B. Corkum, Theory of high-harmonic generation by low-frequency laser fields, *Physical Review A* **49**, pp. 2117–2132 (1994).

[45] G. Vampa and T. Brabec, Merge of high harmonic generation from gases and solids and its implications for attosecond science, *Journal of Physics B: Atomic, Molecular and Optical Physics* **50**, 8, p. 083001 (2017).

https://doi.org/10.1142/9789811279560_0005

# Chapter 5

# High-Harmonic Spectroscopy of Coherently Vibrating Solids

Navdeep Rana and Gopal Dixit

*Department of Physics, Indian Institute of Technology Bombay, Powai, Mumbai 400076, India*

## 5.1 Introduction

Several important optical processes occur when an intense laser pulse interacts with matter [1, 2]. Strong field-driven high-order harmonic generation (HHG) is one such process. HHG is a highly non-perturbative frequency up-conversion process, which emits radiation at integer multiples of incident laser frequency and converts infrared light into bright, coherent, ultrashort bursts of extreme ultraviolet and soft X-ray radiation [3, 4]. In addition, HHG forms the foundation of present attosecond technology: the generation of isolated attosecond pulse [5] and pulse trains [6], and enables time-resolved spectroscopy of laser-driven attosecond electronic motion in matter [7–12]. Owing to its immense importance, the Nobel Prize in Physics for 2023 was awarded to attosecond physics and its applications in interrogating electron dynamics in atoms and molecules.

The success of HHG from gases over the past decades has also stimulated attempts to generate strong field-driven non-perturbative high-harmonics from solids [13]. However, due to their low damage threshold, HHG from solids was not feasible until 2010, when technological advancement in mid-infrared and terahertz laser sources enabled the study of strong field-driven processes in solids, including semiconductors and dielectrics. After the pioneering work of Ghimire *et al.* [14], HHG in solids has developed into a novel method for probing different aspects of static and dynamic properties of solids [15–30]. On the technological side, HHG from solids offers an attractive possibility of fascinating all-solid-state compact optical devices to generate bright, coherent attosecond pulses in extreme ultraviolet energy range [31, 32]. Despite the significant progress in high-harmonic spectroscopy of solids, how the coherent atomic vibrations in solids affect HHG still needs to be better understood. Thus, the role of coherent atomic vibrations in solid HHG is less explored territory, except in recent reports [33, 34]. Contrary to that, the role of nuclear motion in various gaseous molecules has been extensively explored by high-harmonic spectroscopy [35–37]. Note that coherent atomic vibrations influence a diverse range of physical and chemical processes in solids. Thus, probing coherent atomic vibrations is essential to comprehend several ubiquitous phenomena in solids, such as heat capacity and thermal conductivity [38–40], as well as structural aspects including crystalline arrangement characterization and the manifestation of phase transitions arising from external stimulus [41, 42].

The present chapter presents a detailed discussion of how coherent atomic vibrations affect solid HHG. We will analyze how different properties of the vibrating solid, such as amplitude of the vibration, role of anharmonic vibration, interference of different symmetries of the vibrations, and the mass of the vibrating atoms impact the harmonic generation. In addition, the influence of the role of different laser parameters like pulse duration, intensity, and polarization direction on HHG will be presented in depth. We choose monolayer graphene to demonstrate the sensitivity of the coherent atomic vibrations on HHG. Strong field-driven HHG from graphene without atomic vibrations has been discussed experimentally and theoretically in detail [43–59]. Owing to $\mathbf{D}_{6h}$ point group symmetry, monolayer graphene exhibits three acoustic and three optical phonon branches, in which two optical phonon modes are planar in nature as shown in Fig. 5.1. Both the in-plane phonon modes are Raman active and degenerate at $\mathbf{\Gamma}$ point with 194 meV phonon frequency [60]. These in-plane optical phonon modes are also known as $\mathbf{E}_{2g}$ or **G** mode. We will focus on these two in-plane optical phonon modes in the following.

## 5.2 Theoretical Method

The interaction of an intense laser with coherent electron and lattice dynamics in graphene is described within the density-matrix framework. The temporal evolution of the density matrix is solved using semiconductor

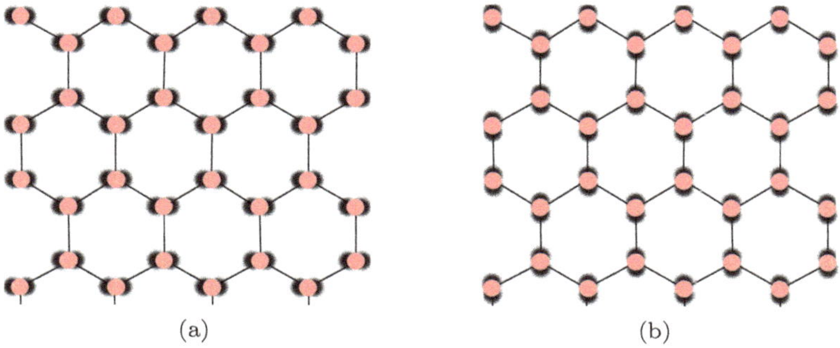

Fig. 5.1 Real-space representations of the coherent atomic vibrations associated with the degenerate $\mathbf{E}_{2g}$ phonon modes of monolayer graphene at the $\mathbf{\Gamma}$ point. Coherent atomic vibrations associated with the in-plane (a) longitudinal optical (**iLO**) and (b) transverse optical (**iTO**) phonon modes. The red solid dots represent carbon atoms at equilibrium in graphene, whereas vibrating black solid dots correspond to the in-plane **iTO** and **iLO** phonon modes.

Bloch equations in Houston basis as [61–63]

$$i\dot{\rho}_{mn}^{\mathbf{k}} = \mathcal{E}_{mn}^{\mathbf{k}_t}\rho_{mn}^{\mathbf{k}} + i\frac{\tilde{\delta}_{mn}}{\mathrm{T}_2}\rho_{mn}^{\mathbf{k}} - \mathbf{F}(t)\cdot\sum_{l}\left(\mathbf{d}_{ml}^{\mathbf{k}_t}\rho_{ln}^{\mathbf{k}} - \mathbf{d}_{ln}^{\mathbf{k}_t}\rho_{ml}^{\mathbf{k}}\right). \tag{5.1}$$

Here, $\rho_{mn}$ is the density matrix between eigenstates $|m\rangle$ and $|n\rangle$, $\mathcal{E}_{mn}^{\mathbf{k}}$ and $\mathbf{d}_{mn}^{\mathbf{k}}$ are the band-gap energy and the dipole-matrix element between $m$ and $n$ bands at crystal momentum $\mathbf{k}$, respectively. The dipole-matrix element is related to the periodic part of the wavefunction ($|u_{m,\mathbf{k}}\rangle$) as $\mathbf{d}_{mn} = -i\langle u_{m,\mathbf{k}}|\nabla_{\mathbf{k}}|u_{n,\mathbf{k}}\rangle$. In the presence of an intense laser field, the crystal momentum $\mathbf{k}$ changes to $\mathbf{k}_t = \mathbf{k} + \mathbf{A}(t)$, where $\mathbf{A}(t)$ is the vector potential of the laser field, which is related to the electric field as $\mathbf{F}(t) = -d\mathbf{A}(t)/dt$. In the above expression, we have defined $\tilde{\delta}_{mn}$ as $(1\text{-}\delta_{mn})$. A phenomenological constant dephasing time $\mathrm{T}_2$ is introduced in the semiconductor Bloch equations to account for decoherence between electrons and holes [61].

To account for the coherent lattice vibrations, tight-binding Hamiltonian within nearest-neighbor interaction for graphene is extended from static to time domain as [64, 65]

$$\mathcal{H}_t = -\sum_{i\in nn}\gamma(t)e^{i\mathbf{k}\cdot\mathbf{d}_i(t)}\hat{a}_{\mathbf{k}}^{\dagger}\hat{b}_{\mathbf{k}} + \mathrm{H.c.} \tag{5.2}$$

where H.c. stands for Hermitian conjugate. To account for the temporal variations in the relative distance between carbon atoms, the hopping term is modeled as $\gamma(t) = \gamma_0 \, e^{-(|\mathbf{d}_i(t)|-a)/\delta}$, where $\delta = 0.184a_0$ with $a_0 = 2.46$ Å as the lattice parameter, $|\mathbf{d}_i| = a = 1.42$ Å is the inter-atomic distance and $\gamma_0 = 2.7$ eV is the nearest-neighbor hopping energy [66]. Here, $\hat{a}_k^\dagger$ ($\hat{b}_k$) is the creation (annihilation) operator for atom A (B) in the unit cell. At each time step during the high-harmonic probe pulse and coherent lattice vibration, time-dependent tight-binding Hamiltonian $\mathcal{H}_t$ is diagonalized and updated eigenstates and eigenenergies are obtained.

In this chapter, we assume that the coherent atomic vibrations is induced by a pump pulse and consequently probed by the high-harmonic generating pulse. As the pump pulse triggers coherent vibrations, the relative distance between carbon atoms in graphene changes. As a result of this change, dipole matrix elements become time-dependent. At each instant during the coherent vibrations, dipole matrix elements are calculated with the updated eigenstates at each time step. The dipole-matrix elements and the eigenenergies are smoothly updated in semiconductor-Bloch equations at each consecutive time step as long as the maximum displacement of the atoms is small and the time step is too small compared to the lattice oscillation period. The semiconductor Bloch equations are solved using the fourth-order Runge-Kutta method with a time step of 0.01 fs. We sampled the Brillouin zone with 251 × 251 grid. The total current at any **k**-point in the entire Brillouin zone is simulated as $\mathbf{J}(\mathbf{k},t) = \sum_{m,n} \rho_{mn}^{\mathbf{k}}(t)\ \mathbf{p}_{nm}^{\mathbf{k}_t}$ with $\mathbf{p}_{nm}^{\mathbf{k}}$ as the momentum matrix element. The high-harmonic spectrum is determined from the Fourier transform of the time derivative of the current as

$$\mathcal{I}(\omega) = \left| \mathcal{FT}\left( \frac{d}{dt}\left[ \int_{BZ} \mathbf{J}(\mathbf{k},t)\ \mathrm{d}\mathbf{k} \right] \right) \right|^2 . \tag{5.3}$$

Here, $\mathcal{FT}$ stands for the Fourier transform. In the present chapter, we employ a linearly polarized pulse with an eight-cycle duration having a sin-squared envelope. The pulse has a wavelength of 2.0 $\mu$m and an intensity of $1 \times 10^{11}$ W/cm$^2$. We have also considered $\mathrm{T}_2 = 10$ fs and maximum displacement of atoms as $0.03a_0$ from their equilibrium positions during coherent lattice dynamics. The same parameters are used throughout this chapter unless stated otherwise.

## 5.3 Results and Discussion

Figure 5.2 shows that the coherent atomic vibrations in monolayer graphene lead to the generation of equally spaced higher-order sidebands along with

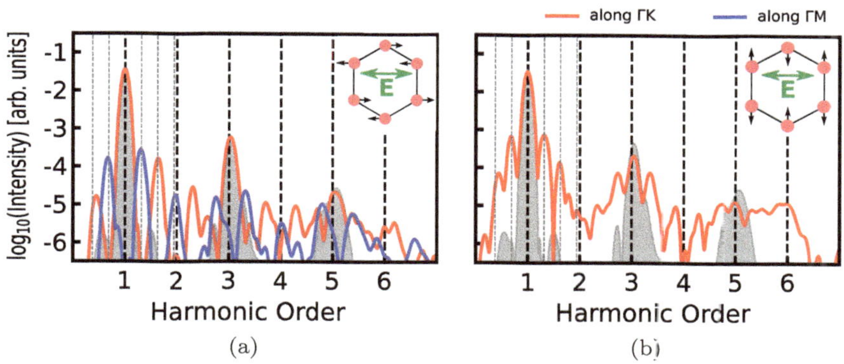


Fig. 5.2 High-harmonic spectra of graphene with and without coherent in-plane atomic vibrations. The harmonic spectra corresponding to the coherent (a) **iLO** and (b) **iTO** phonon mode, while the probe harmonic pulse is polarized along $\mathbf{\Gamma} - \mathbf{K}$ direction. The unit cell of the graphene with the corresponding lattice eigenvector and polarization of the harmonic generating probe pulse are shown in the respective insets. The red (blue) color corresponds to the polarization of emitted radiation parallel (perpendicular) to the polarization of the harmonic generating probe pulse. In both scenarios, sidebands corresponding to the first harmonic are marked at frequencies ($\omega_0 \pm m\omega_{\rm ph}$), where $\omega_0$ is the frequency of the harmonic generating probe pulse, and $\omega_{\rm ph}$ is the phonon frequency. The gray-shaded region serves as the reference, representing the spectra of graphene without atomic vibrations. The probe HHG pulse is a linearly polarized pulse with a wavelength of 2.0 $\mu$m and an intensity of $1 \times 10^{11}$ W/cm$^2$. In addition, the spectra correspond to a maximum 0.03a$_0$ displacement of atoms from their equilibrium positions and T$_2$ = 10 fs. Adapted with permission from [67].

the main harmonic peaks. The frequency of the coherent vibrations matches the sidebands' positions in the spectra. As evident from Fig. 5.2(a), the in-plane **iLO** lattice excitation results in odd- and even-order sidebands polarized perpendicular and parallel to the probe harmonic pulse's polarization, respectively [67]. On the other hand, in-plane **iTO** lattice excitation leads all sidebands polarized along the polarization of the probe harmonic pulse [see Fig. 5.2(b)].

Let us analyze the time-frequency map of the harmonic spectra to know how high-harmonic spectroscopy is sensitive to an interplay of the coherent atomic and sub-cycle electron dynamics. The time-frequency maps are obtained by performing the Gabor transformation. Two bursts of the harmonics per cycle, at the extrema of the electric field, are evident from the time-frequency map of graphene without lattice dynamics [Fig. 5.3(a)]. In this case, the Gabor profile does not vary periodically with $\omega_{\rm ph}$. When **iLO** phonon mode is coherently excited, currents in both

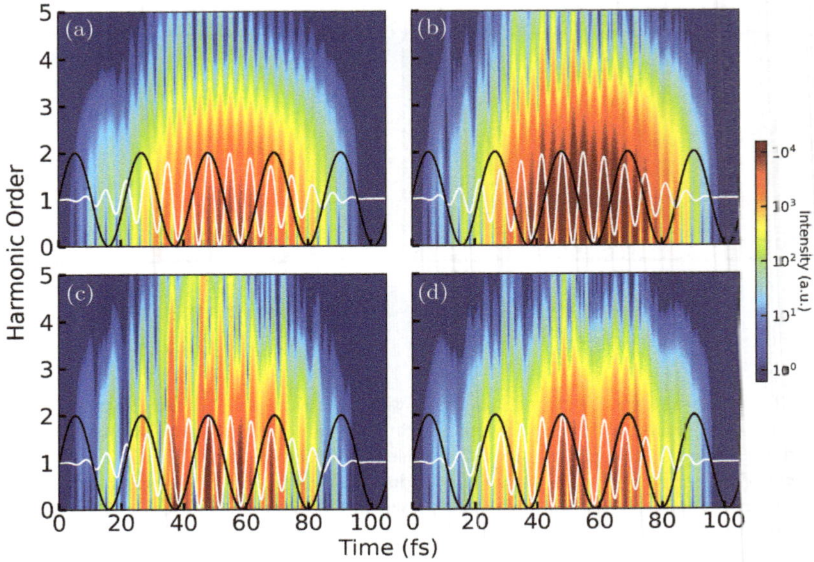


Fig. 5.3 Time-frequency representation of the high-harmonic generation. Time-frequency map of the current corresponds to (a) graphene without atomic vibrations, (b) parallel (**X**), and (c) perpendicular (**Y**) components in graphene with the **iLO** coherent phonon mode. (d) Time-frequency map of the current in graphene with the **iTO** coherent phonon mode. The electric field of the probe pulse is shown by a white curve, which is along the **X** direction. The black curve represents the periodic displacement of atoms from the equilibrium position with phonon frequency $\omega_{\text{ph}}$.

parallel and perpendicular directions, with respect to the probe pulse, are generated [Fig. 5.2(a)]. The Gabor profile of the parallel (**X** component) and the perpendicular (**Y** component) are presented in Figs. 5.3(b) and 5.3(c), respectively. The profile of the perpendicular component exhibits modulation with a time period equal to 21.3 fs, which matches with $\omega_{\text{ph}}$. However, the parallel profile does not show any periodic modulation of frequency $\omega_{\text{ph}}$ (see black curve). Moreover, the harmonic bursts are not exactly at the extremum of the probe electric field but are slightly shifted, which could be attributed to the possible electron-phonon interaction. In the case of the coherent **iTO** phonon mode, current along the probe pulse's direction is generated [Fig. 5.2(b)]. Thus, the resultant Gabor profile of the current display periodic modulation with frequency $\omega_{\text{ph}}$. The reason behind the periodic modulation of the Gabor profile can be attributed to the time-periodic Hamiltonian as given in Eq. (5.2).

One phonon period corresponds to approximately three optical cycles of the probe pulse, and two harmonic bursts per laser cycle are evident from the Gabor profile. It means that there are six harmonic bursts within one phonon period: three at the maxima of the cycle and three at the minima. This explains the generation of three frequency up ($\omega_0 + m\omega_{\rm ph}$) and three frequency down sidebands ($\omega_0 - m\omega_{\rm ph}$), which are consistent with our observations made in Fig. 5.2. It has been shown that the Floquet formalism of the dynamical symmetries of the combined system, including the coherent atomic vibrations in graphene and probe harmonic pulse, determine the selection rules for the polarization of the sidebands associated with the symmetry of the coherently excited atomic vibration; detailed discussion is reported in Ref. [67].

### 5.3.1 *Role of System Properties*

So far, we have established that high-harmonic spectroscopy is responsive to the coherent atomic vibrations in graphene. Moreover, we have also examined how the system's dynamical symmetries provide valuable insights into the symmetry of the induced atomic vibrations [67]. At this juncture, it is natural to wonder how different properties of the coherently vibrating graphene affect the high-harmonic spectrum. In the following, we will investigate various aspects of the system's properties on HHG from vibrating graphene. The high-harmonic spectra are normalized with respect to the first harmonic throughout this section unless stated otherwise.

Figure 5.4 presents high-harmonic spectra corresponding to the coherently excited **iLO** phonon mode for different amplitudes of the atomic displacement with respect to the equilibrium position. The intensity of the sidebands increases monotonically with respect to the displacement amplitude, whereas the intensity of the main harmonic peaks remains unaffected, as evident from the figure. An increase in the sideband intensity with the displacement amplitude can be attributed to the extended degree of symmetry breaking caused by the larger displacement of the atoms within graphene. The extended symmetry breaking leads to an overall increase in the intensity of the sidebands. Remarkably, the polarization properties of the sidebands remain unaffected despite the "boost" in the symmetry breaking.

A similar trend is observed for the coherently excited **iTO** phonon mode with the displacement amplitude ranging from $0.01a_0$ to $0.05a_0$ relative to the equilibrium positions. It is important to emphasize that our findings

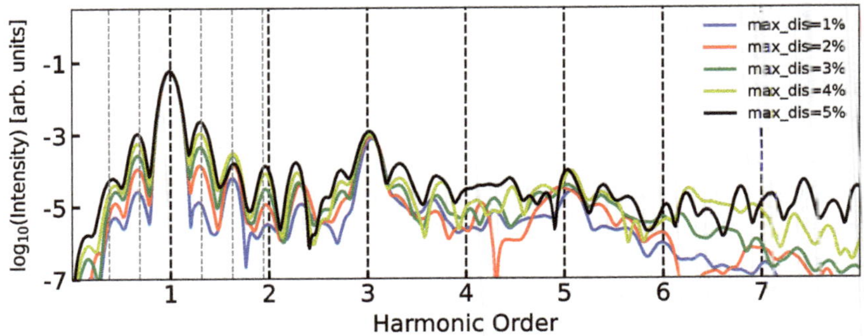


Fig. 5.4 Sensitivity of the high-harmonic spectra with respect to different amplitudes of the atomic vibrations. The spectra correspond to graphene with the coherent **iLO** phonon mode. The parameters of the probe laser pulse are the same as in Fig. 5.2.

are valid as long as the atomic vibrations with different amplitudes are within the harmonic approximation. Thus, it is pertinent to ask how the anharmonicity in the atomic vibrations affects the harmonic spectra.

Until now, we have limited our discussion of the lattice dynamics within harmonic approximation as the time-dependent displacement vector for a particular phonon mode is linear in amplitude as $\mathbf{q}(t) = \mathrm{q}_0 \,\mathrm{Re}\left\{\exp(i\omega_{\mathrm{ph}}t)\right\} \hat{\mathbf{e}}$, where $\mathrm{q}_0$ is the maximum displacement of an atom from its equilibrium position and $\hat{\mathbf{e}}$ is the normalized eigenvector for a particular phonon mode [67]. Let us extend our discussion beyond the harmonic approximation by adding a non-linear term in the displacement vector. The updated equation of motion for the lattice dynamics reads as

$$\mathbf{q}(t) = \left[\mathrm{q}_0 \;\sin(\omega_{\mathrm{ph}}t) - \frac{\mathrm{q}_0^2}{2}\left\{\cos(2\omega_{\mathrm{ph}}t) - 1\right\}\right] \hat{\mathbf{e}}. \tag{5.4}$$

Here, $\mathrm{q}_0^2$-term on the right-hand side includes anharmonicity in our simulation. High-harmonic spectra corresponding to the coherent **iLO** and **iTO** phonon modes for two displacement amplitudes are shown in Fig. 5.5. As evident from the figure, there is no meaningful effect of the anharmonicity on the generated spectra for both phonon modes, even for an amplitude equal to 0.05 of the lattice constant. Thus, high-harmonic spectroscopy is not appropriate to capture anharmonic effects in graphene within our theoretical framework.

During an excitation process of a coherent phonon mode, there is a possibility that other phonon modes can also get excited with different

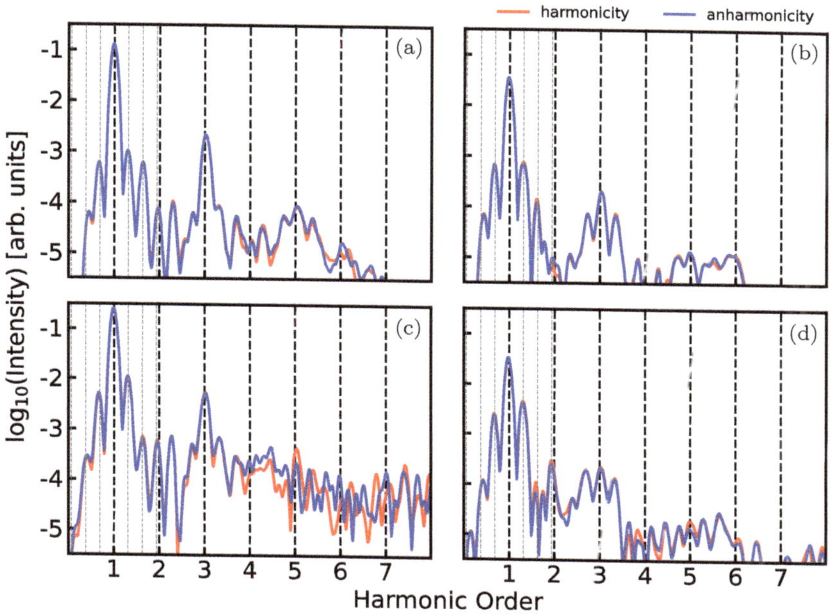


Fig. 5.5 Effect of the anharmonicity in the high-harmonic spectroscopy of the coherent lattice dynamics in graphene. High-harmonic spectra corresponding to the coherent **iLO** phonon mode with the maximum amplitude equal to (a) 0.03 and (c) 0.05 of the lattice constant. (b) and (d) are the same as (a) and (c) except **iTO** phonon mode is coherently excited. The blue (red) color corresponds to the lattice dynamics with (without) an anharmonic effect. The probe harmonic pulse is polarized along $\mathbf{\Gamma} - \mathbf{K}$ direction. The parameters of the probe laser pulse are the same as in Fig. 5.2.

amplitudes as both in-plane phonon modes are degenerate at $\mathbf{\Gamma}$-point. Thus, it is interesting to see how the coupling between the two modes affects the harmonic spectra. Let us introduce the coupling of the **iTO** phonon mode with **iLO** phonon mode as

$$\begin{aligned} \mathbf{q}_{\mathbf{iLO}}(t) &= \mathrm{q}_0 \ \mathrm{Re}\{\exp(i\omega_{\mathrm{ph}}t)\} \ \hat{\mathbf{e}}_{\mathbf{iLO}}, \\ \text{and} \quad \mathbf{q}_{\mathbf{iTO}}(t) &= \alpha \mathrm{q}_0 \ \mathrm{Re}\{\exp(i\omega_{\mathrm{ph}}t)\} \ \hat{\mathbf{e}}_{\mathbf{iTO}}, \end{aligned} \tag{5.5}$$

where $\alpha$ is a real number, which determines the strength of the coupling between two phonon modes. We primarily focus on the coherently excited **iLO** phonon mode and its coupling with the **iTO** phonon mode.

Figure 5.6 presents high-harmonic spectra for different coupling strengths ranging from 0.1 to 1.0, i.e., weak coupling to strong coupling regime. It is established that when the **iLO** phonon is coherently excited,

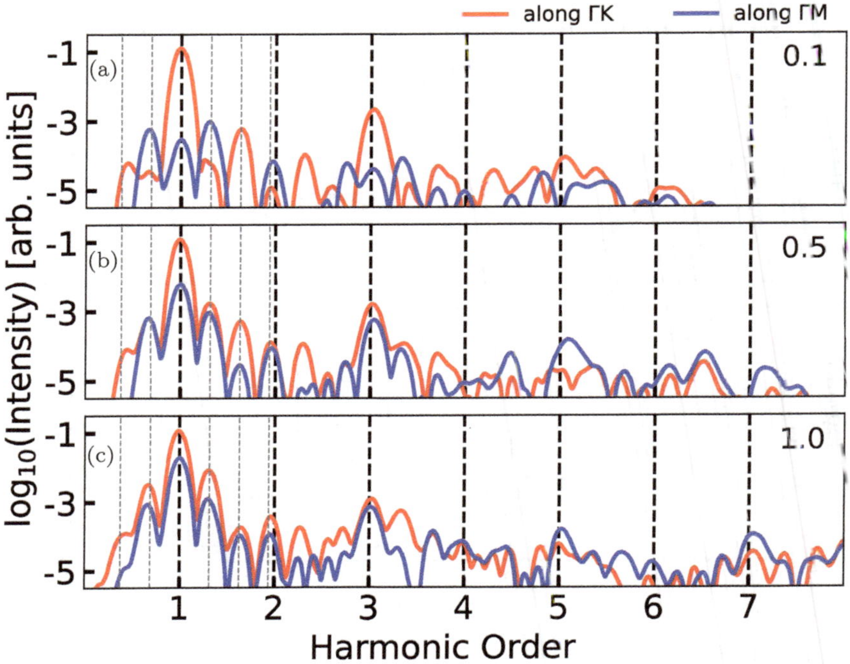


Fig. 5.6 High-harmonic spectra corresponding to graphene with the coherent **iLO** phonon mode coupled with $\alpha$ times **iTO** phonon mode strength. The spectra shown in (a), (b), and (c) correspond to $\alpha$ = 0.1, 0.5, and 1.0, respectively. The red (blue) color corresponds to the polarization of emitted radiation parallel (perpendicular) to the polarization of the harmonic generating probe pulse. The probe harmonic pulse is polarized along $\mathbf{\Gamma}-\mathbf{K}$ direction. The parameters of the probe laser pulse are the same as in Fig. 5.2.

odd-order sidebands are polarized perpendicular to the probe pulse ($\mathbf{\Gamma}-\mathbf{M}$) direction, whereas the even-order sidebands are polarized parallel to the probe pulse ($\mathbf{\Gamma}-\mathbf{K}$) direction. However, additional symmetry restrictions apply to the polarization of the sidebands when the **iTO** phonon mode is coupled with the **iLO** mode. In this scenario, reflection-plane symmetry perpendicular to the probe pulse is absent, resulting in the generation of harmonics along $\mathbf{\Gamma}-\mathbf{K}$ as well as $\mathbf{\Gamma}-\mathbf{M}$ directions as evident from the figure. Moreover, the polarization properties of the sidebands also alter as the coupling strength increases. This means that the polarization of the emitted harmonics are elliptical in nature, and the ellipticity increases as the coupling strength increases. Thus, the coupling between the phonon modes can be extracted by analyzing the polarization properties of the main harmonics as well as their sidebands.

After analyzing how different properties of the excited coherent phonon mode affect the generated harmonic spectra, it is important to know how HHG is sensitive to a change in the vibration frequency. For this purpose, we replace $^{12}$C by $^{13}$C in graphene and the corresponding change in the frequency can be estimated as

$$\omega_{\rm ph}^{13_{\rm C}} = \omega_{\rm ph}^{12_{\rm C}} \sqrt{\frac{m_{12_{\rm C}}}{n_{12} m_{12_{\rm C}} + n_{13} m_{13_{\rm C}}}}, \tag{5.6}$$

where $n_{12}$ and $n_{13}$ represent the atomic fraction of $^{12}$C and $^{13}$C carbon atoms, respectively [68]. Similarly, $m_{12_{\rm C}}$ and $m_{13_{\rm C}}$ represent the atomic mass of $^{12}$C and $^{13}$C carbon atoms, respectively. We consider graphene entirely composed of $^{13}$C, and the corresponding frequency of the in-plane phonon mode comes out as 186.4 meV. In this case, it is assumed that the coherent **iLO** phonon mode with a frequency of 186.4 meV is excited by the pump pulse, and the resulting high-harmonic spectra is presented in Fig. 5.7. The main harmonic peak is unaffected by the isotope substitution, as expected. However, the sidebands exhibit sensitivity toward a change in the isotope. The shift in the peak of a sideband increases as the order of the sideband increases, i.e., the third sideband is most affected by the isotope substitution. It is instructive to use a probe laser pulse with a longer wavelength to separate the peaks of the sidebands associated with

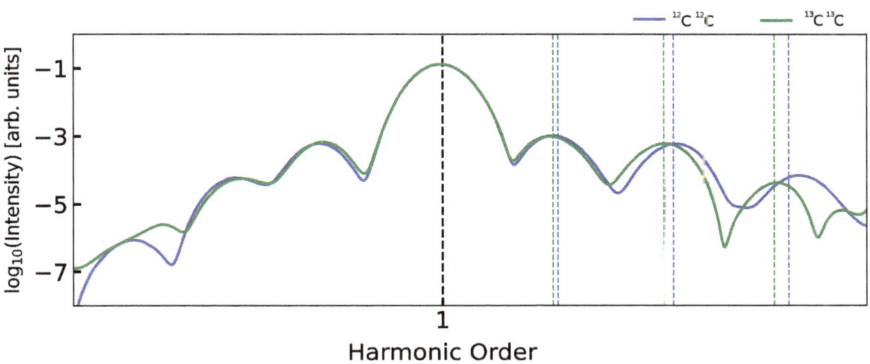


Fig. 5.7 Effect of the carbon isotope on high-harmonic spectroscopy of the coherent lattice dynamics corresponding to the in-plane **iLO** phonon mode in graphene. The parameters of the probe pulse are the same as in Fig. 5.2.

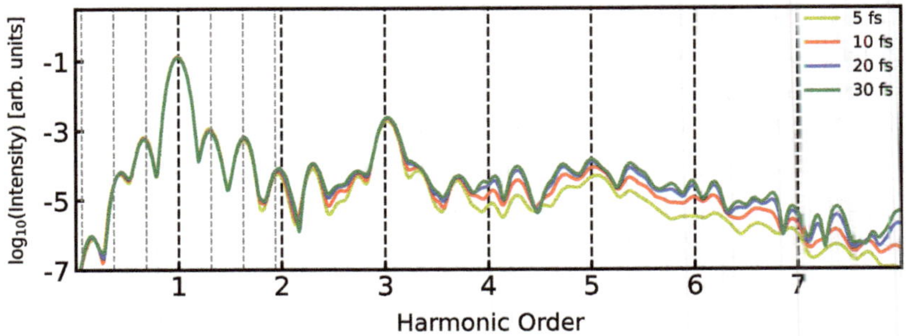


Fig. 5.8 Variations in the high-harmonic spectra corresponding to graphene with the coherent **iLO** phonon mode for the different dephasing time, $T_2$. The parameters of the probe laser pulse are the same as in Fig. 5.2.

graphene composed of $^{13}C$ and $^{12}C$ atoms and to avoid the effect of the peak broadening to discern isotopes.

Before we start our discussion on how the HHG is responsive to different parameters of the probe laser pulse, let us investigate how the emitted harmonics and sidebands are sensitive to the dephasing time — a phenomenological term accounting for the decoherence between electrons and holes in the semiconductor Bloch equations [see Eq. (5.2)]. Figure 5.8 presents the harmonic spectra corresponding to the coherent **iLO** phonon mode for different dephasing times. The low-order harmonics are insensitivity to the change in the dephasing time as the underlying mechanism for HHG from vibrating graphene is dominated by the intraband process. However, the intensity of higher-order harmonics are boosted as the dephasing time increases. It is known that the population in the conduction band increases as the dephasing time increases, which leads to an enhancement in the harmonic yield [69]. Note that the increase in the harmonic yield does not alter the polarization properties of the emitted harmonics as well as the sidebands.

### 5.3.2 *Impact of Laser Parameters*

In the previous section, we have examined how different properties of the vibrating graphene, such as displacement amplitude, anharmonicity, coupling between phonon modes, isotope-induced frequency change, and dephasing time affect the harmonic spectra. In this section, we will investigate how various parameters of the probe pulse influence HHG from the vibrating graphene. Along this line, high-harmonic spectra of graphene

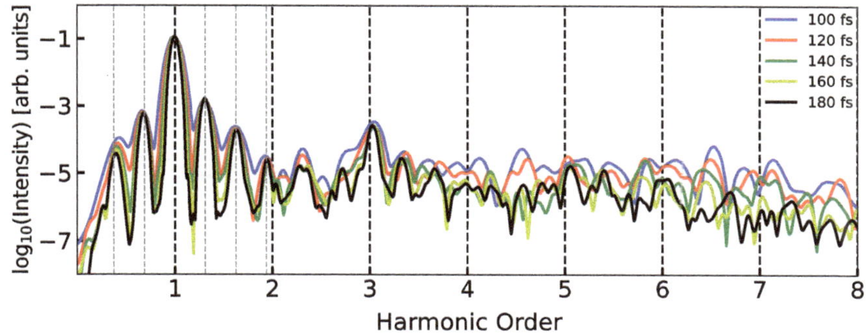


Fig. 5.9 High-harmonic spectra corresponding to the coherent **iLO** phonon mode for the different probe pulse duration. The rest of the probe pulse parameters are same as in Fig. 5.2.

corresponding to the coherent **iLO** phonon mode for different duration of the probe pulse is shown in Fig. 5.9. It is evident that the generation of the sidebands is insensitive to the pulse duration. The main harmonic peaks and the sidebands become sharper as the pulse duration increases. Owing to the recombination of an electron with a hole, a short burst of radiation is emitted during each half-cycle of the laser pulse. As the pulse duration becomes larger, multi-cycle accumulations of the radiation burst cause sharper peaks in the spectra. Note that the polarization properties of the main harmonics and the sidebands remain unaffected as the pulse duration increases.

How the intensity of the probe pulse changes the harmonic generation from the graphene with coherent **iLO** phonon mode is addressed in Fig. 5.10. The yield of the main harmonics and the sidebands increases as the intensity increases from $10^{10}$ to $5\times10^{11}$ W/cm$^2$. The drastic boost in the yield can be attributed to the significant increase in the conduction band population as the excitation probability increases with intensity. However, an interesting observation can be made when the harmonics are normalized with respect to the first harmonic corresponding to $10^{10}$ W/cm$^2$ intensity. The intensity of the sidebands decreases as the intensity increases. At higher intensities, the effect of the electron dynamics becomes more dominant, suppressing the contributions from the coherent lattice dynamics. In other words, the intensity of the main harmonic peaks increases due to a higher population of electrons in the conduction band. However, the influence of the lattice dynamics does not increase at the same rate at higher intensities.

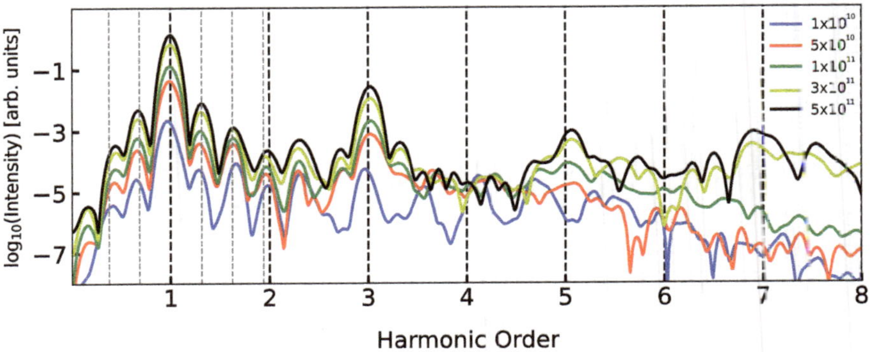


Fig. 5.10 High-harmonic spectra corresponding to the coherent **iLO** phonon mode for the different intensity of the probe pulse. The rest of the probe pulse parameters are same as in Fig. 5.2.

The intensity of the probe pulse does not alter other properties of the spectra.

After analyzing the effects of the pulse duration and intensity of the probe pulse, it is interesting to see how variation in the polarization of the probe pulse impacts the harmonic spectra. For this purpose, we tune the linear polarization of the probe pulse from $\mathbf{\Gamma} - \mathbf{K}$ to $\mathbf{\Gamma} - \mathbf{M}$ direction and the corresponding high-harmonic spectra associated with the coherent **iLO** and **iTO** phonon modes are depicted in Figs. 5.11 and 5.12, respectively. As discussed earlier, the six-fold symmetry of graphene reduces to the two-fold symmetry when atoms vibrate along the **iLO** phonon mode, and there is no reflection plane symmetry. Moreover, the even- (odd-) order sidebands are polarized parallel (perpendicular) to the polarization of the probe pulse, whereas the main harmonics are always polarized along the probe pulse. This means that the harmonics and the sidebands are emitted along $\mathbf{\Gamma} - \mathbf{K}$ and $\mathbf{\Gamma} - \mathbf{M}$ directions when the polarization of the probe pulse is not aligned along $\mathbf{\Gamma} - \mathbf{K}$ direction. Let's consider a situation where the probe pulse is polarized at 45° with respect to the $\mathbf{\Gamma} - \mathbf{K}$ direction. This can be seen as two orthogonally polarized probe pulses along $\mathbf{\Gamma} - \mathbf{K}$ and $\mathbf{\Gamma} - \mathbf{M}$ directions with an amplitude equal to $1/\sqrt{2}$ times of the original probe pulse along 45°. It means that the emitted harmonics and their sidebands have both components with the same magnitude, resulting in Epsilon equal to zero, as evident from Fig. 5.11. A similar analysis is applicable to the harmonic spectra corresponding to the **iTO** phonon mode, as shown in Fig. 5.12.

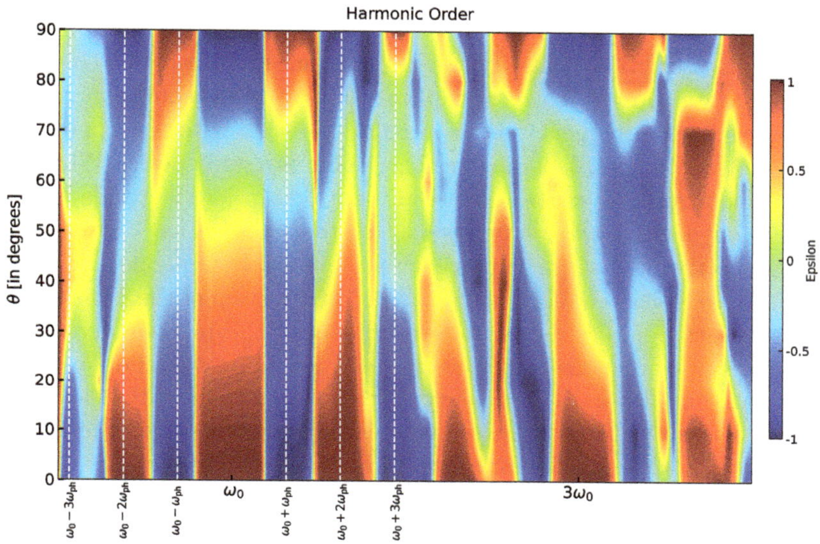


Fig. 5.11 Sensitivity of the harmonic generation for polarization direction of the probe pulse. Polarization of the main harmonics and the sidebands along $\mathbf{\Gamma - K}$ ($\mathbf{\Gamma - M}$) is represented by Epsilon equal to 1 ($-1$) in the colorbar. The rest of the probe pulse parameters are same as in Fig. 5.2. The harmonic spectrum corresponds to the coherent **iLO** phonon mode.

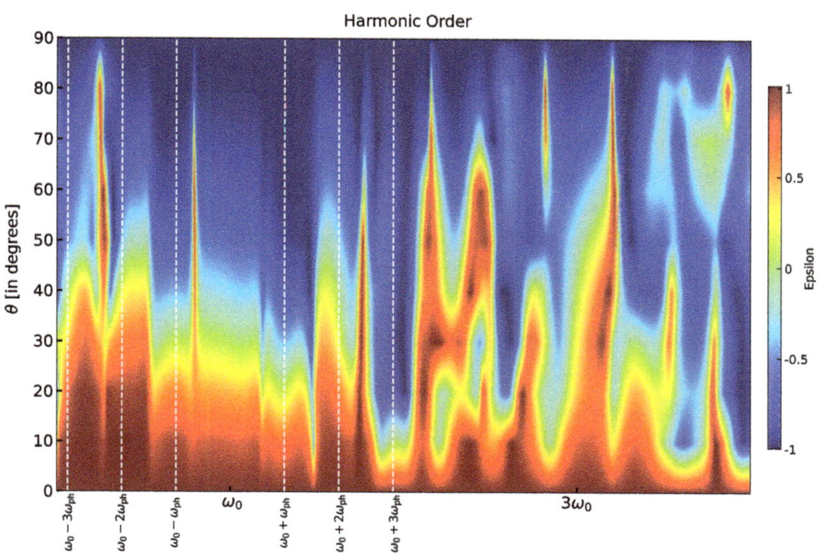


Fig. 5.12 Same as Fig. 5.11 for the coherent **iTO** phonon mode.

## 5.4 Summary

We have discussed the sensitivity of high-harmonic spectroscopy to coherent atomic vibrations in monolayer graphene. The high-harmonic spectrum is modulated by the frequency of the atomic vibration with the appearance of higher-order sidebands along with the main harmonic peaks. When the **iLO** phonon mode is excited, the even- and odd-order sidebands exhibit polarization parallel and perpendicular to the polarization of the probe pulse, respectively. However, upon excitation of the **iTO** phonon mode, all sidebands are polarized along the polarization direction of the probe pulse. These distinct polarization properties arise from the inherent dynamical symmetries governing the combined system, including the phonon modes and the probing laser pulse. Therefore, the polarization of the sidebands serves as a sensitive probe of these dynamical symmetries. Also, the six-fold symmetry of graphene reduces to two-fold dynamically due to the coherent atomic vibrations corresponding to a single phonon mode. This symmetry alteration generates symmetry-forbidden harmonics of circularly polarized probe pulses [70].

Our comprehensive investigation delved into various aspects of the vibrating graphene and its interaction with the probe laser pulse. Notably, we observed a direct correlation between the displacement amplitude and the intensity of sidebands, with the sidebands consistently amplifying as the displacement amplitude increased. In addition, we acknowledged the intrinsic limitations of our theoretical framework in capturing the role of anharmonicity. Another intriguing facet we explored was the impact of the coupling of two phonon modes, and it was found that analysis of the polarization of the emitted sidebands provides the coupling information. Furthermore, we have observed that the high-harmonic spectra of vibrating graphene is sensitive to the isotope substitution. Also, the properties of the sideband's polarization are robust with respect to the dephasing time.

We have also extended our study to explore the influence of various laser parameters on HHG from the vibrating graphene. As we prolonged the pulse duration, the polarization properties of both main harmonics and sidebands remained unaffected. Additionally, we have shown that the polarization of sidebands remains the same on increasing the intensity of the probe pulse. However, the intensity of the main harmonic peak and the sidebands increases. Finally, by changing the direction of the probe pulse from $\mathbf{\Gamma}-\mathbf{K}$ to $\mathbf{\Gamma}-\mathbf{M}$, we unveiled the ability to tune the polarization properties of the sidebands with the main harmonics. Overall, our findings demonstrate the potential of high-harmonic spectroscopy as a powerful tool for studying coherent lattice dynamics and its influence on electron dynamics in solids, paving the way for further advancements in the solid HHG.

## Acknowledgments

We acknowledge fruitful discussions with Misha Ivanov (MBI, Berlin) and M. S. Mrudul (Uppsala University, Sweden). G. D. acknowledges financial support from SERB India (Project No. MTR/2021/000138).

## References

[1] R. W. Boyd, *Nonlinear Optics*. Academic Press (2020).
[2] T. Brabec and F. Krausz, Intense few-cycle laser fields: Frontiers of nonlinear optics, *Reviews of Modern Physics* **72**, 2, p. 545 (2000).
[3] A. McPherson, G. Gibson, H. Jara, U. Johann, T. S. Luk, I. McIntyre, K. Boyer and C. K. Rhodes, Studies of multiphoton production of vacuum-ultraviolet radiation in the rare gases, *Journal of the Optical Society of America B* **4**, 4, pp. 595–601 (1987).
[4] M. Ferray, A. L'Huillier, X. F. Li, L. A. Lompre, G. Mainfray and C. Manus, Multiple-harmonic conversion of 1064 nm radiation in rare gases, *Journal of Physics B* **21**, 3, p. L31 (1988).
[5] M. Hentschel, R. Kienberger, C. Spielmann, G. A. Reider, N. Milosevic, T. Brabec, P. Corkum, U. Heinzmann, M. Drescher and F. Krausz, Attosecond metrology, *Nature* **414**, 6863, pp. 509–513 (2001).
[6] P.-M. Paul, E. S. Toma, P. Breger, G. Mullot, F. Augé, P. Balcou, H. G. Muller and P. Agostini, Observation of a train of attosecond pulses from high harmonic generation, *Science* **292**, 5522, pp. 1689–1692 (2001).
[7] F. Krausz and M. Ivanov, Attosecond physics, *Reviews of Modern Physics* **81**, 1, p. 163 (2009).
[8] G. Dixit, O. Vendrell and R. Santra, Imaging electronic quantum motion with light, *Proceedings of the National Academy of Sciences* **109**, 29, pp. 11636–11640 (2012).
[9] T. Bredtmann, M. Ivanov and G. Dixit, X-ray imaging of chemically active valence electrons during a pericyclic reaction, *Nature Communications* **5**, 1, p. 5589 (2014).
[10] K. Midorikawa, Progress on table-top isolated attosecond light sources, *Nature Photonics* **16**, 4, pp. 267–278 (2022).
[11] R. Borrego-Varillas, M. Lucchini and M. Nisoli, Attosecond spectroscopy for the investigation of ultrafast dynamics in atomic, molecular and solid-state physics, *Reports on Progress in Physics* **85**, 6, p. 066401 (2022).
[12] F. Lépine, M. Y. Ivanov and M. J. J. Vrakking, Attosecond molecular dynamics: Fact or fiction? *Nature Photonics* **8**, 3, pp. 195–204 (2014).

[13] S. Y. Kruchinin, F. Krausz and V. S. Yakovlev, Colloquium: Strong-field phenomena in periodic systems, *Reviews of Modern Physics* **90**, 2, p. 021002 (2018).

[14] S. Ghimire, A. D. DiChiara, E. Sistrunk, P. Agostini, L. F. DiMauro and D. A. Reis, Observation of high-order harmonic generation in a bulk crystal, *Nature Physics* **7**, 2, pp. 138–141 (2011).

[15] S. Ghimire and D. A. Reis, High-harmonic generation from solids, *Nature Physics* **15**, 6, pp. 10–16 (2019).

[16] M. Mrudul, Á. Jiménez-Galán, M. Ivanov and G. Dixit, Light-induced valleytronics in pristine graphene, *Optica* **8**, 3, pp. 422–427 (2021).

[17] M. Mrudul, N. Tancogne-Dejean, A. Rubio and G. Dixit, High-harmonic generation from spin-polarised defects in solids, *npj Computational Materials* **6**, 1, p. 10 (2020).

[18] A. Pattanayak, M. S. Mrudul and G. Dixit, Influence of vacancy defects in solid high-order harmonic generation, *Physical Review A* **101**, 1, p. 013404 (2020).

[19] M. Mrudul and G. Dixit, Controlling valley-polarisation in graphene via tailored light pulses, *Journal of Physics B* **54**, 22, p. 224001 (2021).

[20] E. Goulielmakis and T. Brabec, High harmonic generation in condensed matter, *Nature Photonics* **16**, 6, pp. 411–421 (2022).

[21] A. Bharti, M. Mrudul and G. Dixit, High-harmonic spectroscopy of light-driven nonlinear anisotropic anomalous Hall effect in a Weyl semimetal, *Physical Review B* **105**, 15, p. 155140 (2022).

[22] A. Pattanayak, S. Pujari and G. Dixit, Role of Majorana fermions in high-harmonic generation from Kitaev chain, *Scientific Reports* **12**, 1, p. 6722 (2022).

[23] M. S. Mrudul, A. Pattanayak, M. Ivanov and G. Dixit, Direct numerical observation of real-space recollision in high-order harmonic generation from solids, *Physical Review A* **100**, 4, p. 043420 (2019).

[24] G. Vampa, T. J. Hammond, N. Thiré, B. E. Schmidt, F. Légaré, C. R. McDonald, T. Brabec and P. B. Corkum, Linking high harmonics from gases and solids, *Nature* **522**, 7557, pp. 462–464 (2015).

[25] G. Vampa, T. J. Hammond, N. Thiré, B. E. Schmidt, F. Légaré, C. R. McDonald, T. Brabec, D. D. Klug and P. B. Corkum, All-optical reconstruction of crystal band structure, *Physical Review Letters* **115**, 19, p. 193603 (2015).

[26] F. Langer, C. P. Schmid, S. Schlauderer, M. Gmitra, J. Fabian, P. Nagler, C. Schüller, T. Korn, P. G. Hawkins, J. T. Steiner *et al.*, Lightwave valleytronics in a monolayer of tungsten diselenide, *Nature* **557**, 7703, p. 76 (2018).

[27] T. T. Luu and H. J. Wörner, Measurement of the Berry curvature of solids using high-harmonic spectroscopy, *Nature Communications* **9**, 1, p. 916 (2018).

[28] A. Bharti and G. Dixit, Role of topological charges in the nonlinear optical response from Weyl semimetals, *Physical Review B* **107**, 22, p. 224308 (2023).

[29] R. E. F. Silva, I. V. Blinov, A. N. Rubtsov, O. Smirnova and M. Ivanov, High-harmonic spectroscopy of ultrafast many-body dynamics in strongly correlated systems, *Nature Photonics* **12**, 5, p. 266 (2018).

[30] A. Chacón, W. Zhu, S. P. Kelly, A. Dauphin, E. Pisarty, A. Picón, C. Ticknor, M. F. Ciappina, A. Saxena and M. Lewenstein, Observing topological phase transitions with high harmonic generation, *arXiv preprint arXiv:1807.01616* (2018).

[31] O. Schubert, M. Hohenleutner, F. Langer, B. Urbanek, C. Lange, U. Huttner, D. Golde, T. Meier, M. Kira, S. W. Koch and R. Huber, Sub-cycle control of terahertz high-harmonic generation by dynamical Bloch oscillations, *Nature Photonics* **8**, 2, p. 119 (2014).

[32] T. T. Luu, M. Garg, S. Y. Kruchinin, A. Moulet, M. T. Hassan and E. Goulielmakis, Extreme ultraviolet high-harmonic spectroscopy of solids, *Nature* **521**, 7553, p. 498 (2015).

[33] O. Neufeld, J. Zhang, U. D. Giovannini, H. Hübener and A. Rubio, Probing phonon dynamics with multi-dimensional high harmonic carrier envelope phase spectroscopy, *Proceedings of the National Academy of Sciences* **119**, 25, p. e2204219119 (2022).

[34] S.-Q. Hu, H. Zhao, X.-B. Liu, D.-Q. Chen and S. Meng, Probing phonon dynamics and electron-phonon coupling by high harmonic generation in solids, *arXiv preprint arXiv:2304.09459* (2023).

[35] S. Patchkovskii, Nuclear dynamics in polyatomic molecules and high-order harmonic generation, *Physical Review Letters* **102**, 25, p. 253602 (2009).

[36] N. L. Wagner, A. Wüest, I. P. Christov, T. Popmintchev, X. Zhou, M. M. Murnane and H. C. Kapteyn, Monitoring molecular dynamics using coherent electrons from high harmonic generation, *Proceedings of the National Academy of Sciences* **103**, 36, pp. 13279–13285 (2006).

[37] A.-T. Le, T. Morishita, R. Lucchese and C. D. Lin, Theory of high harmonic generation for probing time-resolved large-amplitude molecular vibrations with ultrashort intense lasers, *Physical Review Letters* **109**, 20, p. 203004 (2012).

[38] B. Fultz, Vibrational thermodynamics of materials, *Progress in Materials Science* **55**, 4, pp. 247–352 (2010).

[39] J. Niedziela, D. Bansal, A. May, J. Ding, T. L.-Atkins, G. Ehlers, D. Abernathy, A. Said and O. Delaire, Selective breakdown of phonon quasiparticles across superionic transition in $CuCrSe_2$, *Nature Physics* **15**, pp. 73–78 (2019).

[40] L. Lindsay and C. A. Polanco, Thermal transport by first-principles anharmonic lattice dynamics, in W. Andreoni and S. Yip (Eds.), *Handbook of Materials Modeling: Applications: Current and Emerging Materials*, Springer, pp. 735–765 (2020).

[41] M. Dove, *Introduction to Lattice Dynamics.* Cambridge University Press (1993).

[42] N. Rana, A. P. Roy, D. Bansal and G. Dixit, Four-dimensional imaging of lattice dynamics using ab-initio simulation, *npj Computational Materials* **7**, 1, p. 7 (2021).

[43] N. Yoshikawa, T. Tamaya and K. Tanaka, High-harmonic generation in graphene enhanced by elliptically polarized light excitation, *Science* **356**, 6339, pp. 736–738 (2017).
[44] M. Taucer, T. J. Hammond, P. B. Corkum, G. Vampa, C. Couture, N. Thiré, B. E. Schmidt, F. Légaré, H. Selvi, N. Unsuree *et al.*, Nonperturbative harmonic generation in graphene from intense midinfrared pulsed light, *Physical Review B* **96**, 19, p. 195420 (2017).
[45] S. Cha, M. Kim, Y. Kim, S. Choi, S. Kang, H. Kim, S. Yoon, G. Moon, T. Kim, Y. W. Lee *et al.*, Gate-tunable quantum pathways of high harmonic generation in graphene, *Nature Communications* **13**, 1, p. 6630 (2022).
[46] H. K. Avetissian and G. F. Mkrtchian, Impact of electron-electron Coulomb interaction on the high harmonic generation process in graphene, *Physical Review B* **97**, 11, p. 115454 (2018).
[47] H. K. Avetissian, G. F. Mkrtchian, K. G. Batrakov, S. A. Maksimenko and A. Hoffmann, Multiphoton resonant excitations and high-harmonic generation in bilayer graphene, *Physical Review B* **88**, 16, p. 165411 (2013).
[48] M. Mrudul and G. Dixit, High-harmonic generation from monolayer and bilayer graphene, *Physical Review B* **103**, 9, p. 094308 (2021).
[49] I. Al-Naib, J. E. Sipe and M. M. Dignam, High harmonic generation in undoped graphene: Interplay of inter-and intraband dynamics, *Physical Review B* **90**, 24, p. 245423 (2014).
[50] L. A. Chizhova, F. Libisch and J. Burgdörfer, High-harmonic generation in graphene: Interband response and the harmonic cutoff, *Physical Review B* **95**, 8, p. 085436 (2017).
[51] H. A. Hafez, S. Kovalev, J. C. Deinert, Z. Mics, B. Green, N. Awari, M. Chen, S. Germanskiy, U. Lehnert, J. Teichert *et al.*, Extremely efficient terahertz high-harmonic generation in graphene by hot Dirac fermions, *Nature* **561**, 7724, pp. 507–511 (2018).
[52] Ó. Zurrón, A. Picón and L. Plaja, Theory of high-order harmonic generation for gapless graphene, *New Journal of Physics* **20**, 5, p. 053033 (2018).
[53] Ó. Zurrón-Cifuentes, R. Boyero-García, C. Hernández-García, A. Picón and L. Plaja, Optical anisotropy of non-perturbative high-order harmonic generation in gapless graphene, *Optics Express* **27**, 5, pp. 7776–7786 (2019).
[54] S. A. Sato, H. Hirori, Y. Sanari, Y. Kanemitsu and A. Rubio, High-order harmonic generation in graphene: Nonlinear coupling of intraband and interband transitions, *Physical Review B* **103**, 4, p. L041408 (2021).
[55] R. Boyero-García, A. García-Cabrera, O. Zurrón-Cifuentes, C. Hernández-García and L. Plaja, Non-classical high harmonic generation in graphene driven by linearly-polarized laser pulses, *Optics Express* **30**, 9, pp. 15546–15555 (2022).
[56] Y. Zhang, L. Li, J. Li, T. Huang, P. Lan and P. Lu, Orientation dependence of high-order harmonic generation in graphene, *Physical Review A* **104**, 3, p. 033110 (2021).
[57] C. Liu, Y. Zheng, Z. Zeng and R. Li, Driving-laser ellipticity dependence of high-order harmonic generation in graphene, *Physical Review A* **97**, 6, p. 063412 (2018).

[58] F. Dong, Q. Xia and J. Liu, Ellipticity of the harmonic emission from graphene irradiated by a linearly polarized laser, *Physical Review A* **104**, 3, p. 033119 (2021).

[59] N. Rana, M. Mrudul and G. Dixit, Generation of circularly polarized high harmonics with identical helicity in two-dimensional materials, *Physical Review Applied* **18**, 6, p. 064049 (2022).

[60] J.-H. Kim, A. Nugraha, L. Booshehri, E. Hároz, K. Sato, G. Sanders, K.-J. Yee, Y.-S. Lim, C. Stanton, R. Saito *et al.*, Coherent phonons in carbon nanotubes and graphene, *Chemical Physics* **413**, pp. 55–80 (2013).

[61] J. Wilhelm, P. Grössing, A. Seith, J. Crewse, M. Nitsch, L. Weigl, C. Schmid and F. Evers, Semiconductor Bloch-equations formalism: Derivation and application to high-harmonic generation from Dirac fermions, *Physical Review B* **103**, 12, p. 125419 (2021).

[62] N. Rana and G. Dixit, All-optical ultrafast valley switching in two-dimensional materials, *Physical Review Applied* **19**, 3, p. 034056 (2023).

[63] M. Wu, S. Ghimire, D. A. Reis, K. J. Schafer and M. B. Gaarde, High-harmonic generation from Bloch electrons in solids, *Physical Review A* **91**, 4, p. 043839 (2015).

[64] V. Mohanty and E. J. Heller, Lazy electrons in graphene, *Proceedings of the National Academy of Sciences* **116**, 37, pp. 18316–18321 (2019).

[65] M. Rodriguez-Vega, M. Vogl and G. A. Fiete, Direct driving of electronic and phononic degrees of freedom in a honeycomb bilayer with infrared light, *Physical Review B* **104**, 24, p. 245135 (2021).

[66] P. Moon and M. Koshino, Optical absorption in twisted bilayer graphene, *Physical Review B* **87**, 20, p. 205404 (2013).

[67] N. Rana, M. Mrudul, D. Kartashov, M. Ivanov and G. Dixit, High-harmonic spectroscopy of coherent lattice dynamics in graphene, *Physical Review B* **106**, 6, p. 064303 (2022).

[68] C. Zhang, Q. Li, B. Tian, Z. Huang, W. Lin, H. Li, D. He, Y. Zhou and W. Cai, Isotope effect of the phonons mean free path in graphene by micro-Raman measurement, *Science China Physics, Mechanics & Astronomy* **57**, pp. 1817–1821 (2014).

[69] C. Heide, Y. Kobayashi, A. C. Johnson, F. Liu, T. F. Heinz, D. A. Reis and S. Ghimire, Probing electron-hole coherence in strongly driven 2D materials using high-harmonic generation, *Optica* **9**, 5, pp. 512–516 (2022).

[70] N. Rana and G. Dixit, Probing phonon-driven symmetry alterations in graphene via high-order-harmonic spectroscopy, *Physical Review A* **106**, 5, p. 053116 (2022).

https://doi.org/10.1142/9789811279560_0006

# Chapter 6

# Analyzing High-Order Harmonic Generation in Solids Based on Semi-Classical Recollision Models

Hongdan Zhang[1], Ruixin Zuo[2], Shidong Yang[3], Alexander Trautmann[2], Xiaohong Song[1], Torsten Meier[2] and Weifeng Yang[1,4]

[1] *School of Physics and Optoelectronic Engineering, Hainan University, Haikou 570228, China*
[2] *Department of Physics, Paderborn University, Warburger Str. 100, D-33098 Paderborn, Germany*
[3] *Institute of Mathematics, Shantou University, Shantou, Guangdong 515063, China*
[4] *Center for Theoretical Physics, Hainan University, Haikou 570228, China*

## 6.1 Introduction

Rapid advances in the development of highly intense lasers have paved the way for the study of nonlinear optics, i.e., interactions between light and matter, in very extreme regimes which resulted in the development of powerful approaches for controlling and probing ultrafast processes in matter [1–6]. High-order harmonic generation (HHG) is one of the most important laser-induced nonlinear phenomena and is closely related to the so-called collision dynamics, which is vividly being described by the three-step model or semi-classical recollision model which was introduced for gaseous atoms [7, 8]. First, under the influence of a strong laser field, the atom is ionized due to electron tunneling. Then the unbound electron is accelerated by the laser field and moves like a free charged particle away from and then back to the parent ion when the field changes its polarity. Finally, the electron recollides with the parent ion, which leads to the emission of high-energy radiation in the form of high harmonics of the laser frequency (see Fig. 6.1(c)). The emitted high-energy photons encode

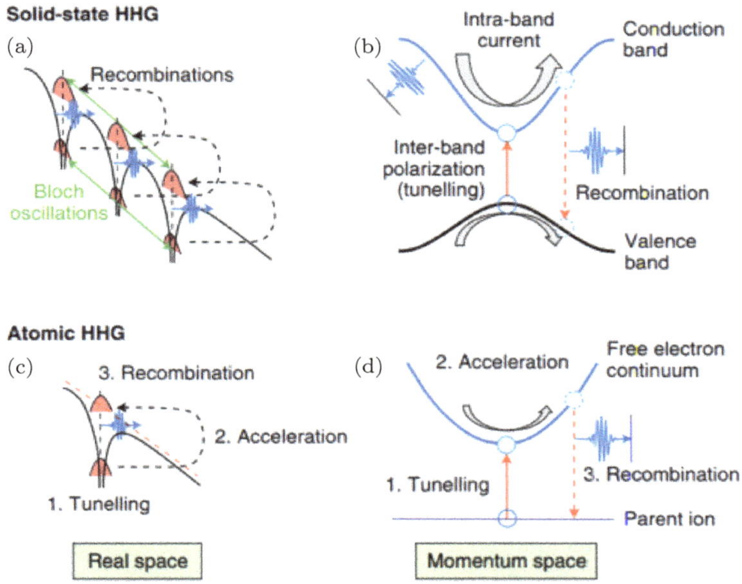


Fig. 6.1 (a), (c) Visualization of solid and atomic HHG in real space. In (c) the three-step recollision model is described, while the recollision in (a) for solids is, in general, not limited to the parent ion but may occur at other ions. (b), (d) Sketches for solid and atomic HHG in momentum space, in correspondence with (a) and (c), respectively. Reproduced with permission from [17].

a lot of important structural and dynamic information about the atoms or molecules they leave behind. This semi-classical recollision model successfully explains the main features of atomic HHG offering a fundamental insight into the microscopic HHG processes, and predictions based on this model have received numerous experimental validations [9–14].

Initially, the generation of HHG and studies of other high-field effects were performed on atoms and small molecules. More recently, experiments have demonstrated that HHG can also be generated by subjecting solid-state systems to strong laser fields. Following the first pioneering experiment which demonstrated HHG from a bulk ZnO crystal [15], many experimental and theoretical studies on solid HHG have emerged [16–91]. Ghimire *et al.* [15] reported the first non-perturbative HHG experiment in solids. Here, a ZnO sample is driven by a mid-infrared laser and harmonics up to the $25^{\text{th}}$ order were obtained. It is demonstrated that in the non-perturbative regime the high harmonic plateau scales linearly with the amplitude of the driving laser field and the wavelength. Afterwards, solid HHG experiments were also performed with THz fields of longer wavelengths [69, 92] and on more materials with different band gaps and other material properties, such as, e.g., monolayer graphene, GaSe, MgO, and $MoS_2$ [69, 93–98]. Various theoretical approaches have been adopted to simulate solid HHG, such as the semiconductor Bloch equations [24, 67, 99–102], numerical solutions of the time-dependent Schrödinger equation [53, 103–106], and time-dependent density functional theory [107, 108]. Besides, several effects contribute to and views can be used to analyze solid HHG, such as dynamic and interband Bloch oscillations [109], the influence of the Berry curvature [43, 102], and a strong-field perspective provided by a Wannier-Stark state-based approach [110] and so on.

Generally, two mechanisms leading to HHG in solids have to be considered, the interband and the intraband currents [99] as is shown in Fig. 6.1(b). Interband current is a rough analogy to the ionization and recombination involved in the atomic recollision model. As early as fifty years ago, Keldysh introduced a crucial idea. Treating both gases and solids on equal footing, he showed that under appropriate conditions the valence to conduction band transitions in solids can be approximated by tunneling as the strong-field ionization in gases (i.e., the first step) [111]. Based on this approach, the electron in the conduction band and the corresponding hole in the valence band, both being accelerated by the driving laser field, are in some kind of correspondence to the electron that has tunneled away from the atom and is then accelerated by the laser field (i.e., the second step). When the conduction band electron recombines with its hole in the valence band by stimulated interband recombination, a high-energy photon is emitted (i.e., the third step). Here, the prefix 'inter' in the interband mechanism highlights the transition from conduction to valence band in the final recombination step. Figure 6.1(b) nicely illustrates solid HHG in momentum space, in comparison with atomic HHG in Fig. 6.1(d).

The second mechanism is known as intraband current, which is different from the typical recollision model in gases. After the 'tunneling step', which is still the transition from the valence to conduction band, the electron is driven by the laser field in the anharmonic conduction band. The hole is similarly driven by the laser in the same manner in the anharmonic valence band. The strongly anharmonic motion of the electrons and holes in respective bands leads to an overall anharmonic current, which is the so-called intraband current (see Fig. 6.1(a) in real space). Likewise, the prefix 'intra' depicts the current originating within separate bands.

Though the above two mechanisms can qualitatively explain the solid HHG, remarkable differences exist between atomic and solid HHG. These originate from the complexity of crystalline systems which represent a dense periodic arrangement of many atoms that are bound together by interatomic forces. Since the internuclear distance between the atoms in gases is very large, each atom can be regarded as an independent source for HHG. For periodically arranged atoms in crystals the internuclear distances are much shorter. Correspondingly, interactions between atoms are much stronger than those in atomic gases. These strong interatomic interactions are responsible for the existence of stable crystals and give rise to their distinct properties, many of which are manifested in the electronic band structure. Therefore, on the one hand, understanding solid HHG on a microscopic level is very complex; however, on the other hand, solid HHG offers a large variety of phenomena and design possibilities which do not exist in atoms.

Since HHG emitted from solid crystals was observed [107, 108], it has frequently been compared with atomic HHG. There are two main reasons for this: The first is historical, since the research on HHG in atomic gases laid the foundation for attosecond science, which was later extended to bulk crystals. The second is that the three-step recollision model in gaseous media has similarities to the interband current in solid HHG. That is the reason why some characteristics that are similar between solid and atomic HHG can be interpreted as the dominance of the interband current over the intraband current [23, 51]. However, as we will discuss subsequently there are distinct discrepancies between solid and atomic HHG due to different electronic dynamics and emission mechanisms.

At first sight, the difference of atomic gas and solid crystal lies in the arrangement of atoms. The density of solid crystals is much higher than that of atomic gases, which should certainly enhance the HHG signal. Considering that higher laser intensity can lead to enhancement of atomic HHG yield, it seems that the HHG yield of solid crystals should be higher than that of gaseous atoms also for higher laser intensities. However, solid crystal can suffer damage when exposed to too intense laser fields beyond their damage threshold [17, 94, 100]. Typically, damage occurs when the conduction bands are too highly populated. The energetically excited electrons and the generated holes reduce the attractive interatomic forces which hold the crystal together and when too much energy is deposited in the crystal its structure will be damaged. Choosing materials with large band gaps reduces the level of excitations from the valence band to the conduction band for fixed laser parameters. For a chosen material one will use photon energies and intensities of driving laser to stay below the respective damage threshold. This explains the usual choices of driving laser with rather long wavelengths and low intensities in solid HHG.

In this chapter, we will present an overview of the comparison of HHG in solids and gases from the perspective of semi-classical three-step models. Further comparisons are presented following the research development of solid HHG. Finally, we summarize and provide an outlook on the potential of future studies.

## 6.2 Solid-State High-Order Harmonic Generation

### 6.2.1 *General Comparison of HHG in Gases and Solids*

In Ref. [15], HHG from a ZnO crystal is obtained by driving it with a quite long wavelength centered at 3.25 $\mu$m. It is found that for a high driving laser field strength the high-energy cutoff position can be as high at the 25$^{\text{th}}$ harmonic, as shown in Fig. 6.2(a). Besides, the high-energy

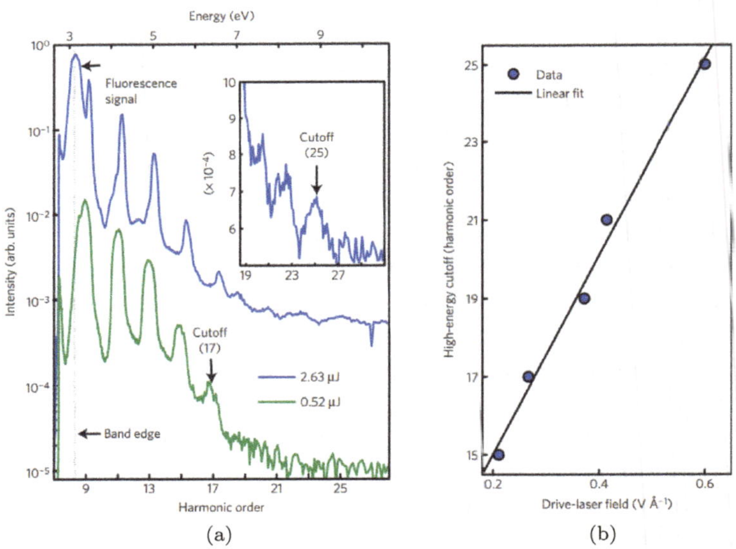


Fig. 6.2 (a) Measured high-order harmonic spectra for laser field energies of 0.52 $\mu$J (solid green line) and 2.63 $\mu$J (solid blue line), respectively. (b) The field dependence of the high-energy cutoff, which scales linearly with the laser field amplitude. Reproduced with permission from [15].

cutoff scales linearly with driving laser field amplitude, which can be clearly seen in Fig. 6.2(b). Theoretically, a quasi-classical model was introduced to interpret the linear dependence of the cutoff on both the wavelength and field strength of laser pulse [53]. The linear dependence of the cutoff on the laser field amplitude is different to the case of gaseous atoms, where the dependence on the laser field strength is quadratic [51].

In general, gas-phase HHG yields depend on the ellipticity of the driving laser fields [112,113]. Budil *et al.* [112] conducted experiments on noble gases (Ne, Xe, and Ar) driven by a laser with a wavelength centered at 825 nm and studied the dependence of HHG on ellipticities. As the ellipticity increases, the intensity of every high-order harmonic decreases. This could be interpreted by the three-step recollision model with ionization, acceleration, and recombination of electrons. When the ellipticity increases to 1 (i.e., circular polarization), the ionized electron hardly recombines with its parent nucleus because the trajectory of electron greatly drifts away. Solid HHG contains much more rich ellipticity-dependent features compared to atomic HHG, which is directly related to the more complex structure of materials. For example, HHG in rare-gas solids exhibits an atomic-like dependence on the ellipticity [51]. In contrast, the emission is not that sensitive to ellipticity in ZnO, where the HHG yields decrease only by a small extent. The decreases are more pronounced in the higher-energy region [15]. However, the HHG shows an anomalous and anisotropic ellipticity dependence in single-crystal MgO [94]. Furthermore, the situation is much more complex in two-dimensional crystals. For graphene the HHG yield is enhanced when driven by elliptically polarized laser fields [97], but it is suppressed with increasing ellipticity in monolayer $MoS_2$ [98]. The origin for these differences lies in the bonding character between the atoms which plays an important role for the strong-field nonlinear response.

Besides the ellipticity-dependent characteristics, the solid HHG of bulk crystals is strongly anisotropic. Actually, many earlier studies on solid HHG adopt one-dimensional models [45, 51, 93, 114]. This is not only computationally convenient, but also provides a simple and intuitive picture of solid HHG. The physical mechanisms can be well described including even the energy band reconstruction. For example, Ref. [114] presented that the energy dispersion of the lowest conduction band of $SiO_2$ has been successfully retrieved. To achieve this, the authors of [112] assumed that intraband currents are the dominant source for HHG. The dependence of solid HHG on the interaction between multiple bands has been confirmed in $GaSe_9$ [51]. Already these few selected results clearly demonstrate the huge potential of solid HHG in probing the structure of energy bands, even when limited to a one-dimensional analysis. However, clearly one-dimensional models cannot fully describe the complete dynamics following extremely nonlinear light-matter interactions.

To better describe and understand solid HHG, higher dimensional models describing the periodic potentials of bulk crystals are indispensable. In multi-dimensional solid crystals, the atoms have a periodic geometrical arrangement and the energy band structure landscapes depend on the direction in reciprocal space. In Ref. [94], You *et al.* measured the orientation-dependent HHG of a (001)-cut MgO sample driven by a linearly polarized laser with wavelength centered at 1.3 $\mu$m. The signal for different orientations are obtained by rotating the MgO sample around the axis along the laser propagation direction, as shown in Fig. 6.3(a). Figure 6.3(b) displays the measured HHG covering harmonic orders from $13^{\text{th}}$ to $21^{\text{st}}$ for various orientations showing a four-fold symmetry. The integrated HHG signal from $13^{\text{th}}$ to $21^{\text{st}}$ order shown in Fig. 6.3(c) confirms the four-fold periodicity which reflects the symmetry of the crystal. Moreover, further experiments were conducted on single crystal sapphire prepared by cutting along the M-plane and the C-plane which have a fourfold and a sixfold symmetry, respectively [115]. The wavelength of driven laser field is 800 nm and the HHG yield also presents orientation dependence, corresponding to the symmetry of the sapphire single crystal [115]. Actually, this result can be understood by considering one-dimensional cuts of the band structure which include the band gap and are oriented along the polarization of the driving laser. With such models the n-fold symmetry of the HHG signal can clearly be reproduced [116].

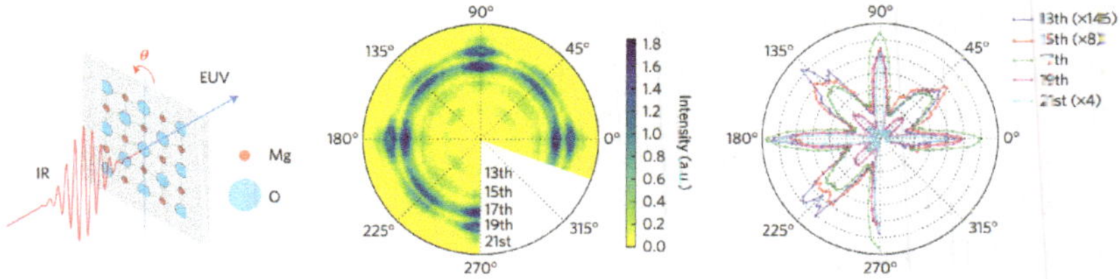


(a) (b) (c)

Fig. 6.3 (a) A strong infrared laser with wavelength centered at 1.3 $\mu$m is focused on a (001)-cut MgO crystal. The orientation dependence of HHG is realized by rotating the crystal with respect to the fixed laser polarization. (b) Measured high-harmonic spectra for various orientations. (c) Integrated intensity for different orientations for harmonic orders from $13^{\text{th}}$ to $21^{\text{st}}$. Reproduced with permission from [94].

### 6.2.2 *Detailed Comparison from the Perspective of the Three-Step Model*

In this subsection, we focus on the difference between the HHG from gaseous atoms and solid crystals using the three-step model. Following the essence of HHG in atoms [7], Vampa *et al.* [23, 24, 45] proposed a generalized three-step recollision model mapping from atomic to solid HHG. With this model, the main mechanism of solid HHG in the off-resonant limit can be well understood, when mid-infrared driving laser frequencies are much smaller than the band gap. Besides, they found that the interband current dominates in solid HHG for mid-infrared driving pulses by adding a weak second-harmonic field.

From the perspective of the three-step model, the above differences between solid and gaseous HHG are mainly reflected in three aspects: First, the ionization process of atomic HHG is not affected by electronic motion in the acceleration process, since the electron tunnels from a bound state. However, in semiconducting materials, the electrons occupy delocalized states in both the valence and the conduction bands, and it has been demonstrated that the above interband and intraband currents are not independent but closely coupled [99]. The influence of the complex coupled dynamics of the two mechanisms is not limited to the carrier injection from the valence band into the conduction band, but affects also the motion of the excited electron wave packet [117–119]. Second, different from atomic HHG, the acceleration step in the three-step model for solids contributes an additional source to solid HHG, in the form of an intraband current [98, 114, 120]. Third, unlike HHG in atoms, in solids the hole generated by the interband tunneling also moves in the excitation process since it is accelerated in its band. Thus, the recombination of the electron is not limited to occur at its original site. It can instead recollide with neighboring atomic sites of the crystal [28, 55, 94], as shown by the black dashed line in Fig. 6.1(a).

The semi-classical theory has successfully explained spectral structures of HHG with an intuitive picture and is widely applied [17, 46, 51, 53, 105, 121]. However, it is not able to provide a satisfactory quantitative description. In Ref. [29] an abnormal ellipticity dependence of the HHG yield of ZnO driven by elliptically or circularly polarized laser field was found. The HHG yield under elliptically polarized or circularly polarized laser fields turned out to be comparable to that under linearly polarized excitation. For some harmonic orders, the former is even higher than the latter. This ellipticity dependence is counterintuitive from the perspective of the three-step recollision model. To explain this abnormal phenomenon, Li *et al.* [29] developed a reciprocal-space-trajectory method based on accelerated Bloch states and proposed a four-step model to describe the relevant physical mechanisms. In this model, the solid HHG is obtained by considering an ensemble of trajectories with different ionization times as well as initial states in reciprocal space.

Figure 6.4 is a sketch which describes the four-step model in reciprocal space considering one valence band and one conduction band. The dynamical processes denoted by '0' to '3' are the four processes of solid HHG. The dynamics from '1' to '3' are the processes described by the three-step model adapted to solids [22, 24, 53, 103]. In this model, the electron is initially located at the top of the valence band, then it is vertically excited into the conduction band at the same crystal momentum and later on it is accelerated. However, the '0' process is different not only from the initial state of electrons in atomic gases, but also from the three-step model for solids. In the former case, the electron is bound in the ground state before ionization. In the latter case, the electron is located at the top of the valence band when ionization occurs. Prior to interband tunnelling the electron is accelerated in the valence band and driven by the laser field to the top of the band. This acceleration process before ionization is named as preacceleration. The following processes from '1' to '3' are the same as the three-step model for solids.

The reciprocal-space-trajectory analysis method enables us to decompose HHG which originates from different trajectories, and further depict

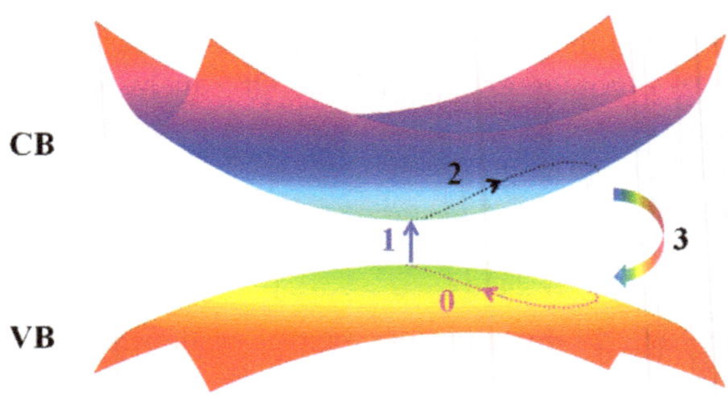


Fig. 6.4 Sketch of electron trajectories in two-dimensional reciprocal space. The red dashed line denotes the preacceleration in valence band, and the black dashed line denotes the acceleration in conduction band. Reproduced with permission from [29].

a clear picture of solid HHG [29]. By comparing the total HHG yield with separate spectra with and without preacceleration, they further confirm the importance of the preacceleration. It is for example found that the farther away from the top of the valence band the initial location of each electron trajectory is, the higher is the energy cutoff of HHG. It should be noted that with the added preacceleration the four-step model should be viewed as an improved version of the three-step model and that taking preacceleration into consideration is now widely accepted [28, 35, 43, 122]. Song *et al.* [35] confirmed the validity of the four-step model also for ZnO driven by multi-color linearly polarized laser fields. It is demonstrated in Ref. [35] that the preacceleration process prior to ionization results in a robust enhancement of both the HHG yield and the population in the conduction band.

Recently, Lang *et al.* [90] presented an extended three-step model considering three bands for describing solid HHG driven by both infrared and extreme ultraviolet pulses. In the absence of the extreme ultraviolet pulse, the illustration in Fig. 6.5(b) is the common three-step process. A more complex recollision process is depicted in Fig. 6.5(c). With the assistance of extreme ultraviolet pulse, the electrons that are far from the center of the Brillouin zone can be accelerated to the energetically highest point of the valence band V1 at $t_1$, then excited to the energetically lowest

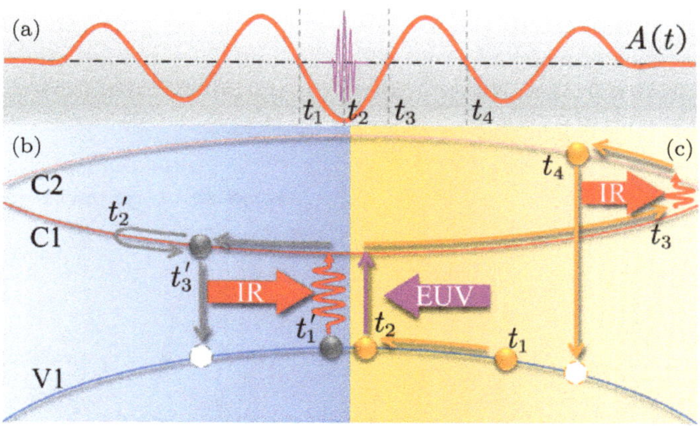


Fig. 6.5 Illustration of the extended three-step model of solid HHG driven by both infrared and extreme ultraviolet pulses. (a) Vector potential of the incident infrared (red) and extreme ultraviolet (purple) pulses. (b) Electron trajectory in the conventional three-step model for HHG in solids. (c) Electron trajectory in the extended three-step model. Reproduced with permission from [90].

point of the conduction band C1 through one-photon absorption at $t_2$. After being accelerated on lowest conduction band C1, the electrons have possibility to tunnel [87] to higher conduction band C2 near the Brillouin zone boundary driven by infrared pulse at $t_3$. When electrons recombine with holes at valence band V1, higher-energy photons are irradiated at $t_4$. The extended three-step model well describes the extension of high-energy cutoff, providing a clear physical process.

Yu *et al.* [50] investigated bilayers of two-dimensional materials and explored the interlayer scattering in bilayer hexagonal boron nitride ($h$-BN) with large interlayer spacing. In contrast to HHG spectra from monolayer $h$-BN, the HHG spectra from bilayer $h$-BN show a double-plateau structure, with a cutoff being highly extended. For monolayer $h$-BN, the HHG can be depicted by an atomic-like three-step recollision model, as can be seen in Fig. 6.6(a). For distant bilayer $h$-BN, another four-step model was proposed to understand the increase of the cutoff energy. As shown in Fig. 6.6(b), the backscattering of the electrons from the neighboring layer is taken into account. First, electrons are ionized from monolayer $h$-BN. Second, confined by the interlayer spacing, the electrons are backscattered by the other layer before being slowed down to zero velocity, which constitutes an additional step relative to the three-step model. Third, the scattered electrons are accelerated and obtain non-zero velocity after scattering. Finally, the electrons recombine in the original layer, emitting high-energy photons. Apparently, compared to the four-step model taking preacceleration into account in the 'ionization' step [29], the four-step model of [50] takes

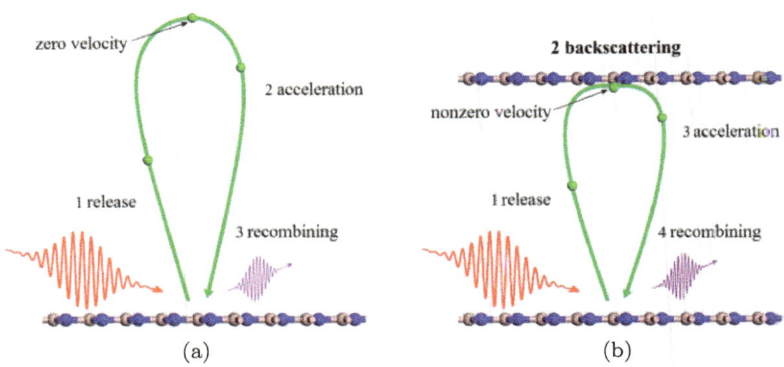


Fig. 6.6 Schematic HHG emission described by (a) three-step model and (b) four-step model adapted to describe monolayer and distant bilayers of hexagonal boron nitride, respectively. Reproduced with permission from [50].

backscattering into consideration in the 'acceleration' step to interpret the HHG from bilayer $h$-BN.

Hansen *et al.* [73] extended the original semi-classical recollision model from two bands to multiple bands. Furthermore, Fu *et al.* [25] not only verified the validity of the recollision model in real space, but adopted a three-band recollision model to analyze the HHG in solids. They vividly presented the physical mechanisms connected to different kinds of electron and hole trajectories. The first type of trajectories (T1) in k space (shown in Fig. 6.7(a)) and real space (shown in Fig. 6.7(b)) involves only one conduction band and the high-energy photon is emitted when the hole and electron encounter each other. The recollision position is labeled by the red solid arrow in Figs. 6.7(a) and (b). The second type (T2) involves both the first and the second conduction bands. The formation mechanism of T2 is similar to that of T1 at the first recollision. However, in T2 the electron continues to move in the conduction band after the first recollision. The electron moves away from the center of the first Brillouin zone and is transferred to the second conduction band when it reaches the position in

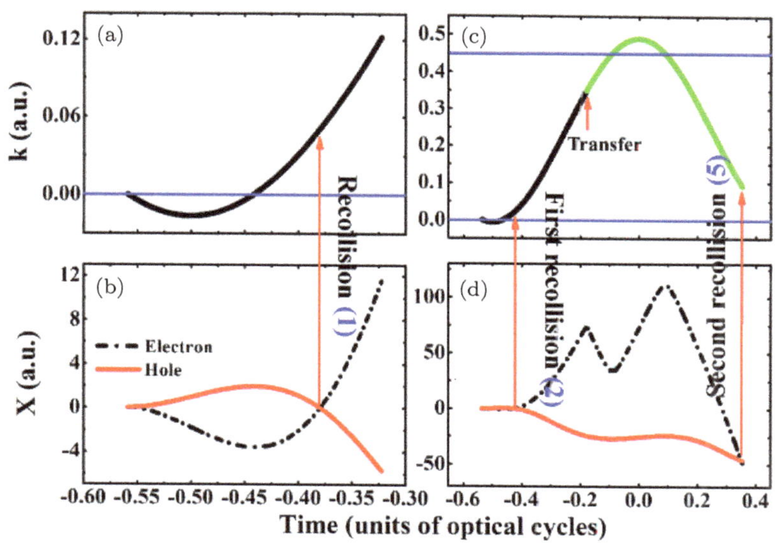


Fig. 6.7 Two kinds of typical trajectories of electrons and holds in (a, c) $k$ space and (b, d) real space. The first and second conduction bands are labeled by black and green solid line, respectively. The boundary of the first Brillouin zone and the position of $k = 0$ are marked with blue lines. Reproduced with permission from [25].

k-space where the two bands cross. After velocity reversals happening three times, the electron finally recollides with the hole again, which is labeled by the red solid arrow in the second recollision in Figs. 6.7(c) and (d). Note that only two typical trajectories are presented here; additional kinds of trajectories can be found in Ref. [25] which also provides additional details and visualizations of recollisions in three-band systems.

The intraband current is an additional source for HHG in solids with no counterpart in HHG of gaseous atoms. It is still debated which of the two currents, i.e., interband or intraband, is dominant for solid HHG [15, 24, 45, 48, 67, 69, 93, 94, 114, 123–125]. Clearly there is no simple answer to this question as it depends on the investigated material, the experimental conditions, and which emission frequencies are considered. A large amount of experimental and theoretical studies have demonstrated that the interband current dominates above the band gap for GaAs, MgO, and ZnO driven by mid-infrared fields [15, 22, 24, 95, 99, 120, 126]. However, Ref. [48] mentioned that in almost all semiconductors, the Rabi frequency is usually much lower than the Bloch frequency, when they are driven by the same laser, which actually gives an intuitive physical picture. A considerable amount of electrons traverses the Brillouin zone in a much shorter time scale than the interband transition process. Thus, although the interband current dominates the emission in many studies, the intraband acceleration cannot be neglected as it influences the carrier dynamic and thus the interband emission. It has also been reported in Ref. [124] that the carrier excitation from the valence band to the conduction band does strongly depend on the coupling between the interband and intraband mechanisms. Though the intraband current is constrained only in one separate energy band, an interband excitation is the starting point of the nonlinear dynamics without which no intraband dynamics and no corresponding emission of HHG would exist. Reference [36] demonstrated that the carrier excitation from the valence to the conduction band is limited to an extremely short time scale due to the intraband motion. It was further confirmed that the existence of the intraband motion has great influence on interband polarization and, in fact, contributes to making the interband current the dominant mechanism.

For the generation of high harmonics in gaseous atoms the electrons have to return to the original site, i.e., their parent ion, in real space, which is different to the situation in bulk crystals [28, 43, 44, 55, 105]. In 2017, Liu *et al.* [55] found that the electron can recollide with neighboring atomic sites and that the cutoff energy is linearly related to the migration distance of the electron. In Ref. [105], Osika *et al.* developed an analytic model using a mixed representation: delocalized Bloch functions and localized Wannier wave functions in the conduction band and valence band, respectively. Based on this model, the authors provide a clear physical picture, which is visualized in Fig. 6.8. First, the localized electron (green solid area) located at the atomic site $j_0$ tunnels from the valence to the conduction band at time $t'$. Later, the excited electron is accelerated in the conduction band under the influence of the driving laser from $t'$ to $t$. At the same time, the initially localized electron wave function (blue solid area) spreads along the periodic lattice. Finally, the electron recombines at the $j^{\text{th}}$ site (original site or other site), which is denoted by red arrows in Fig. 6.8. The two violet solid lines depict the high-energy photon emission upon different recollision

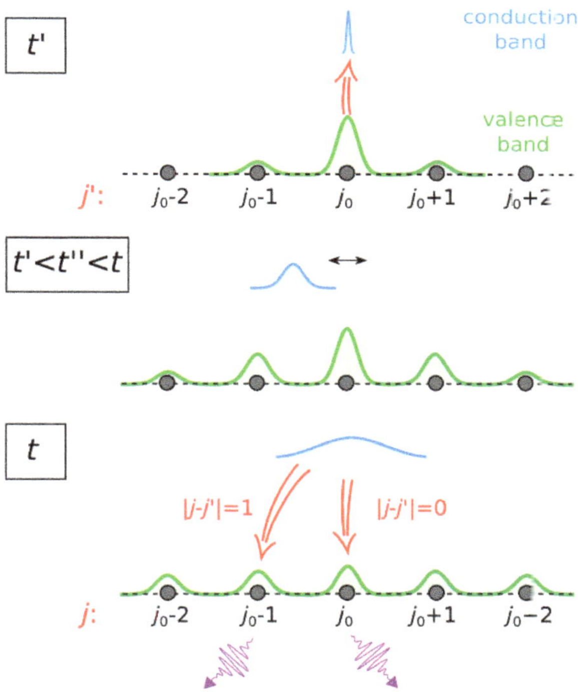


Fig. 6.8 Schematic illustration of the electron excitation-recollision process for HHG originating from a crystal with a periodic lattice. The positions of the atoms of the periodic lattice are represented by gray circles along the black dashed line. The localized Wannier wave functions are depicted in green. The wave packet excited at time $t'$ from the $j_0^{\text{th}}$ site to the conduction band is denoted by the blue area. Later on this wave packet is accelerated within the conduction band. Reproduced with permission from [105].

sites. Since the Bloch wave functions are delocalized in coordinate space, the electron and hole wave packets can overlap even if their centers do not coincide. In such situations, an additional source contributes HHG which arises from the electron-hole pair polarization energy, which was named imperfect recollision [43]. By developing an extended recollision model, Yue *et al.* [43] confirmed that the coherent radiation of a delocalized electron wave packet originates from recollisions with its correlated hole [22, 23].

Recently, Zuo *et al.* [28] presented another neighboring atom scattering-assisted electron-hole recombination mechanism. They found that the electron does not directly recombine at the neighboring atom but is scattered by it. When the driving pulse is linearly polarized, the recollision probability with its associated hole can increase by backward scattering with neighboring atoms, which provides an interpretation for the anisotropy of HHG from MgO. Figures 6.9(a) and (b) present the two-dimensional conduction and valence bands, respectively. The deep pink, green, and

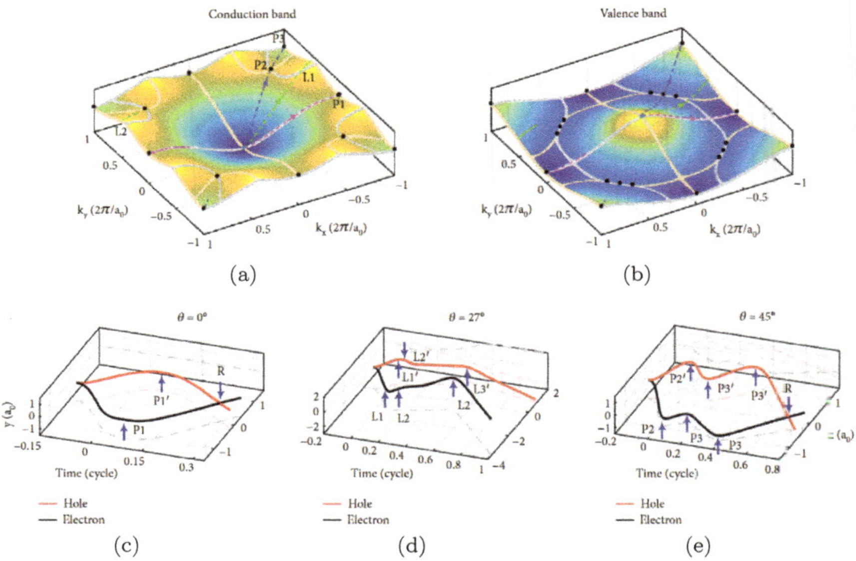


Fig. 6.9 Two-dimensional (a) conduction and (b) valence band structures of MgO. The black dots indicate the van Hove singularities. The time-dependent position of the electron (black) and hole (red) trajectories that ionize before the field crest for (c) $\theta = 0°$, (d) $\theta = 27°$, and (e) $\theta = 45°$, respectively. Here, $\theta$ is the angle between the linear polarization direction of the driving laser field and the nearest-neighbor bonding direction of MgO. Reproduced with permission from [28].

yellow dashed arrow lines are electron trajectories at $\theta = 0°$, $27°$, and $45°$, respectively, in momentum space. Figures 6.9(c), (d), and (e) show the corresponding trajectories of electrons (black solid lines) and holes (red solid lines) for laser polarizations along three directions. The black dots in Figs. 6.9(a) and (b) represent the van Hove singularities with $\nabla_k E_{c/v}(\mathrm{k})=0$ in momentum space. Comparing the electron/hole trajectories in momentum space (dashed arrow lines in Figs. 6.9(a) and (b)) to that in real space (Figs. 6.9(d) and (e)), it can be clearly seen that the electron and hole reverse the directions of their motion in real space once they get to the critical points/van Hove singularities. Such an inversion is attributed to collisions which originates from the backscattering of the electron/hole with neighboring atoms. For $\theta = 0°$, one critical point exists in this direction. Accordingly, the electron trajectory labeled by P1 in Fig. 6.9(c) undergoes direction reversion just once. After the backscattering by neighboring atoms, the electron and hole finally reencounter with each other. The physical picture for $\theta = 45°$ is similar to that for $\theta = 0°$. The only difference is that there are three critical points along the trajectory for this angle, thus, correspondingly the electrons and holes reverse their movement directions three times (P2, P3 trajectories). This mechanism can also explain the delocalization mechanism recently reported in HHG from ZnO, i.e., that an electron ionized in the periodic lattice may recombine at another site [105]. In addition to the backward scattering, there also exists forward scattering during recollision. For $\theta = 27°$, the electron and its associated hole that are generated before the laser crest pass through critical lines which are depicted by the green dashed arrows in Figs. 6.9(a) and (b). By further comparing the electron and hole trajectories in momentum and real space, it was demonstrated that whenever the electron or hole crosses a critical line in momentum space, the electron or hole trajectory in real space, labeled by L1 and L2 in Fig. 6.9(d), experiences a forward scattering at neighboring atoms. After several side collisions of forward scattering, the electron gets much farther away from its associated hole. The forward scattering is not limited to linearly polarized laser fields. Figure 6.10 depicts the trajectories containing forward scattering of electrons (black line) and holes (magenta line) with ellipticity of 1 and 0.65, respectively. As can be seen in Fig. 6.9, when the driving pulse is circularly or elliptically polarized, forward scatterings by collisions with neighboring atoms can help the electron to reencounter with its associated hole. This result is in correspondence with the abnormal ellipticity dependence of HHG in MgO [94]. It should be noted that collisions with other atoms can only

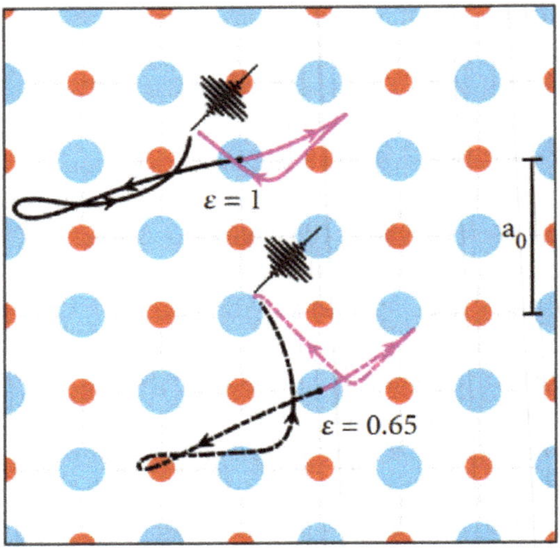


Fig. 6.10 Electron (black) and hole (magenta) trajectories for circular $|\varepsilon| = 1$ (solid lines) and elliptical polarization $|\varepsilon| = 0.65$ (dot-dashed lines). Reproduced with permission from [28].

occur when the de Broglie wavelength of electron or hole is small enough and comparable to atomic scale. This viewpoint provides a novel perspective of the recollision process in terms of wavelike property. In brief, this electron-hole recombination picture provides an unified interpretation and new perspective to further inspect previous experimental and theoretical studies [22, 94, 105].

In Ref. [30] it was demonstrated that the wavelike behavior of electrons plays an essential role in solid HHG. Within collision models the electrons are treated as particles. However, the wave-particle duality is an elementary effect which is built into quantum mechanics and has to be considered. By adopting the analogy to the Huygens-Fresnel principle, the authors of [30] simulated solid HHG from a wavelike perspective, as shown in Fig. 6.11. At time $t'$, in the perspective of Huygens-Fresnel principle, the electron wave packet ionized by the laser field is considered to be a composition of wavelets. While for the harmonic emission at time $t_r$, it can be modeled by the interference of the contributions from all wavelets. Under the condition that diffraction during the propagation can be neglected, one thus can regard the electron as a particle that moves along trajectories which is similar to light rays in geometrical optics. However, due to the complex

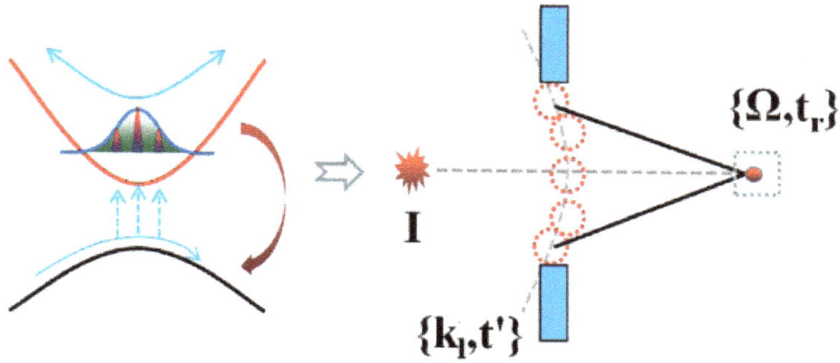


Fig. 6.11 Sketch for the HHG analogy to the Huygens-Fresnel principle. Reproduced with permission from [30].

band structure of solids and delocalization of the electron wave packet, the wave packet will be distorted during its dynamical evolution. Therefore, it is necessary to consider electronic dynamics as a composition of a series of wavelets. In Ref. [30], the deviation of the cutoff energy present from a particle-like perspective disappears and the spectrotemporal characteristics were well reproduced based on the wave-like recollision picture.

## 6.3 Conclusion

In this chapter, we compare HHG in the gas and solid phases and provide an overview over several distinct features present in HHG from solids. The fundamental steps for HHG in gases and solids are similar. However, significant differences exist which originate from the much more rich and complex structure of solid bulk crystals. From the perspective of semi-classical recollision models, we briefly describe the history of the study of solid HHG. We discuss the extra preacceleration process in the ionization step, the additional HHG source provided by the intraband current in the acceleration step, and the various developments related to the final recollision step. These distinct new features of solid HHG go strongly beyond atomic HHG and bring more opportunities and ways to study and control the nonlinear interaction between lasers and bulk crystals. Solid HHG has a huge potential for the further development of attosecond science. Examples for applications include imaging the dynamics in crystals with attosecond temporal and subnanometer spatial resolution, measuring field-induced charge flows, and extracting information on physical properties of bulk crystals as well as attosecond pulse generation. It should be noted that solid HHG can still be considered as a burgeoning field and several important questions are still open and under current debate. It can be expected that in the near future the underlying mechanisms that govern HHG from solids will be resolved in more detail on a microscopic basis which will provide a route to novel applications of solid HHG in attosecond science.

## Acknowledgements

This work was supported by the National Natural Science Foundation of China (Grants Nos. 12074240, 12264013, 12204136, 12204135 and 91950101), Hainan Provincial Natural Science Foundation of China (Grant Nos. 122CXTD504, 123MS002, 123QN180, 123QN179 and 122QN217), the Deutsche Forschungsgemeinschaft (DFG, German Research Foundation) (project number 231447078 TRR 142/3) (project A10), and the Sino-German Mobility Programme (Grant No. M-0031). We thank the PC$^2$ (Paderborn Center for Parallel Computing) for a computing time grant.

## References

[1] M. Uiberacker, T. Uphues, M. Schultze *et al.*, Attosecond real-time observation of electron tunnelling in atoms, *Nature* **446**, pp. 627–632 (2007).

[2] S. Baker, J. S. Robinson, C. A. Haworth *et al.*, Probing proton dynamics in molecules on an attosecond time scale, *Science* **312**, pp. 424–427 (2006).

[3] C. I. Blaga, J. Xu, A. D. DiChiara *et al.*, Imaging ultrafast molecular dynamics with laser-induced electron diffraction, *Nature* **483**, pp. 194–197 (2012).

[4] P. M. Paul, E. S. Toma, P. Breger *et al.*, Observation of a train of attosecond pulses from high harmonic generation, *Science* **292**, pp. 1689–1692 (2001).

[5] M. Hentschel, R. Kienberger, C. Spielmann *et al.*, Attosecond metrology, *Nature* **414**, pp. 509–513 (2001).

[6] Y. Pertot, C. Schmidt, M. Matthews *et al.*, Time-resolved X-ray absorption spectroscopy with a water window high-harmonic source, *Science* **355**, pp. 264–267 (2017).

[7] P. B. Corkum, Plasma perspective on strong field multiphoton ionization, *Phys. Rev. Lett.* **71**, pp. 1994–1997 (1993).

[8] P. B. Corkum, Recollision physics, *Phys. Today* **64**, pp. 36–41 (2011).

[9] P. B. Corkum and F. Krausz, Attosecond science, *Nat. Phys.* **3**, pp. 381–387 (2007).

[10] W. Becker, S. P. Goreslavski, D. B. Milosevic and G. G. Paulus, The plateau in above-threshold ionization: The keystone of rescattering physics, *J. Phys. B* **51**, p. 162002 (2018).

[11] M. Nisoli and G. Sansone, New frontiers in attosecond science, *Prog. Quant. Electron* **33**, pp. 17–59 (2009).

[12] R. Borrego-Varillas, M. Lucchini and M. Nisoli, Attosecond spectroscopy for the investigation of ultrafast dynamics in atomic, molecular and solid-state physics, *Rep. Prog. Phys.* **85**, p. 066401 (2022).

[13] F. Krausz and M. Ivanov, Attosecond physics, *Rev. Mod. Phys.* **81**, pp. 163–234 (2009).

[14] M. Lewenstein, P. Balcou, M. Y. Ivanov, A. L'Huillier and P. B. Corkum, Theory of high-harmonic generation by low-frequency laser fields, *Phys. Rev. A* **49**, pp. 2117–2132 (1994).

[15] S. Ghimire, A. D. Dichiara, E. Sistrunk, P. Agostini, L. F. Dimauro and D. A. Reis, Observation of high-order harmonic generation in a bulk crystal, *Nat. Phys.* **7**, pp. 138–141 (2010).

[16] K. Amini, J. Biegert, F. Calegari *et al.*, Symphony on strong field approximation, *Rep. Prog. Phys.* **82**, p. 116001 (2019).

[17] S. Ghimire and D. A. Reis, High-harmonic generation from solids, *Nat. Phys.* **15**, pp. 10–16 (2019).

[18] S. Ghimire, G. Ndabashimiye, A. D. DiChiara *et al.*, Strong-field and attosecond physics in solids, *J. Phys. B* **47**, p. 204030 (2014).

[19] K. K. Hansen, D. Bauer and L. B. Madsen, Finite-system effects on high-order harmonic generation: From atoms to solids, *Phys. Rev. A* **97**, p. 043424 (2018).

[20] U. Huttner, K. Schuh, J. Moloney and S. W. Koch, Similarities and differences between high-harmonic generation in atoms and solids [Invited], *J. Opt. Soc. Am. B* **33**, pp. C22–C29 (2016).

[21] N. Tancogne-Dejean, M. Sentef and A. Rubio, Ultrafast modification of Hubbard U in a strongly correlated material: ab initio high-harmonic generation in NiO, *Phys. Rev. Lett.* **121**, p. 097402 (2018).

[22] G. Vampa, T. J. Hammond, N. Thiré *et al.*, Linking high harmonics from gases and solids, *Nature* **522**, pp. 462–464 (2015).

[23] G. Vampa, C. McDonald, G. Orlando, P. Corkum and T. Brabec, Semi-classical analysis of high harmonic generation in bulk crystals, *Phys. Rev. B* **91**, p. 064302 (2015).

[24] G. Vampa, C. McDonald, G. Orlando, D. Klug, P. Corkum and T. Brabec, Theoretical analysis of high-harmonic generation in solids, *Phys. Rev. Lett.* **113**, p. 073901 (2014).

[25] S. Fu, Y. Feng, J. Li, S. Yue, X. Zhang, B. Hu and H. Du, Recollision dynamics analysis of high-order harmonic generation in solids, *Phys. Rev. A* **101**, p. 023402 (2020).

[26] C. Liu, Y. Zheng, Z. Zeng and R. Li, Driving-laser ellipticity dependence of high-order harmonic generation in graphene, *Phys. Rev. A* **97**, p. 063412 (2018).

[27] H. Wang, Y. Feng, S. Fu, J. Li, X. Zhang and H. Du, Complex carrier-envelope-phase effect of solid harmonics under nonadiabatic conditions, *Phys. Rev. A* **99**, p. 023406 (2019).

[28] R. Zuo, A. Trautmann, G. Wang *et al.*, Neighboring atom collisions in solid-state high harmonic generation, *Ultrafast Science* **2021**, pp. 1–8 (2021).

[29] L. Li, P. Lan, X. Zhu, T. Huang, Q. Zhang, M. Lein and P. Lu, Reciprocal-space-trajectory perspective on high-harmonic generation in solids, *Phys. Rev. Lett.* **122**, p. 193901 (2019).

[30] L. Li, P. Lan, X. Zhu and P. Lu, Huygens-Fresnel picture for high harmonic generation in solids, *Phys. Rev. Lett.* **127**, p. 223201 (2021).

[31] L. Li, Y. Zhang, P. Lan *et al.*, Dynamic core polarization in high harmonic generation from solids: The example of MgO crystals, *Phys. Rev. Lett.* **126**, p. 187401 (2021).

[32] T. Du, Z. Guan, X. Zhou and X. Bian, Enhanced high-order harmonic generation from periodic potentials in inhomogeneous laser fields, *Phys. Rev. A* **94**, p. 023419 (2016).

[33] T. Du, X. Huang and X. Bian, High-order-harmonic generation from solids: The contributions of the Bloch wave packets moving at the group and phase velocities, *Phys. Rev. A* **97**, p. 013403 (2018).

[34] T. Shao, L. Lv, J. Liu and X. Bian, Quantum path interferences and selection in interband solid high-order harmonic generation in MgO crystals, *Phys. Rev. A* **101**, p. 053421 (2020).

[35] X. Song, S. Yang, R. Zuo, T. Meier and W. Yang, Enhanced high-order harmonic generation in semiconductors by excitation with multicolor pulses, *Phys. Rev. A* **101**, p. 033410 (2020).

[36] X. Song, R. Zuo, S. Yang, P. Li, T. Meier and W. Yang, Attosecond temporal confinement of interband excitation by intraband motion, *Opt. Express* **27**, p. 2225 (2019).

[37] R. Zuo, X. Song, X. Liu, S. Yang and W. Yang, Influence of intraband motion on the interband excitation and high harmonic generation, *Chin. Phys. B* **28**, p. 094208 (2019).

[38] J. Jin, H. Liang, X. Xiao *et al.*, Contribution of Floquet-Bloch states to high-order harmonic generation in solids, *Phys. Rev. A* **100**, p. 013412 (2019).

[39] X. Wu, H. Liang, X. Kong, Q. Gong and L. Peng, Enhancement of high-order harmonic generation in two-dimensional materials by plasmonic fields, *Phys. Rev. A* **103**, p. 043117 (2021).

[40] X. Kong, H. Liang, X. Wu, L. Geng, W. Yu and L. Peng, Manipulation of the high-order harmonic generation in monolayer hexagonal boron nitride by two-color laser field, *J. Chem. Phys.* **156**, p. 074701 (2022).

[41] Y. He, J. Guo, F. Gao *et al.*, Interference between harmonics of different crystal momentum channels in solid high-order harmonic generation, *Phys. Rev. A* **104**, p. 013104 (2021).

[42] Y. He, J. Guo, F. Gao *et al.*, Interference between harmonics of different crystal momentum channels in solid high-order harmonic generation, *Phys. Rev. A* **104**, p. 013104 (2021).

[43] L. Yue and M. Gaarde, Imperfect recollisions in high-harmonic generation in solids, *Phys. Rev. Lett.* **124**, p. 153204 (2020).

[44] L. Yue and M. Gaarde, Expanded view of electron-hole recollisions in solid-state high-order harmonic generation: Full-Brillouin-zone tunneling and imperfect recollisions, *Phys. Rev. A* **103**, p. 063105 (2021).

[45] G. Vampa, T. Hammond, N. Thiré *et al.*, All-optical reconstruction of crystal band structure, *Phys. Rev. Lett.* **115**, p. 193603 (2015).

[46] M. Hassan, T. Luu, A. Moulet *et al.*, Optical attosecond pulses and tracking the nonlinear response of bound electrons, *Nature* **530**, pp. 66–70 (2016).

[47] W. Yang, Y. Lin, X. Chen *et al.*, Wave mixing and high-harmonic generation enhancement by a two-color field driven dielectric metasurface [Invited], *Chin. Opt. Lett.* **19**, p. 123202 (2021).

[48] S. Kruchinin, F. Krausz and V. Yakovlev, Strong-field phenomena in periodic systems, *Rev. Mod. Phys.* **90**, p. 021002 (2018).

[49] C. Qian, C. Yu, S. Jiang *et al.*, Role of shift vector in high-harmonic generation from noncentrosymmetric topological insulators under strong laser fields, *Phys. Rev. X* **12**, p. 021030 (2022).

[50] C. Yu, S. Jiang, T. Wu, G. Yuan, Y. Peng, C. Jin and R. Lu, Higher harmonic generation from bilayer nanostructures assisted by electron backscattering, *Phys. Rev. B* **102**, p. 241407(R) (2020).

[51] G. Ndabashimiye, S. Ghimire, M. Wu *et al.*, Solid-state harmonics beyond the atomic limit, *Nature* **534**, pp. 520–523 (2016).

[52] J. Park, A. Subramani, S. Kim and M. Ciappina, Recent trends in high-order harmonic generation in solids, *Adv. Phys.-X* **7**, p. 2003244 (2022).

[53] T. Du and X. Bian, Quasi-classical analysis of the dynamics of the high-order harmonic generation from solids, *Opt. Express* **25**, p. 151 (2017).

[54] J. Li, X. Zhang, S. Yue, H. Wu, B. Hu and H. Du, Enhancement of the second plateau in solid high-order harmonic spectra by the two-color fields, *Opt. Express* **25**, p. 18603 (2017).

[55] L. Liu, J. Zhao, W. Dong, J. Liu, Y. Huang and Z. Zhao, Spatial coherence in high-order-harmonic generation from periodic solid structures, *Phys. Rev. A* **96**, p. 053403 (2017).

[56] Z. Guan, X. Zhou and X. Bian, High-order-harmonic generation from periodic potentials driven by few-cycle laser pulses, *Phys. Rev. A* **93**, p. 033852 (2016).

[57] T. Du, D. Tang, X. Huang and X. Bian, Multichannel high-order harmonic generation from solids, *Phys. Rev. A* **97**, p. 043413 (2018).

[58] L. Lv and X. Bian, Multielectron interference of intraband harmonics in solids, *Phys. Rev. B* **100**, p. 214312 (2019).

[59] T. Shao, Y. Xu, X. Huang and X. Bian, Strain effects on high-order harmonic generation in solids, *Phys. Rev. A* **99**, p. 013432 (2019).

[60] T. Shao, Y. Xu, X. Huang and X. Bian, Subcycle interference in high-order harmonic generation from solids, *Phys. Rev. A* **98**, p. 063416 (2018).

[61] S. Jiang, C. Yu, J. Chen, Y. Huang, R. Lu and C. Lin, Smooth periodic gauge satisfying crystal symmetry and periodicity to study high-harmonic generation in solids, *Phys. Rev. B* **102**, p. 155201 (2020).

[62] J. Liu and X. Bian, Multichannel high-order harmonic generation from fractal bands in fibonacci quasicrystals, *Phys. Rev. Lett.* **127**, p. 213901 (2021).
[63] N. Tancogne-Dejean, O. Mucke, F. Kartner and A. Rubio, Impact of the electronic band structure in high-harmonic generation spectra of solids, *Phys. Rev. Lett.* **118**, p. 087403 (2017).
[64] X. Wang, Y. Xu, X. Huang and X. Bian, Interference between inter- and intraband currents in high-order harmonic generation in solids, *Phys Rev. A* **98**, p. 023427 (2018).
[65] D. Bauer and K. Hansen, High-harmonic generation in solids with and without topological edge states, *Phys. Rev. Lett.* **120**, p. 177401 (2018).
[66] J. Li, X. Zhang, S. Fu, Y. Feng, B. Hu and H. Du, Phase invariance of the semiconductor Bloch equations, *Phys. Rev. A* **100**, p. 043404 (2019).
[67] S. Jiang, J. Chen, H. Wei, C. Yu, R. Lu and C. Lin, Role of the transition dipole amplitude and phase on the generation of odd and even high-order harmonics in crystals, *Phys. Rev. Lett.* **120**, p. 253201 (2018).
[68] X. Liu, X. Zhu, L. Li, Y. Li, Q. Zhang, P. Lan and P. Lu, Selection rules of high-order-harmonic generation: Symmetries of molecules and laser fields, *Phys. Rev. A* **94**, p. 033410 (2016).
[69] O. Schubert, M. Hohenleutner, F. Langer *et al.*, Sub-cycle control of terahertz high-harmonic generation by dynamical Bloch oscillations, *Nat. Photonics* **8**, pp. 119–123 (2014).
[70] T. Heinrich, M. Taucer, O. Kfir *et al.*, Chiral high-harmonic generation and spectroscopy on solid surfaces using polarization-tailored strong fields, *Nat. Commun.* **12**, p. 3723 (2021).
[71] T. Du and C. Ma, Temperature-induced dephasing in high-order harmonic generation from solids, *Phys. Rev. A* **105**, p. 053125 (2022).
[72] J. Li, S. Fu, H. Wang, X. Zhang, B. Ding, B. Hu and H. Du, Limitations of the single-active-electron approximation in quantum simulations of solid high-order harmonic generation, *Phys. Rev. A* **98**, p. 043409 (2018).
[73] K. Hansen, T. Deffge and D. Bauer, High-order harmonic generation in solid slabs beyond the single-active-electron approximation, *Phys. Rev. A* **96**, p. 053418 (2017).
[74] X. Wu, H. Liang, X. Kong, Q. Gong and L. Peng, Multiscale numerical tool for studying nonlinear dynamics in solids induced by strong laser pulses, *Phys. Rev. E* **105**, p. 055306 (2022).
[75] Y. Zhang, L. Li, J. Li, T. Hang, P. Lan and P. Lu, Orientation dependence of high-order harmonic generation in graphene, *Phys. Rev. A* **104**, p. 033110 (2021).
[76] C. Liu, Y. Zheng, Z. Zeng and R. Li, Polarization-resolved analysis of high-order harmonic generation in monolayer $MoS_2$, *New J. Phys.* **22**, p. 073046 (2020).
[77] Y. Feng, S. Shi, J. Li, Y. Ren, X. Zhang, J. Chen and H. Du, Semiclassical analysis of ellipticity dependence of harmonic yield in graphene, *Phys. Rev. A* **104**, p. 043525 (2021).

[78] X. Kong, H. Liang, X. Wu and L. Peng, Symmetry analyses of high-order harmonic generation in monolayer hexagonal boron nitride, *J. Phys. B* **54**, p. 124004 (2021).

[79] Z. Guan, L. Liu, G. Wang, S. Zhao, Z. Jiao and X. Zhou, An improved method for the investigation of high-order harmonic generation from graphene, *Chin. Phys. B* **29**, p. 104206 (2020).

[80] J. Jin, H. Liang, X. Xiao, M. Wang, S. Chen, X. Wu, Q. Gong and L. Peng, Contribution of Floquet-Bloch states to high-order harmonic generation in solids, *Phys. Rev. A* **100**, p. 013412 (2019).

[81] J. Jin, X. Xiao, H. Liang, M. Wang, S. Chen, Q. Gong and L. Peng, High-order harmonic generation from a two-dimensional band structure, *Phys. Rev. A* **97**, p. 043420 (2018).

[82] J. Chen, C. Liu, Z. Zeng and R. Li, Control of ultrafast photocurrent in twisted bilayer graphene by circularly polarized few-cycle lasers, *Phys. Rev. B* **105**, p. 014309 (2022).

[83] Z. Lou, Y. Zheng, C. Liu, Z. Zeng, R. Li and Z. Xu, Controlling of the harmonic generation induced by the Berry curvature, *Opt. Express* **29**, p. 37809–37819 (2021).

[84] X. Pan, B. Li, J. Zhang and X. Liu, Exploration of the high-order harmonic generation from periodic potentials by Bohmian trajectories, *J. Phys. B* **54**, p. 025601 (2021).

[85] L. Li, T. Huang, P. Lan, J. Li, Y. Zhang, X. Zhu, L. He, W. Cao and P. Lu, Fingerprint of the interbond electron hopping in second-order harmonic generation, *Phys. Rev. Lett.* **128**, p. 027401 (2022).

[86] X. Liu, Y. Li, D. Liu, X. Zhu, X. Zhang and P. Lu, Effects of quantum interferences among crystal-momentum-resolved electrons in solid high-order harmonic generation, *Phys. Rev. A* **103**, p. 033104 (2021).

[87] A. Uzan, G. Orenstein, A. Jiménez-Galán *et al.*, Attosecond spectral singularities in solid-state high-harmonic generation, *Nat. Photonics* **14**, p. 183–187 (2020).

[88] Y. Kobayashi, C. Heide, H. Kelarde, A. Johnson, F. Li, T. Hein, D. Reis and S. Ghimire, Polarization flipping of even-order harmonics in monolayer transition-metal dichalcogenides, *Ultrafast Science* **2021**, p. 9820716 (2021).

[89] Y. Kobayashi, C. Heide, H. Kelardeh, A. Johnson, F. Liu, T. Heinz, D. Reis and S. Ghimire, Self-consistent Maxwell–Bloch model for high-order harmonic generation in nanostructured semiconductors, *Photonics Res.* **10**, p. 2099 (2022).

[90] Y. Lang, Z. Peng, J. Liu, Z. Zhao and S. Ghimire, Proposal for high-energy cutoff extension of optical harmonics of solid materials using the example of a one-dimensional ZnO crystal, *Phys. Rev. Lett.* **129**, p. 167402 (2022).

[91] X. Song, S. Yang, G. Wang, J. Lin, L. Wang, T. Meier and W. Yang, Control of the electron dynamics in solid-state high harmonic generation on ultrafast time scales by a polarization-skewed laser pulse, *Opt. Express* **31**, pp. 18862–18870 (2023).

[92] B. Zaks, R. Liu and M. Sherwin, Attosecond spectral singularities in solid-state high-harmonic generation, *Nature* **483**, p. 580–583 (2012).

[93] M. Hohenleutner, F. Langer, O. Schubert *et al.*, Attosecond spectral singularities in solid-state high-harmonic generation, *Nature* **523**, pp. 572–575 (2012).

[94] Y. You, D. Reis and S. Ghimire, Anisotropic high-harmonic generation in bulk crystals, *Nat. Phys.* **13**, pp. 345–349 (2017).

[95] Y. You, M. Wu, Y. Yin *et al.*, Anisotropic high-harmonic generation in bulk crystals, *Opt. Lett.* **42**, p. 1816 (2017).

[96] J. Lu, E. Cunningham, Y. You, D. Reis and S. Ghimire, Interferometry of dipole phase in high harmonics from solids, *Opt. Lett.* **13**, pp. 96–100 (2019).

[97] N. Yoshikawa, T. Tamaya and K. Tanaka, High-harmonic generation in graphene enhanced by elliptically polarized light excitation, *Science* **356**, pp. 736–738 (2017).

[98] H. Liu, Y. Li, Y. You, S. Ghimire, T. Heinz and D. Reis, High-harmonic generation from an atomically thin semiconductor, *Nat. Phys.* **13**, pp. 262–265 (2017).

[99] D. Golde, T. Meier and S. W. Koch, High harmonics generated in semiconductor nanostructures by the coupled dynamics of optical inter- and intraband excitations, *Phys. Rev. B* **77**, p. 075330 (2008).

[100] T. T. Luu. and H. Worner, High-order harmonic generation in solids: A unifying approach, *Phys. Rev. B* **94**, p. 115164 (2016).

[101] D. Golde, M. Kira, T. Meier and S. W. Koch, Microscopic theory of the extremely nonlinear terahertz response of semiconductors, *Phys. Stat. Sol. (b)* **248**, p. 863–866 (2011).

[102] L. Thong, C. Ngo, H. T. Duc, X. Song and T. Meier, Microscopic analysis of high harmonic generation in semiconductors with degenerate bands, *Phys. Stat. Sol. (b)* **103**, p. 085201 (2021).

[103] M. Wu, S. Ghimire, D. Reis, K. Schafer and M. Gaarde, High-harmonic generation from Bloch electrons in solids, *Phys. Rev. A* **91**, p. 043839 (2015).

[104] T. Ikemachi, Y. Shinohara, T. Sato *et al.*, Trajectory analysis of high-order-harmonic generation from periodic crystals, *Phys. Rev. A* **95**, p. 043416 (2017).

[105] E. Osika, A. Chacón, L. Ortmann *et al.*, Wannier-bloch approach to localization in high-harmonics generation in solids, *Phys. Rev. X* **7**, p. 021017 (2017).

[106] X. Liu, X. Zhu, X. Zhang, D. Wang, P. Lan and P. Lu, Wavelength scaling of the cutoff energy in the solid high harmonic generation, *Opt. Express* **25**, p. 29216–29224 (2017).

[107] E. Runge and E. K. U. Gross, Density-functional theory for time-dependent systems, *Phys. Rev. Lett.* **52**, p. 997–1000 (1984).

[108] R. van Leeuwen, Density-functional theory for time-dependent systems, *Phys. Rev. Lett.* **80**, p. 1280 (1998).

[109] C. McDonald, G. Vampa, P. Corkum and T. Brabec, Interband Bloch oscillation mechanism for high-harmonic generation in semiconductor crystals, *Phys. Rev. A* **92**, p. 033845 (2015).

[110] T. Higuchi, M. Stockman and P. Hommelhoff, Strong-field perspective on high-harmonic radiation from bulk solids, *Phys. Rev. Lett.* **113**, p. 213901 (2014).

[111] L. Keldysh, Ionization in the field of a strong electromagnetic wave, *Sov. Phys. JETP* **20**, p. 1307 (1965).

[112] K. Budil, P. Salières, A. L'Huillier, T. Dimire and M. Perry, Influence of ellipticity on harmonic generation, *Phys. Rev. A* **48**, p. R3437–R3440 (1993).

[113] N. Burnett, C. Kan and P. Corkum, Ellipticity and polarization effects in harmonic generation in ionizing neon, *Phys. Rev. A* **51**, p. R3418–R3421 (1995).

[114] T. Luu, M. Garg, S. Kruchinin, A. Moulet, M. Hassan and E. Goulielmakis, Extreme ultraviolet high-harmonic spectroscopy of solids, *Nature* **521**, p. 498–502 (2015).

[115] H. Kim, S. Han, Y. Kim, S. Kim and S. Kim, Generation of coherent extreme-ultraviolet radiation from bulk sapphire crystal, *ACS Photonics* **4**, p. 1627–1632 (2017).

[116] F. Langer, M. Hohenleutner, U. Huttner, S. W. Koch, M. Kira and R. Huber, Symmetry-controlled temporal structure of high-harmonic carrier fields from a bulk crystal, *Nat. Photonics* **11**, p. 227–231 (2017).

[117] P. Zhokhov and A. Zheltikov, Field-cycle-resolved photoionization in solids, *Phys. Rev. Lett.* **113**, p. 133903 (2014).

[118] M. Wismer, S. Kruchinin, M. Ciappina, M. Stockman and V. Yakovlev, Strong-field resonant dynamics in semiconductors, *Phys. Rev. Lett.* **116**, p. 197401 (2016).

[119] S. Sato, M. Lucchini, M. Volkov, F. Schlaepfer, L. Gallmann, U. Keller and A. Rubio, Role of intraband transitions in photocarrier generation, *Phys. Rev. B* **98**, p. 035202 (2018).

[120] M. Wu, D. Browne, K. Schafer and M. Gaarde, Multilevel perspective on high-order harmonic generation in solids, *Phys. Rev. A* **94**, p. 063403 (2016).

[121] M. Garg, M. Zhan, T. Luu, H. Lakhotia, T. Klostermann, A. Guggenmos and E. Goulielmakis, Multi-petahertz electronic metrology, *Nature* **538**, p. 359–363 (2016).

[122] L. Li, P. Lan, L. He, W. Cao, Q. Zhang and P. Lu, Determination of electron band structure using temporal interferometry, *Phys. Rev. Lett.* **124**, p. 157403 (2020).

[123] A. Lanin, E. Stepanov, A. Fedotov and A. Zheltikov, Mapping the electron band structure by intraband high-harmonic generation in solids, *Optica* **4**, pp. 516–519 (2017).

[124] F. Schlaepfer, M. Lucchini, S. Sato *et al.*, Attosecond optical-field-enhanced carrier injection into the GaAs conduction band, *Nat. Phys.* **14**, pp. 560–564 (2018).

[125] M. Garg, H. Kim and E. Goulielmakis, Ultimate waveform reproducibility of extreme-ultraviolet pulses by high-harmonic generation in quartz, *Nat. Photonics* **12**, p. 291–296 (2018).
[126] Z. Wang, H. Park, Y. Lai, J. Xu, C. Blaga, F. Yang, P. Agostini and L. DiMauro, The roles of photo-carrier doping and driving wavelength in high harmonic generation from a semiconductor, *Nat. Commun.* **8**, p. 1686 (2017).

https://doi.org/10.1142/9789811279560_0007

# Chapter 7

# Quantum Optical Analysis of High-Order Harmonic Generation in Semiconductors

Javier Rivera-Dean[1], Philipp Stammer[1], Andrew S. Maxwell[2], Theocharis Lamprou[3,4], Andrés F. Ordóñez[5], Emilio Pisanty[6], Paraskevas Tzallas[3,7], Maciej Lewenstein[1,8] and Marcelo F. Ciappina[9,10,11]

[1]*ICFO — Institut de Ciencies Fotoniques, The Barcelona Institute of Science and Technology, 08860 Castelldefels (Barcelona), Spain*

[2]*Department of Physics and Astronomy, Aarhus University, DK-8000 Aarhus C, Denmark*

[3]*Foundation for Research and Technology-Hellas, Institute of Electronic Structure & Laser, GR-70013 Heraklion (Crete), Greece*

[4]*Department of Physics, University of Crete, P.O. Box 2208, GR-70013 Heraklion (Crete), Greece*

[5]*Department of Physics, Imperial College London, SW7 2AZ London, United Kingdom*

[6]*Department of Physics, King's College London, WC2R 2LS London, United Kingdom*

[7]*ELI-ALPS, ELI-Hu Non-Profit Ltd., Dugonics tér 13, H-6720 Szeged, Hungary*

[8]*ICREA, Pg. Lluís Companys 23, 08010 Barcelona, Spain*

[9]*Physics Program, Guangdong Technion–Israel Institute of Technology, Shantou, Guangdong 515063, China*

[10]*Technion — Israel Institute of Technology, Haifa, 32000, Israel*

[11]*Guangdong Provincial Key Laboratory of Materials and Technologies for Energy Conversion, Guangdong Technion — Israel Institute of Technology, Shantou, Guangdong 515063, China*

## 7.1 Introduction

Quantum optics stands as one of the most promising platforms for practical applications in quantum information science, with light serving as the primary carrier of information [1–3]. To advance such applications, non-classical states of light are necessary — states inherently described by quantum electrodynamics tools [4]. Some of the available non-classical light sources rely on nonlinear interactions between light and matter [5], and one of the most striking examples of these interactions is high-order harmonic generation (HHG) [6, 7]. In HHG, a strong, pulsed laser field interacts with matter, resulting in the emission of radiation, emitted as a periodic series of ultrashort harmonics of the driving laser, spanning numerous harmonic orders [8]. While semiclassical theoretical descriptions of HHG processes — where light is treated classically while the matter system is treated quantum mechanically — suffice to replicate most experimental results, they leave untapped the potential of strong-field physics in quantum optics.

Recent years have witnessed renewed attraction in analyzing strong-field processes from a quantum optical perspective [9, 10]. This surge in interest corresponds to the observation of strong laser-matter interactions being useful for creating non-classical states. For example, by means of HHG processes in atomic systems it is possible to generate non-classical states of light [11–16] with intensities enough to drive nonlinear processes in matter [17], as well as massively frequency-entangled states [18, 19], light-matter entangled states [20, 21], and high-photon-number squeezed states [22, 23]. Moreover, it has been demonstrated that the use of non-classical light sources can significantly impact electron dynamics [24–26], resulting in extended and better-resolved harmonic spectra [27], as well as leading to new paradigms regarding the semiclassical interpretation of HHG processes [28, 29].

In this chapter, we delve into recent findings in solid-state systems, specifically semiconductor materials, along these lines where the quantum optical behavior is brought to attention. These systems have been recently analyzed in Ref. [30], demonstrating that the band structure of solids lead to non-trivial effects on the quantum optical state of the system. However, the content presented in this chapter builds upon Refs. [31, 32], which provide a quantum optical description of strongly driven light-matter interactions using two distinct basis sets to characterize the electronic system. Although yielding similar outcomes, these different approaches offer diverse insights into electron dynamics and provide more suitable frameworks for computing specific quantum information measures.

## 7.2 Semiclassical Analysis of the Light-Matter Interaction

In this section, we delve into the interaction between a laser field and a solid-state system from a semiclassical perspective. Within this framework, the solid-state system is treated quantum mechanically, while the field is regarded as a linearly polarized classical field. Therefore, under the length gauge and within the dipole approximation, the interaction can be described under a single-active electron picture by the following Hamiltonian

$$\hat{H} = \hat{H}_{\text{cr}} + \mathsf{e}\hat{r}_i E_{\text{cl}}(t), \tag{7.1}$$

where $\mathsf{e}$ denotes the electron charge, $\hat{r}_i$ represents the position operator along the polarization direction $\varepsilon_i$ with $i$ denoting the direction ($i \in \{x, y, z\}$), and $E_{\text{cl}}(t)$ stands for a classical electric field. It is worth noting that the dipole approximation is applicable in this context since, as in typical experimental implementations [33, 34], the laser fields used have frequencies within the mid-infrared regime ($\lambda_L \sim$ 1–5 $\mu$m), while typical crystal lattices are on the order of 1 Å.

In the expression above, $\hat{H}_{\text{cr}}$ denotes the crystal Hamiltonian. We will model this Hamiltonian using a two-band, tight-binding model, with the band-dispersion relations for both the valence ($v$) and conduction ($c$) bands satisfying

$$E_v(\mathbf{k}) = \sum_{i=x,y,z} \sum_{j=0}^{\infty} \alpha_{v,i}^j \cos(jk_i a_i), \tag{7.2}$$

$$E_c(\mathbf{k}) = E_g + \sum_{i=x,y,z} \sum_{j=0}^{\infty} \alpha_{c,i}^j \cos(jk_i a_i), \tag{7.3}$$

where $\mathbf{k}$ denotes the crystal momentum, $a_i$ represents the lattice constant along the $i$ direction, $E_g$ corresponds to the band gap energy at the $\Gamma$ point, and $\alpha_{m,i}^j$ the coefficients arising from expanding the bands within the tight-binding approximation [35].

In the following subsections, we will address the dynamics associated with the Hamiltonian in Eq. (7.1) by employing two basis sets to expand the electronic wavefunctions. First, we adopt a Bloch-based description [36, 37], which are eigenstates of the crystal Hamiltonian and possess a well-defined crystal momentum. Subsequently, we shift to a Wannier-based description as discussed in Ref. [38] which, in contrast, uses eigenstates localized in space while being entirely delocalized in reciprocal space. Despite the approaches being expected to yield similar results, given that unitary transformations should not impact the physics of the process they offer distinct insights into the electron dynamics inside the crystal [39].

### 7.2.1 *Bloch-Bloch Picture*

In this subsection, we focus on a description in which the quantum state of the electron in both the valence and conduction bands is represented using the Bloch basis $\{|\phi_{\mathbf{k},m}\rangle\}$, where $m$ denotes the respective band. As previously mentioned, these states are eigenstates of the crystal Hamiltonian with eigenvalue $E_m(\mathbf{k})$, that is

$$\hat{H}_{\rm cr}\,|\phi_{\mathbf{k},m}\rangle = E_m(\mathbf{k})\,|\phi_{\mathbf{k},m}\rangle, \tag{7.4}$$

while the matrix elements for the corresponding position operator $\hat{r}_i$ have both diagonal and non-diagonal contributions. In the following, we express this operator in terms of its intraband ($\hat{r}_{i,\rm tra}$) and interband ($\hat{r}_{i,\rm ter}$) components, that is

$$\hat{r}_i = \hat{r}_{i,\rm tra} + \hat{r}_{i,\rm ter}. \tag{7.5}$$

As their name suggests, intraband terms involve matrix elements that pertain to Bloch states within the same band, while interband terms relate to Bloch states in different bands. More specifically, we have the following expressions for these matrix elements

$$\langle\phi_{\mathbf{k},m}|\hat{r}_{i,\rm tra}|\phi_{\mathbf{k}',l}\rangle = i\hbar\delta_{m,l}\frac{\partial}{\partial k_i}\delta(\mathbf{k}-\mathbf{k}'), \tag{7.6}$$

$$\langle\phi_{\mathbf{k},m}|\hat{r}_{i,\rm ter}|\phi_{\mathbf{k}',l}\rangle = \frac{1}{\mathsf{e}}d_{ml}^{(i)}(\mathbf{k})\delta(\mathbf{k}-\mathbf{k}'), \tag{7.7}$$

with $d_{ml}^{(i)}(\mathbf{k})$ the dipole moment matrix element between the two bands, which can be generally expressed as [35–37]

$$d_{ml}^{(i)}(\mathbf{k}) = \sqrt{\frac{\mathfrak{E}_{p,i}}{[2(E_l(\mathbf{k})-E_c(\mathbf{k}))^2]}}, \tag{7.8}$$

where $\mathfrak{E}_{p,i}$ is the so-called Kane parameter, whose values are tabulated for many materials.

In terms of these intraband and interband operators, we can express the Hamiltonian describing the dynamics of this system as

$$i\hbar\frac{\partial\,|\psi(t)\rangle}{\partial t}=\Big[\hat{H}_{\text{cr}}+\mathsf{e}\hat{r}_{i,\text{tra}}E_{\text{cl}}(t)+\mathsf{e}\hat{r}_{i,\text{ter}}E_{\text{cl}}(t)\Big]\,|\psi(t)\rangle. \tag{7.9}$$

Hereupon, we work in the interaction picture with respect to the intraband excitation component. In more explicit mathematical terms, we introduce in our equations $|\psi(t)\rangle=\hat{U}_{\text{tra}}(t)|\bar{\psi}(t)\rangle$. In this expression, $\hat{U}_{\text{tra}}(t)=e^{i\mathsf{e}A_{\text{cl}}(t)\hat{r}_{i,\text{tra}}/\hbar}$ with $A_{\text{cl}}(t)$ the classical vector potential related to the electric field via $E_{\text{cl}}(t)=-\partial A_{\text{cl}}(t)/\partial t$. The rationale behind this transformation arises from the fact that the intraband term, as presented in Eq. (7.6), couples Bloch states with different values of $k$, leading to a set of coupled differential equations when projecting Eq. (7.9) onto the Bloch basis. By transitioning to the interaction picture with respect to this term, we effectively shift our frame of reference to that of the oscillating electron within the field, and work with the canonical momentum $\mathbf{K}=\mathbf{k}-\mathsf{e}\boldsymbol{\varepsilon}_{i}A_{\text{cl}}(t)$, where $\boldsymbol{\varepsilon}_{i}$ denotes the polarization direction.

Within this frame, the Schrödinger equation in Eq. (7.9) transforms into

$$i\hbar\frac{\partial|\bar{\psi}(t)\rangle}{\partial t}=\Big[\hat{H}_{\text{cr}}(t)+\mathsf{e}\hat{r}_{i,\text{ter}}(t)E_{\text{cl}}(t)\Big]|\bar{\psi}(t)\rangle, \tag{7.10}$$

where we have defined

$$\hat{H}_{\text{cr}}(t)\equiv\hat{U}^{\dagger}_{\text{tra}}(t)\hat{H}_{\text{cr}}\hat{U}_{\text{tra}}(t),\quad \hat{r}_{\text{ter}}(t)\equiv\hat{U}^{\dagger}_{\text{tra}}(t)\hat{r}_{\text{ter}}\hat{U}_{\text{tra}}(t), \tag{7.11}$$

and by projecting Eq. (7.10) with respect to the basis set spanned by the canonical crystal momentum $\{|\mathbf{K},m\rangle\}$, we can rewrite this differential equation as

$$\begin{aligned} i\hbar\frac{\partial b_m(\mathbf{K},t)}{\partial t}&=E_m\big(\mathbf{K}+\mathsf{e}\varepsilon_i A_{\text{cl}}(t)\big)b_m(\mathbf{K},t)\\ &\quad+\sum_{l=c,v}d^{(i)}_{ml}\big(\mathbf{K}+\mathsf{e}\varepsilon_i A_{\text{cl}}(t)\big)b_l(\mathbf{K},t), \end{aligned} \tag{7.12}$$

defined for both the valence and conduction bands, leading to a set of coupled differential equations. In this expression, the $b_m(\mathbf{K},t)\equiv\big\langle\mathbf{K}\big|\bar{\psi}(t)\big\rangle$ are the probability amplitudes of finding the electron with canonical crystal momentum $\mathbf{K}$ in band $m$, with $E_m(\mathbf{k})$ the dispersion relation of the $m$th band, given in Eqs. (7.2) and (7.3).

From this system of differential equations, where as we can see has become uncoupled with respect to $\mathbf{K}$, one can derive what are commonly referred to as semiconductor Bloch equations [39, 40]. These equations govern the dynamics of the band populations and coherences, denoted here as $n_m(\mathbf{K},t) \equiv b_m^*(\mathbf{K},t)b_m(\mathbf{K},t)$ and $\pi(\mathbf{K},t) \equiv b_v^*(\mathbf{K},t)b_c(\mathbf{K},t)$, respectively. The beneficial aspect about this matrix formulation is that one can straightforwardly include environmental effects into the solid dynamics by adding the corresponding terms affecting either the populations and/or the coherences. Here, we focus on dephasing effects due to electron-electron and electron-phonon couplings, which can be introduced in terms of what is commonly known as a dephasing time $T_2$ into our equations. With this, the differential equations describing the population and coherence dynamics are given by

$$i\hbar\frac{\partial n_m(\mathbf{K},t)}{\partial t} = \theta_m E_{\text{cl}}(t)\left[d_{cv}^{(i)}\left(\mathbf{K}+\mathrm{e}\varepsilon_i A_{\text{cl}}(t)\right)\right]^*\pi(\mathbf{K},t) + \text{c.c.}, \tag{7.13}$$

$$\begin{aligned} i\hbar\frac{\partial \pi(\mathbf{K},t)}{\partial t} &= \left[\mathcal{E}_g\left(\mathbf{K}+\mathrm{e}\varepsilon_i A_{\text{cl}}(t)\right) + E_{\text{cl}}(t)\xi_g^{(i)}\left(\mathbf{K}+\mathrm{e}\varepsilon_i A_{\text{cl}}(t)\right) - \frac{i}{T_2}\right] \\ &\quad \times \pi(\mathbf{K},t) + E_{\text{cl}}(t)d_{cv}^{(i)}(t)\left(\mathbf{K}+\mathrm{e}\varepsilon_i A_{\text{cl}}(t)\right)w(\mathbf{K},t), \end{aligned} \tag{7.14}$$

where we have defined $\theta_v \equiv 1$ and $\theta_c \equiv -1$, the band gap energy as $\mathcal{E}_g(\mathbf{k}) \equiv E_c(\mathbf{k}) - E_v(\mathbf{k})$, the difference between the Berry connection $d_{mm}^{(i)}(\mathbf{k})$ of each band as $\xi_g^{(i)} \equiv d_{cc}^{(i)}(\mathbf{k}) - d_{vv}^{(i)}(\mathbf{k})$ and the population difference between bands as $w(\mathbf{K},t) \equiv n_v(\mathbf{K},t) - n_c(\mathbf{K},t)$.

These equations allow us to compute one of the most characteristic observables associated to HHG, commonly referred to as the HHG spectrum. This quantity is given as $\omega^2|\text{FT}[j_{i,\text{ter}}(t) + j_{i,\text{tra}}(t)]|^2$, where $j_{i,\text{ter}}(t)$ and $j_{i,\text{tra}}(t)$ represent the time-dependent inter and intraband terms which are, specially for centrosymmetric materials in which the Berry connection is zero [41], the dominant contributions to the HHG process [34, 36, 37, 40] and FT denotes the Fourier transform.

On the one hand, interband dynamics provide an analogous mechanism to the HHG process in atomic and molecular systems, governed by a three step-like model [42–45]. Specifically, for solid-state systems (1) an electron-hole pair is created due to the promotion of the electron to the conduction band, (2) the pair gets accelerated by the field in their respective bands, and (3) the pair meets each other in the valence band by emitting a photon with energy equal to the energy difference between the bands [36, 37]. On the other hand, intraband dynamics provide a mechanism for the emission of radiation related to electron scattering within the energy dispersion profile of the specific band in which it is located. This mechanism is often referred to as Bloch oscillations [46].

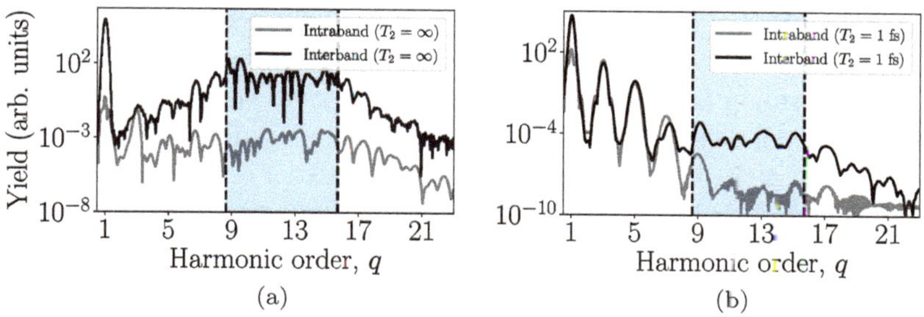


Fig. 7.1 Contributions of interband and intraband currents to the HHG spectra. The spectra have been computed for a ZnO semiconductor after interaction with a linearly polarized field along the $\Gamma - A$ direction, with a peak field strength $0.5\,\mathrm{V/\AA}$, central wavelength $\lambda_L = 3.25\,\mu\mathrm{m}$ and duration of $\Delta t \approx 96\,\mathrm{fs}$ ($\approx$ 9 total cycles). In panel (a), an ideal solid is considered where dephasing effects do not affect the electron dynamics ($T_2 \to \infty$). In contrast, panel (b) considers the case where dephasing effects are indeed included. The dashed blue region span the minimum and maximum energy band gaps for the considered crystal direction.

These current components can be respectively computed from Eqs. (7.13) and (7.14) as

$$j_{i,\mathrm{ter}}(t) = \mathrm{e}\frac{\mathrm{d}}{\mathrm{d}t}\left[\mathrm{tr}(\hat{r}_{i,\mathrm{ter}}\hat{\rho}(t))\right], \tag{7.15}$$

$$j_{i,\mathrm{tra}}(t) = -\mathrm{e}\,\mathrm{tr}\Big(\big[\hat{r}_{i,\mathrm{tra}}, \hat{H}_{\mathrm{cr}}]\hat{\rho}(t)\big]\Big), \tag{7.16}$$

where $\hat{\rho}(t)$ is the density matrix with populations and coherences given by $n_m(\mathbf{K}, t)$ and $\pi(\mathbf{K}, t)$, when expanded in the canonical crystal momentum basis.

In Fig. 7.1, we present the spectrum calculated for ZnO when excited with a linearly polarized field along the $\Gamma - A$ direction, with field strength 0.5 V/Å, central wavelength $\lambda_L = 3.25\,\mu\mathrm{m}$, duration of $\Delta t \approx 96\,\mathrm{fs}$ ($\approx$ 9 total cycles). Specifically, in panel (a) the dephasing time has been set $T_2 \to \infty$, while in panel (b) it is set to $T_2 = 1\,\mathrm{fs}$. As observed, in both cases, interband contributions dominate over the intraband one within the perturbative part of the spectrum, which is presented in the figure with the blue region delimited by the minimum and maximum energy band gaps for this solid direction. For energies smaller than the minimum energy band gap, i.e., within the perturbative region, no clear dominance is found between them, and the difference becomes less evident for smaller dephasing times. However, dephasing effects have another important consequence, which is that of making the harmonic peaks more evident than in the case without dephasing ($T_2 \to \infty$). While the use of these small dephasing times to obtain well-resolved theoretical spectra has been common practice in theories describing HHG spectra in solids, their small values have also been a subject of controversy within the attoscience community [47]. And this is mainly because the used values are on the order of a quarter of an optical driving field cycle, while experimental evidence suggest they should be likely on the order of 15~50 fs at least [48, 49].

In this context, some works justify their values by suggesting that electrons with high values of crystal momentum could decohere at this scales [36]. Others opposing to the presence of these small values propose additional mechanisms beyond electron-photon dephasing. For instance, in Ref. [50], the importance of propagation effects is highlighted in this regard. On the other hand, in Ref. [51], through extensive numerical computations, they observe that issues regarding numerical convergence could potentially be a concern.

### 7.2.2 *Wannier-Bloch Picture*

Unlike the previous subsection, here we approach the same problem but from a slightly different perspective. Instead of working with Bloch states for both valence and conduction bands, in this subsection, we characterize valence band states using a Wannier basis $\{|w_{j,v}\rangle\}$, with $j$ denoting the Wannier site and $v$ the band. For the conduction band, we utilize Bloch states $\{|\phi_{\mathbf{k},c}\rangle\}$. The main distinction between Wannier and Bloch states is that the former can be defined to be maximally localized in space, while the latter are completely localized in reciprocal space. Thus, this representation in terms of localized states in the valence band provides an atomistic-like interpretation of the HHG process in solids [38]. Specifically, within the single-electron picture, one could understand the electron as initially being localized in a given Wannier site and recombining with a potentially different site, somewhat resembling the localization of an electron in an atomic nucleus.

However, unlike Bloch states, Wannier states are not eigenstates of the crystal Hamiltonian. This highlights an additional degree of freedom electrons have in solid-state systems when compared to atomic ensembles: electrons can hop from one Wannier site to another — a process that does not occur between different atomic nuclei, which are far apart and barely interact with each other. Specifically, the matrix elements of the crystal Hamiltonian under Wannier basis are given, under the tight-binding approximation, by

$$\langle w_{j,v}|\hat{H}_{\text{cr}}|w_{j',v}\rangle = -I_v\delta_{|j-j'|,1}, \tag{7.17}$$

where $I_v < 0$ corresponds to the hopping parameter. This relation implies that we consider only nearest-neighbor interactions between the different sites. However, depending on the specific system that is used, other neighbors may need to be included. It is worth highlighting that, as we are working with linearly polarized fields, in the following we consider a one-dimensional description for the Wannier states such that $j$ is a scalar and not a vector representing the spatial location of the Wannier site in real space.

Initially, the system's state corresponds to a completely filled Fermi sea, indicating that all Wannier states are occupied. However, here we consider a single-active electron description, wherein we solve the equation for a single electron and then combine all results at the end. Thus, under these considerations and in terms of the mixed Wannier-Bloch basis representation, we propose the following ansatz to solve the Schrödinger equation in Eq. (7.9)

$$|\psi(t)\rangle = \sum_j a_j(t)\,|w_{j,v}\rangle + \int \mathrm{d}\mathbf{k} a_c(\mathbf{k},t)\,|\phi_{\mathbf{k},c}\rangle, \tag{7.18}$$

under the initial condition $a_j(0) = \delta_{j,j_0}$. Introducing this ansatz into Eq. (7.9) and projecting it with respect to both a Wannier state $|w_{j,v}\rangle$ and a Bloch state $|\phi_{\mathbf{k},c}\rangle$, we arrive at the following set of coupled differential equations

$$\begin{aligned} i\hbar\frac{\partial a_j(t)}{\partial t} &= -I_v a_{j-1}(t) - I_v a_{j+1}(t) + \mathsf{e} r_j E_{\text{cl}}(t) a_j(t) \\ &\quad + E_{\text{cl}}(t)\int \mathrm{d}\mathbf{k}\; d_{jc}(\mathbf{k}) a_c(\mathbf{k},t), \end{aligned} \tag{7.19}$$

$$i\hbar\frac{\partial a_c(\mathbf{k},t)}{\partial t} = E_c(\mathbf{k})a_c(\mathbf{k},t) + i\hbar E_{\text{cl}}(t)\frac{\partial a_c(\mathbf{k},t)}{\partial k_i} + E_{\text{cl}}(t)\sum_j d^*_{jc}(\mathbf{k})a_j(t), \tag{7.20}$$

where $r_j$ represents the position of the $j$th site, and $d_{jc}(\mathbf{k}) \equiv \mathsf{e}\,\langle w_{j,v}|\,\hat{r}_i\,|\phi_{\mathbf{k},v}\rangle$. This term corresponds to the transition matrix element between an electron in site $j$ of the valence band to a Bloch state with crystal momentum $\mathbf{k}$ in the conduction band.

It is worth noting that the two main dynamics of HHG processes in solids are accounted for in Eqs. (7.19) and (7.20). Specifically, intraband transitions within the valence band are represented in Eq. (7.19) through the electron hopping between the neighboring sites. Similarly, intraband transition within the conduction band are characterized by the crystal momentum derivatives. Finally, interband transitions between the valence and conduction bands are represented by the terms including the transition dipole matrix elements $d_{jc}(\mathbf{k})$.

For the first equation, and in line with the strong-field approximation (SFA) assumptions [45], we consider the depletion due to conduction band excitations to be very weak in comparison to intraband transitions. Consequently, we approximate Eq. (7.19) as

$$i\hbar\frac{\partial a_j(t)}{\partial t} \approx -I_v a_{j-1}(t) - I_v a_{j+1}(t) + \mathsf{e} r_j E_{\text{cl}}(t) a_j(t), \tag{7.21}$$

and perform a discrete Fourier transform over the site label $j$ by multiplying the entire equation by $e^{-iqj}$ and summing over all sites $j$. Hence, if we define

$$a_q(t) = \sum_j a_j(t) e^{-iqj}, \tag{7.22}$$

the Schrödinger equation in Eq. (7.21) transforms into

$$i\hbar\frac{\partial a_q(t)}{\partial t} = -2I_v\cos(q)a_q(t) + i\hbar\mathsf{e}\mathsf{a}E_{\text{cl}}(t)\frac{\partial a_q(t)}{\partial q}, \tag{7.23}$$

with $\mathsf{a}$ denoting the lattice constant. This differential equation can also be written as

$$i\hbar\frac{\mathrm{d}a_{\tilde{q}}(t)}{\mathrm{d}t} = -2I_v\cos(\tilde{q} + \mathsf{e}\mathsf{a}E_{\text{cl}}(t))a_{\tilde{q}}(t) \tag{7.24}$$

where $\tilde{q} = q - \mathsf{e}\mathsf{a}A_{\text{cl}}(t)$, and whose solution is straightforwardly given by

$$a_{\tilde{q}}(t) = \exp\left[\frac{i}{\hbar}2I_v\int_{t_0}^{t}\mathrm{d}\tau\cos(\tilde{q} + \mathsf{e}\mathsf{a}A_{\text{cl}}(t))\right]a_{\tilde{q}}(t_0). \tag{7.25}$$

Consequently, a solution to Eq. (7.21) can be obtained by inverting the discrete Fourier transform. Thus, we get

$$a_j(t) = \sum_q e^{iq(j-j_0)}\exp\left[\frac{i}{\hbar}2I_v\int_{t_0}^{t}\mathrm{d}\tau\cos(\tilde{q} + \mathsf{e}\mathsf{a}A_{\text{cl}}(t))\right], \tag{7.26}$$

where the initial condition $a_j(t_0) = \delta_{j-j_0}$ has already been implemented.

On the other hand, the solution of Eq. (7.20) is a first-order inhomogeneous differential equation with well-defined homogeneous and inhomogeneous parts. Hence, its solution can be written as a linear combination of the solution to the homogeneous part and a solution to the inhomogeneous component. In this case, after introducing the initial conditions, we arrive at [38, 52]

$$a_c(\mathbf{K}, t) = -\frac{i}{\hbar}\sum_j \int_{t_0}^{t} \mathrm{d}t_1 \exp\left[-\frac{i}{\hbar}\int_{t_0}^{t_1} \mathrm{d}\tau E_c\big(\mathbf{K} + \mathsf{e}\varepsilon_i A_{\mathrm{cl}}(\tau)\big)\right] \times E_{\mathrm{cl}}(t_1) d^*_{jc}\big(\mathbf{K} + \mathsf{e}\varepsilon_i A_{\mathrm{cl}}(t_1)\big) a_j(t_1), \tag{7.27}$$

which bears some resemblance to the atomic ionization process. More specifically, at time $t_1$ the electron is found at site $j$, which does not necessarily have to be the initial $j_0$ according to Eq. (7.26). Then, it undergoes a transition to the conduction band driven by the field at time $t_1$, represented by the dipole matrix element $d_{jc}(\mathbf{k})$. Finally, it accelerates in the conduction band until time $t$, with the state gaining an additional phase factor accounting for this process.

Taking into account that [38]

$$d_{jc}(\mathbf{k}) = d_{vc}(\mathbf{k})\tilde{w}_v e^{ikr_j/\hbar}, \tag{7.28}$$

where $\tilde{w}_m$ is a normalization constant for Wannier states which for one-dimensional lattices is k-independent [53], we can write the transition dipole matrix element as

$$\begin{aligned} D(t) &= \langle\psi(t)|\hat{r}|\psi(t)\rangle \\ &\approx -\frac{i}{\hbar}|\tilde{w}_v|^2 \sum_{j,j'}\int_{t_0}^{t}\mathrm{d}t_1 \int \mathrm{d}\mathbf{K} a^*_j(t) d_{vc}\big(\mathbf{K} + \mathsf{e}\varepsilon_i A_{\mathrm{cl}}(t)\big) e^{i(\mathbf{K}+\mathsf{e}\varepsilon_i A_{\mathrm{cl}}(t))r_j/\hbar} \\ &\quad \times e^{-i\varphi(p,t,t_1)/\hbar} E_{\mathrm{cl}}(t_1) d^*_{vc}\big(\mathbf{K} + \mathsf{e}\varepsilon_i A_{\mathrm{cl}}(t_1)\big) \\ &\quad + e^{-i(\mathbf{K}+\mathsf{e}\varepsilon_i A_{\mathrm{cl}}(t_1))r_{j'}/\hbar} a_{j'}(t_1) + \text{c.c.}, \end{aligned} \tag{7.29}$$

where we have neglected the contribution of intraband transitions with the aim of accounting only for the dominant interband transitions in the non-perturbative part of the HHG spectrum. In this expression, we have further defined $\varphi(p, t, t_1) \equiv \int_{t_1}^{t} \mathrm{d}\tau E_c(\mathbf{K} + \mathsf{e}\boldsymbol{E}_i A_{\mathrm{cl}}(\tau))$. This time-dependent dipole moment, whose Fourier transform leads to the HHG spectrum, yields similar dynamics to those occurring in atomic systems. First, at time $t_1$ the electron gets ionized from the Wannier site $j'$ to the conduction band. Within the conduction band, it accelerates gaining a phase $\varphi(p, t, t_1)$, and recombines at time $t$ with the $j$th site.

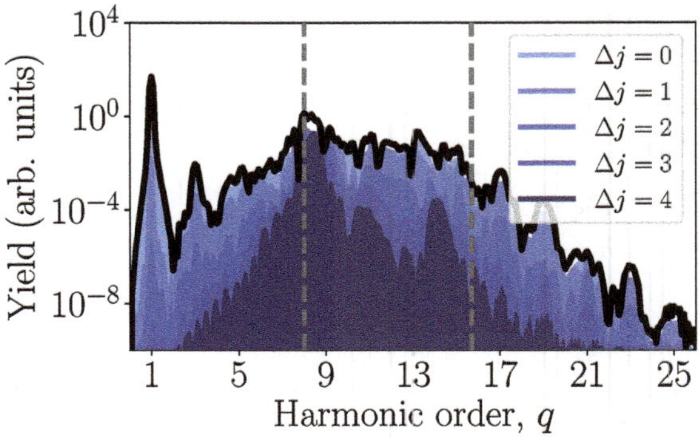


Fig. 7.2 The interband contribution to the HHG spectrum, following the Wannier-Bloch approach presented in the text, is computed for a ZnO material that has been excited with a linearly polarized field along the $\Gamma - A$ direction. In this case, the field strength considered is $E_0 \approx 0.2\,\text{V/Å}$, central wavelength $\lambda_L = 3.0\,\mu\text{m}$, and duration $\Delta t \approx 100\,\text{fs}$ ($\approx$ 10 optical cycles). The different colors represent the different contributions to the spectra depending on the distance between the initial and final Wannier sites. The black curve represents the total contribution. This figure is based on Fig. 3 in Ref. [38].

Given that in general $j \neq j'$, one can already observe one of the main distinctions between HHG in atoms and solid-state systems. While for atoms the electron performs closed trajectories, meaning that ionization and recombination take place at the same real-space position, for solids this is not necessarily the case. This could be a potential explanation for the reduced dependence of harmonic generation on the field's ellipticity found in solid systems [34]. However, depending on the material considered and how the excitation process takes place (for instance, the crystal direction along which the laser is polarized), the influence of increasing values $\Delta j = |j - j'|$ on the HHG spectra might differ. In Fig. 7.2, this is shown for a ZnO material excited by a linearly polarized field parallel to the $\Gamma - A$ direction, with field strength $E_0 \approx 0.2\,\text{V/Å}$, central wavelength $\lambda_L = 3.0\,\mu\text{m}$ and duration of $\Delta t \approx 100\,\text{fs}$ ($\approx$ 10 optical cycles). As observed, in this case, the contribution to the HHG spectra starts to decay already for $\Delta j \geq 2$, meaning that for this particular case, the HHG process is highly localized.

It is worth noting that the HHG spectra obtained here recovers the characteristic features shown in Fig. 7.1 (a) (black curve), with a plateau structure defined within the lowest and maximum values of the energetic band gap. Furthermore, in this figure we also observe that the absence of dephasing effects leads to poorly resolved harmonic peaks. In this direction, it has been recently observed in Ref. [54] that the destructive interference of long trajectories, i.e., those with $\Delta j$ being large, recovers well-resolved harmonic peaks without having to resort to extremely small dephasing times. Similar to atomic systems, this destructive interference could be related to additional phase factors depending on the intensity profile of the applied laser field.

## 7.3 Quantum Optical Analysis of the Light-Matter Interaction

The discussion thus far on the interaction between strong laser fields and solid-state systems has remained at a semiclassical level: the solid-state system has been studied under a quantum mechanical framework while light has been treated as a classical field. However, as mentioned in the Introduction, delving into the quantum optical regime enables the exploration of concepts such as non-classical light and light-matter entanglement, broadening the spectrum of degrees of freedom in strong-field physics. In the following subsections, after briefly introducing the fundamentals of quantum optics, we expand the previously presented semiclassical analysis to the quantum optical context, drawing from Refs. [31, 32]. We illustrate how the various electron dynamics within the solid affect the joint state of the system following the interaction between light and solid.

### 7.3.1 *A Brief Introduction to Quantum Optics*

Quantum optics explores the properties of light and its interaction with matter under a quantum mechanical perspective. As the aim of this chapter is to delineate the interaction between light and matter considering the quantum mechanical nature of light, we provide an overview of fundamental concepts and tools within quantum optics for clarity and contextual understanding.

The foundations of quantum optics rely on the realization that the energy associated with an electromagnetic field in free space can be represented as a collection of uncoupled harmonic oscillators (see, e.g., Refs. [55–58] and references therein). Within this assembly, each electromagnetic field mode, characterized by a wavevector $\boldsymbol{k}$ and a polarization $\mu$, corresponds to an oscillator. The Hamiltonian operator associated to this oscillator is characterized by

$$\hat{H}_{\mathbf{k},\mu} = \frac{1}{2}\left[\hat{p}_{\boldsymbol{k},\mu} + \omega_k^2 \hat{x}^2_{\boldsymbol{k},\mu}\right], \tag{7.30}$$

where $\hat{p}_{\boldsymbol{k},\mu}$ and $\hat{x}_{\boldsymbol{k},\mu}$ are the canonical operators of the harmonic oscillator, satisfying the set of commutation relations

$$[\hat{x}_{\boldsymbol{k},\mu}, \hat{p}_{\boldsymbol{k},\mu}] = i\hbar, \quad [\hat{x}_{\boldsymbol{k},\mu}, \hat{x}'_{\boldsymbol{k}',\mu}] = [\hat{p}_{\boldsymbol{k},\mu}, \hat{p}'_{\boldsymbol{k}',\mu}] = 0. \tag{7.31}$$

The Hamiltonian introduced in Eq. (7.30) corresponds to a single harmonic oscillator. Hence, for the entire electromagnetic field the Hamiltonian is constructed as a linear combination of these individual oscillators, i.e., $\hat{H}_{\text{field}} = \sum_{\boldsymbol{k},\mu} \hat{H}_{\boldsymbol{k},\mu}$. However, for simplicity in the subsequent discussion within this section, we focus solely on a single electromagnetic field mode $(\boldsymbol{k}, \mu)$. Consequently, we omit the $(\boldsymbol{k}, \mu)$ sub-indexes when referring to the operators to enhance clarity. Furthermore, we introduce a set of creation and annihilation operators, denoted as $\hat{a}^\dagger$ and $\hat{a}$ respectively, linked to the canonical variables $\hat{x}$ and $\hat{p}$ by the following canonical transformations

$$\hat{x} = \sqrt{\frac{\hbar}{2\omega}}[\hat{a}^\dagger + \hat{a}], \quad \hat{p} = i\sqrt{\frac{\hbar\omega}{2}}[\hat{a}^\dagger - \hat{a}], \tag{7.32}$$

which satisfy the following set of commutation relations

$$[\hat{a}, \hat{a}^\dagger] = 1, \quad [\hat{a}, \hat{a}] = [\hat{a}^\dagger, \hat{a}^\dagger] = 0. \tag{7.33}$$

According to these, the Hamiltonian in Eq. (7.30) can be rewritten as

$$\hat{H} = \hbar\omega\left(\hat{a}^\dagger\hat{a} + \frac{1}{2}\right), \tag{7.34}$$

which is diagonalized by the set of Fock states $\{|n\rangle : n \in \mathbb{W}\}$[1] with corresponding eigenvalues $\{\hbar\omega(n + 1/2) : n \in \mathbb{W}\}$. These states are often referred to as *photon-number* states, with $|n = 0\rangle$ usually denoted as the *vacuum* state, and they represent the discrete excitations of the electromagnetic field. The action of the creation and annihilation operators over these states is expressed as

$$\hat{a}\,|n\rangle = \sqrt{n}\,|n-1\rangle, \quad \hat{a}^\dagger\,|n\rangle = \sqrt{n+1}\,|n+1\rangle, \tag{7.35}$$

which respectively destroy and create an electromagnetic field excitation.

In this quantum optical context, the classical electric field $\mathbf{E}_{\text{cl}}(t)$ and vector potential $\mathbf{A}_{\text{cl}}(t)$ are substituted with the quantum mechanical

---

[1] Here, $\mathbb{W}$ represents the set of whole numbers, i.e., the natural numbers including zero.

operators $\hat{\mathbf{A}}(t)$ and $\hat{\mathbf{E}}(t)$. The single-mode components of these operators are given by

$$\hat{A}(t) = \sqrt{\frac{\hbar}{2\omega\varepsilon_0 V}}\left[\hat{a}e^{-i\omega t} + \hat{a}^\dagger e^{i\omega t}\right] \tag{7.36}$$

$$\hat{E}(t) = i\sqrt{\frac{\hbar\omega}{2\varepsilon_0 V}}\left[\hat{a}e^{-i\omega t} - \hat{a}^\dagger e^{i\omega t}\right], \tag{7.37}$$

where $\varepsilon_0$ represents the vacuum permittivity and $V$ stands for the quantization volume arising when expanding the electromagnetic field in terms of discrete modes. Specifically, for Fock states, we get $\langle n|\hat{E}(t)|n\rangle = 0$ $\forall n \in \mathbb{W}$, while their variance scales proportionally with the number of photons $n$, i.e., $\Delta E^2(t) = (\hbar\omega)/(2\varepsilon_0 V)[2n+1]$. Moreover, when examining the uncertainty product between the quadrature operators $\hat{x}$ and $\hat{p}$, which are conjugate variables, it is observed that $\Delta x \Delta p = (\hbar/2)(2n+1)$, generally exceeding Heinseberg's uncertainty limit of $\hbar/2$. Consequently, Fock states do not offer an optimal description of classical descriptions of light, as they generally fail to reach this uncertainty product.

In this context, coherent states [59,60] of light reintroduce the notion of a classically oscillating field that provides a solution to Maxwell's equations in vacuum. The coherent states, denoted as $\{|\alpha\rangle : \alpha \in \mathbb{C}\}$, are described as displaced vacuum states, that is

$$|\alpha\rangle = \hat{D}(\alpha)\,|0\rangle, \tag{7.38}$$

where $\hat{D}(\alpha) \equiv \exp\left[\alpha\hat{a}^\dagger - \alpha^*\hat{a}\right]$ represents the displacement operator. From this definition, and using the displacement operator properties, it follows that coherent states are eigenstates of the creation operator, that is,

$$\hat{a}\,|\alpha\rangle = \alpha\,|\alpha\rangle. \tag{7.39}$$

This last property of coherent states becomes particularly useful when evaluating the mean value of the electric field operator

$$\langle\alpha|\hat{E}(t)|\alpha\rangle = \sqrt{\frac{2\hbar\omega}{\epsilon_0 V}}|\alpha|\sin(\omega t - \theta), \tag{7.40}$$

where we have written $\alpha = |\alpha|e^{i\theta}$. Thus, coherent states of light indeed restore the oscillating nature of the electric field and result in a variance of this operator independent of the number of photons $\langle n\rangle = |\alpha|^2$. Specifically, we obtain $\Delta E^2 = (\hbar\omega)/(2\varepsilon_0 V)$. This behavior arises from coherent states providing an optimal description for the classical description of light, as they saturate the product between the quadrature operators' variance, i.e., $\Delta x \Delta p = (\hbar/\omega)$. This, among other reasons [61], justifies our previous statement: that these states recover a classical description of light. Therefore, we refer to *non-classical* states of light as those states exhibiting features unaccounted for by classical optics theories, Fock states being a particular example.

In general, distinguishing between classical and non-classical states of light often relies on the definition of specific observables. One of the most crucial, although not the sole determinant, is the Wigner function [62, 63], which captures all the information about a quantum state in phase space. Initially proposed in the context of quantum corrections to thermodynamic equilibrum [62], it has found widespread use in quantum optics as an indicator of non-classical states of light [64, 65] by featuring negative values in certain phase space regions. Nonetheless, it is essential to note that the absence of negative values in this function does not inherently imply the *classicality* of the state. Other indicators of non-classical behavior, like sub-Poissonian statistics, might manifest in states characterized by positive-valued Wigner functions.

The Wigner function of a given quantum state $\hat{\rho}$ can be computed as [66]

$$W(\beta) = \mathrm{tr}\Big[\hat{\rho}\hat{D}(\beta)\hat{\Pi}\hat{D}(-\beta)\Big], \tag{7.41}$$

where $\hat{\Pi}$ denotes the parity operator, and $\beta$ a parameter related to the optical quadratures through $x \equiv \mathrm{Re}(\beta)$ and $p \equiv \mathrm{Im}(\beta)$. In Fig. 7.3, we present examples of different light states displaying different Wigner functions. In panel (a), a coherent state with amplitude $\alpha = 0$ is shown, i.e., a vacuum state. These states feature Gaussian Wigner functions without

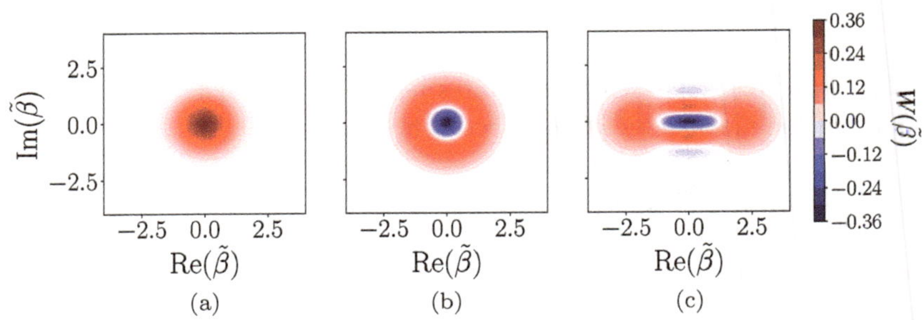


Fig. 7.3 Wigner functions of different states. (a) A vacuum state represented by a coherent state with $|\alpha| = 0$. (b) Wigner function for a Fock state $|n = 1\rangle$. (c) Wigner function of a coherent state superposition $|\alpha\rangle - |-\alpha\rangle$, with $|\alpha| = 1.5$.

negative regions. Conversely, Fock states with $n > 0$ exhibit Wigner functions containing negative regions, as shown in panel (b) for $|n = 1\rangle$. Interestingly, while coherent states themselves lack negative Wigner values in phase space, this is not necessarily true for superpositions composed by different coherent states, i.e., $|\Phi\rangle = \sum_{i=1}^{N} a_i |\alpha_i\rangle$, with $N \geq 2$ and $\alpha_i \neq \alpha_j$ $\forall i \neq j$. In panel (c), we illustrate the case of a coherent state superposition $|\Phi\rangle \propto |\alpha\rangle - |-\alpha\rangle$ with $\alpha = 1.5$. In this instance, quantum interference between the two terms in the superposition, represented by the two maxima at $\text{Re}(\beta) \approx \pm 1.5$, leads to negative regions between these maxima within the Wigner function.

While the Wigner function can indicate non-classical aspects in individual modes of light, in many other situations we have systems composed by multiple subsystems, forming a composite Hilbert space denoted as $\mathcal{H}_T = \bigotimes_{i=1}^{N_s} \mathcal{H}_i$, where $N_s$ represents the number of subsystems. In optics, these Hilbert spaces often represent various modes of light, such as different polarization and/or frequency modes. In this broader context, states $|\psi\rangle \in \mathcal{H}_T$ may exhibit non-classical features beyond the mere presence of negative regions in their Wigner functions. A notable example is the concept of *entanglement*, which pertains to certain quantum states that cannot be expressed as the tensor product of their individual components [67], that is

$$|\psi\rangle \text{ is entangled if } |\psi\rangle \neq \bigotimes_{i=1}^{N_s} |\psi_{\mathcal{H}_i}\rangle, \tag{7.42}$$

where $|\psi_{\mathcal{H}_i}\rangle \in \mathcal{H}_i$.

Determining whether a given state $\hat{\rho}$ is entangled is a problem known to be NP-hard[2] [68]. However, in certain scenarios, such as for bipartite systems involving only two subsystems, it is possible to define mathematical functions that offer a quantitative assessment of entanglement. These functions are commonly known as entanglement measures. Among these measures, one of the most prominent is the von Neumann entropy [69], also referred to as entanglement entropy [70]. For a pure state $|\psi_{AB}\rangle \in \mathcal{H}_A \otimes \mathcal{H}_B$, this quantity is defined as

$$\mathcal{S}(\hat{\rho}) := -\operatorname{tr}[\hat{\rho}_A \log_d(\hat{\rho}_A)], \tag{7.43}$$

[2]Nondeterministic polynomial (NP) problems comprise decision problems for which instances with affirmative answers can have their proofs verified in polynomial time. NP-hard problems, in this context, are as challenging as NP problems but may not necessarily belong to NP.

where $\hat{\rho}_A$ is the reduced density of $|\psi_{AB}\rangle$ with respect to subsystem $B$, defined as $\hat{\rho}_A \equiv \text{tr}_B(|\psi_{AB}\rangle\langle\psi_{AB}|)$. In this expression, $d$ represents the dimension of $\mathcal{H}_A$. Notably, although in this expression we have focused on subsystem $A$, the usage of $A$ and $B$ can be interchanged. If our quantum state is separable, i.e., $|\psi_{AB}\rangle = |\psi_A\rangle \otimes |\psi_B\rangle$, then it can be proven that $\mathcal{S}(\hat{\rho}) = 0$, indicating no entanglement. Conversely, states where $\mathcal{S}(\hat{\rho}) = 1$ are termed maximally entangled. Utilizing entangled states, particularly maximally entangled ones, forms the cornerstone of quantum information science applications such as quantum communication [71, 72] and quantum computation [73, 74], among others [75].

However, in numerous scenarios, both Hilbert spaces $\mathcal{H}_A$ and $\mathcal{H}_B$ can have exceedingly large dimensions. Evaluating the entanglement entropy in Eq. (7.43) under these circumstances becomes numerically challenging, prompting the search for more computationally tractable alternatives. One such example is the linear entropy, defined as

$$\mathcal{S}_{\text{lin}}(\hat{\rho}) := 1 - \text{tr}(\hat{\rho}_A^2), \tag{7.44}$$

where $\hat{\rho}_A^2$ is termed the purity of the state $\hat{\rho}_A$. The linear entropy corresponds to a second-order expansion of the von Neumann entropy in terms of $\hat{\rho}_A$, and akin to the latter, it becomes zero for separable states. In contrast, its maximum value is bounded by $1 - (1/d)$.

In Fig. 7.4, we present both entanglement measures evaluated on a generic two-qubit state of the form $|\Psi_{AB}\rangle = \sqrt{p}\,|0_A\rangle \otimes |0_B\rangle + \sqrt{1-p}\,|1_A\rangle \otimes |1_B\rangle$ as a function of $p$. For $p = 0$ and $p = 1$, corresponding to a separable state, both entanglement measures tend to zero. Conversely, the maximum value for both functions occurs at $p = 0.5$, representing a maximally entangled state $|\psi_{AB}\rangle \propto |0_A\rangle \otimes |0_B\rangle + |1_A\rangle \otimes |1_B\rangle$. The use of these entanglement measures, along with the Wigner function, will be of fundamental interest in the subsequent sections of this chapter.

### 7.3.2 *Bloch-Bloch Picture*

With all the fundamental quantum optics concepts set, let us delve into a quantum optical treatment of the interaction between a solid-state system and a strong laser field, based on an expansion of the electronic states in the Bloch basis. Specifically, we built upon the analysis provided in Sec. 7.2.1 by including in the Hamiltonian presented in Eq. (7.1) the electric field operator. In this context, the Hamiltonian reads

$$\hat{H}(t) = \hat{H}_{\text{cr}} + \mathsf{e}\hat{r}_{i,\text{tra}}\hat{E}(t) + \mathsf{e}\hat{r}_{i,\text{ter}}\hat{E}(t) + \hat{H}_{\text{field}}, \tag{7.45}$$

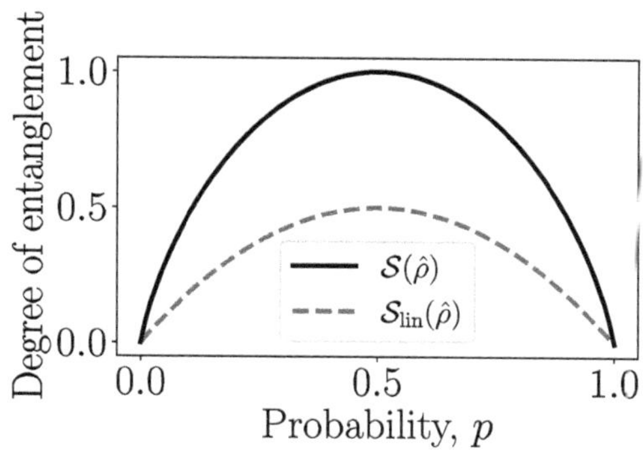


Fig. 7.4 Entanglement measures evaluated for a state in the form $\sqrt{p}\,|0_A\rangle \otimes |0_B\rangle + \sqrt{1-p}\,|1_A\rangle \otimes |1_B\rangle$ plotted as functions of $p$. The black solid curve represents the entanglement entropy, while the grey dashed curve represents the linear entropy. As expected, both measures tend toward zero when the state becomes separable ($p = 0$ and $p = 1$), and they reach a maximum at $p = 0.5$ corresponding to a maximally entangled state.

where $\hat{H}_{\text{field}}$ represents the free-field Hamiltonian and $\hat{E}(t)$ the electric field operator. Our objective is to characterize interactions with lasers of finite duration, which ultimately necessitates the inclusion of a continuum spectrum of light [13]. However, for simplicity, we limit ourselves to a discrete set of modes spanning the harmonic modes from the central frequency of the employed pulse, $\omega_L$, up to the cutoff region of the harmonic spectrum, $\omega_{q_c} = q_c\omega_L$. Thus, we represent the free-field Hamiltonian as $\hat{H}_{\text{field}} \equiv \sum_{q=1}^{q_c} \hbar\omega_q \hat{a}_q^\dagger \hat{a}_q$ where $\omega_q = q\omega_L$ ($\omega_{q=1} \equiv \omega_L$) and $\hat{a}_q^\dagger$ ($\hat{a}_q$) is the creation (annihilation) operator acting on the field mode of frequency $q$. The electric field operator is similarly expressed as

$$\hat{E}(t) = if(t)\sum_{q}^{q_c} \mathrm{g}(\omega_q)\big(\hat{a}_q - \hat{a}_q^\dagger\big), \tag{7.46}$$

where $\mathrm{g}(\omega_q) \equiv \sqrt{\hbar\omega_q/(2\varepsilon_0 V)}$ and $0 \leq f(t) \leq 1$ is a dimensionless function accounting for the pulse envelope.

Within this approximation of considering a discrete set of modes, we can represent the initial state of the system composed by

$$|\Psi(t = t_0)\rangle = |\phi_{\mathbf{k}_0,v}\rangle \otimes |\alpha_L\rangle \bigotimes_{q=2}^{q_c} |0_q\rangle, \tag{7.47}$$

such that the single-active electron is initially in the valence band with crystal momentum $\mathbf{k}_0$, while the driving field mode is in a coherent state

$\alpha_L$ with amplitudes on the order of $10^5$. Meanwhile, as all harmonic modes begin in an unoccupied state, they are initialized as vacuum states. The overall system dynamics are described by the Schrödinger equation

$$i\hbar\frac{\partial\,|\Psi(t)\rangle}{\partial t} = \big[\hat{H}_{\text{cr}} + \mathsf{e}\hat{r}_{i,\text{tra}}\hat{E}(t) + \mathsf{e}\hat{r}_{i,\text{ter}}\hat{E}(t) + \hat{H}_{\text{field}}\big], \tag{7.48}$$

although the analysis can be further simplified by introducing some unitary transformations. Thus, using

$$|\Psi(t)\rangle = \hat{D}_1(\alpha_L)e^{-i\hat{H}_{\text{field}}t/\hbar}\left|\tilde{\Psi}(t)\right\rangle, \tag{7.49}$$

the differential equation in Eq. (7.48) can be rewritten as

$$\begin{aligned} i\hbar\frac{\partial|\tilde{\Psi}(t)\rangle}{\partial t} &= \big[\hat{H}_{\text{cr}} + \mathsf{e}\hat{r}_{i,\text{tra}}\big(E_{\text{cl}}(t) + \hat{E}(t)\big) \\ &\quad + \mathsf{e}\hat{r}_{i,\text{ter}}\big(E_{\text{cl}}(t) + \hat{E}(t)\big)\big]|\tilde{\Psi}(t)\rangle, \end{aligned} \tag{7.50}$$

where $E_{\text{cl}}(t) = \text{tr}(\hat{E}(t)\,|\Psi(t_0)\rangle\langle\Psi(t_0)|)$. These transformations introduce an extra time dependence on the electric field operator, effectively shifting us to the interaction picture with respect to $\hat{H}_{\text{field}}$

$$\hat{E}(t) = if(t)\sum_q^{q_c} \mathrm{g}(\omega_q)\big(\hat{a}_q e^{-i\omega_q t} - \hat{a}_q^\dagger e^{i\omega_q t}\big), \tag{7.51}$$

and displaces the origin of our frame of reference in phase space onto the coherent state $|\alpha_L\rangle$, thus leading to the following initial state for this equation

$$|\tilde{\Psi}(t)\rangle = |\phi_{\mathbf{k}_0,v}\rangle \bigotimes_{q=1}^{q_c} |0_q\rangle. \tag{7.52}$$

The transformations we have outlined are customary in the quantum optical analysis of HHG processes in atomic [11–13, 15] and molecular [21] systems. When examining solid-state targets, along the same lines to what is done in the semiclassical analysis, we introduce an additional transformation akin to the $\hat{U}_{\text{tra}}(t)$ performed in Sec. 7.2.1. Specifically, we consider

$$|\tilde{\Psi}(t)\rangle = e^{i(\mathsf{e}A_{\text{cl}}(t)+\hat{A}(t))\hat{r}_{i,\text{tra}}/\hbar}|\bar{\Psi}(t)\rangle, \tag{7.53}$$

with $\hat{A}(t)$ satisfying $\hat{E}(t) = -\,\partial\hat{A}(t)\big/\partial t$. By substituting this expression into Eq. (7.50) and considering terms up to first order in $\mathrm{g}(\omega_q) \sim 10^{-8}$ a.u., which weighs the quantum optical fluctuations, we can approximate our differential equation as

$$i\hbar\frac{\partial|\bar{\Psi}(t)\rangle}{\partial t} \approx \big[\hat{H}_{\text{cr}}(t) + \mathsf{e}\hat{r}_{i,\text{ter}}(t)\big(E_{\text{cl}}(t) + \hat{E}(t)\big) - i\mathsf{e}\hbar^{-1}\hat{v}_{i,\text{tra}}(t)\hat{A}(t)\big]|\bar{\Psi}(t)\rangle, \tag{7.54}$$

where $\hat{v}_{i,\text{tra}} \equiv [\hat{r}_{i,\text{tra}}, \hat{H}_{\text{cr}}]$. Additionally, the extra time-dependent terms observed in the operators acting on the electron resemble those outlined in Eq. (7.11). Furthermore, these expressions have been derived assuming that inter- and intraband terms are the primary contributors to the HHG in solids. Yet, other factors like the mixture current, whose effects are not fully comprehended from the semiclassical perspective [40], are not considered in this analysis.

The reason for introducing these transformations lies in Eq. (7.54), where the first two terms describe the semiclassical evolution detailed in Sec. 7.2.1. Having solved for these terms as presented in previous subsections, we introduce an additional unitary transformation $|\bar{\Psi}(t)\rangle = \hat{U}_{\text{sc}}(t,t_0)|\tilde{\Psi}'(t)\rangle$ with $\hat{U}_{\text{sc}}(t,t_0) \equiv \hat{\mathcal{T}}\exp\Big[-i\int_{t_0}^{t} \mathrm{d}\tau\hat{\hat{H}}_{\text{sc}}(\tau)/\hbar\Big]$, accounting for the semiclassical dynamics. Here, $\hat{\mathcal{T}}$ denotes the time-ordering operator, and $\hat{\hat{H}}_{\text{sc}}(\tau) \equiv \hat{H}_{\text{cr}}(t) + \mathsf{e}\hat{r}_{i,\text{ter}}(t)E_{\text{cl}}(t)$. As a result, our Schrödinger equation transforms into

$$i\hbar\frac{\partial|\tilde{\Psi}'(t)\rangle}{\partial t} \approx \big[\mathsf{e}\hat{\hat{r}}_{i,\text{ter}}(t)\hat{E}(t) - i\mathsf{e}\hbar^{-1}\hat{\hat{v}}_{i,\text{tra}}(t)\hat{A}(t)\big]|\tilde{\Psi}'(t)\rangle, \tag{7.55}$$

where $\hat{\hat{r}}_{i,\text{ter}} \equiv \hat{U}^{\dagger}_{\text{sc}}(t,t_0)\hat{r}_{i,\text{ter}}(t)\hat{U}_{\text{sc}}(t,t_0)$ and $\hat{\hat{v}}_{i,\text{tra}} \equiv \hat{U}^{\dagger}_{\text{sc}}(t,t_0)\hat{v}_{i,\text{tra}}(t)$ $\hat{U}_{\text{sc}}(t,t_0)$.

With the aim of solving this differential equation, we expand our state in the $\{|\mathbf{K},m\rangle\}$ basis, that is

$$|\tilde{\Psi}'(t)\rangle = \sum_{m=v,c}\int \mathrm{d}\mathbf{K}\,|\mathbf{K},m\rangle \otimes |\Phi_m(\mathbf{K},t)\rangle, \tag{7.56}$$

where $|\Phi_m(\mathbf{K},t)\rangle = \langle\mathbf{K},m|\tilde{\Psi}'(t)\rangle$ corresponds to the quantum optical state conditioned on finding an electron in band $m$ and with canonical crystal momentum $\mathbf{K}$. Introducing this expression into Eq. (7.55) and projecting the whole equation with respect to $|\mathbf{K},m\rangle$, we arrive at

$$i\hbar\frac{\partial}{\partial t}|\Phi_m(\mathbf{K},t)\rangle = \sum_{l=c,v}\int \mathrm{d}\mathbf{K}'\big[\mathsf{e}\,\langle\mathbf{K},m|\,\hat{\hat{r}}_{i,\text{ter}}(t)\,|\mathbf{K}',l\rangle\,\hat{E}(t) - i\mathsf{e}\hbar^{-1}\,\langle\mathbf{K},m|\,\hat{\hat{v}}_{i,\text{tra}}(t)\,|\mathbf{K}',l\rangle\,\hat{A}(t)\big]. \tag{7.57}$$

In order to further proceed in solving the set of differential equations spanned by Eq. (7.57), it is necessary to evaluate the electronic matrix elements appearing in the above expressions. By taking into account the semiclassical evolution described in Eq. (7.12), which leads us to

$$\hat{U}_{\text{sc}}(t,t_0)\,|\mathbf{K},m\rangle = b_v(\mathbf{K},t;m)\,|\mathbf{K},v\rangle + b_c(\mathbf{K},t;m)\,|\mathbf{K},c\rangle, \tag{7.58}$$

and after some algebraic operations, it is possible to show that

$$\begin{aligned}\int \mathrm{d}\mathbf{K}'\,\langle\mathbf{K},m|\,\hat{\bar{r}}_{i,\text{ter}}(t)\,|\mathbf{K}',l\rangle\,|\mathbf{K}',l\rangle \\ &= \sum_{\mathsf{i}=v,c}\sum_{\mathsf{j}=v,c} b_{\mathsf{i}}^*(\mathbf{K},t;m)b_{\mathsf{j}}^*(\mathbf{K},t;l)d_{\mathsf{ij}}\big(\mathbf{K}+\mathsf{e}\varepsilon_i A_{\text{cl}}(t)\big)\,|\mathbf{K},l\rangle \\ &\equiv \mathsf{e}^{-1}M_{m,l}^{(\text{ter})}\,|\mathbf{K},l\rangle,\end{aligned} \tag{7.59}$$

while for the other term we get

$$\begin{aligned}\int \mathrm{d}\mathbf{K}'\,\langle\mathbf{K},m|\,\hat{\bar{v}}_{i,\text{tra}}(t)\,|\mathbf{K}',l\rangle\,|\mathbf{K}',l\rangle \\ &= i\hbar\sum_{\mathsf{i}=v,c} b_{\mathsf{i}}^*(\mathbf{K},t;m)b_{\mathsf{i}}(\mathbf{K},t;l)\frac{\partial}{\partial K_i}\big[E_{\mathsf{i}}\big(\mathbf{K}+\mathsf{e}\varepsilon_i A_{\text{cl}}(t)\big)\big]\,|\Phi_l(\mathbf{K},t)\rangle \\ &\equiv i\mathsf{e}^{-1}\hbar M_{m,l}^{(\text{tra})}(\mathbf{K},t)\,|\Phi_l(\mathbf{K},t)\rangle.\end{aligned} \tag{7.60}$$

These expressions we have just presented are associated with the inter- and intraband currents delineated in Eqs. (7.15) and (7.16). The former corresponds to the time-dependent interband polarization, while the latter deals with the intraband current. However, the quantum optical treatment conducted here has been performed under a noiseless scenario, lacking explicit inclusion of dephasing effects. Nonetheless, these could be incorporated when evaluating the matrix elements $M_{m,l}^{(\text{ter})}(\mathbf{K},t)$ and $M_{m,l}^{(\text{tra})}(\mathbf{K},t)$. It is important to note that this approach is akin to describing the crystal using non-Hermitian Hamiltonians. Similar strategies have been employed in semiclassical analyses [76], yielding analogous outcomes to those presented in Sec. 7.2.1.

Introducing Eqs. (7.15) and (7.16) into Eq. (7.57), and explicitly distinguishing between projections on valence and conduction bands, we obtain the following set of coupled differential equations

$$\begin{aligned}i\hbar\frac{\partial}{\partial t}\,|\Phi_v(\mathbf{K},t)\rangle &= \big[M_{v,v}^{(\text{ter})}(\mathbf{K},t)\hat{E}(t) + M_{v,v}^{(\text{tra})}\hat{A}(t)\big]\,|\Phi_v(\mathbf{K},t)\rangle \\ &\quad + \big[M_{v,c}^{(\text{ter})}(\mathbf{K},t)\hat{E}(t) + M_{v,c}^{(\text{tra})}\hat{A}(t)\big]\,|\Phi_c(\mathbf{K},t)\rangle,\end{aligned} \tag{7.61}$$

$$i\hbar\frac{\partial}{\partial t}\left|\Phi_c(\mathbf{K},t)\right\rangle = \left[M_{c,v}^{(\text{ter})}(\mathbf{K},t)\hat{E}(t) + M_{c,v}^{(\text{tra})}\hat{A}(t)\right]\left|\Phi_v(\mathbf{K},t)\right\rangle \\ + \left[M_{c,c}^{(\text{ter})}(\mathbf{K},t)\hat{E}(t) + M_{c,c}^{(\text{tra})}\hat{A}(t)\right]\left|\Phi_c(\mathbf{K},t)\right\rangle, \quad (7.62)$$

where each of the equations have both a homogeneous and an inhomogeneous component. Thus, their solution can be expressed as the sum of the solution to the homogeneous equation plus a particular solution to the inhomogeneous one. Let us start by the former, for which we have

$$i\hbar\frac{\partial}{\partial t}\left|\Phi_{\text{i,hom}}(\mathbf{K},t)\right\rangle \\ = \left[M_{\text{i,i}}^{(\text{ter})}(\mathbf{K},t)\hat{E}(t) + M_{\text{i,i}}^{(\text{tra})}\hat{A}(t)\right]\left|\Phi_{\text{i,hom}}(\mathbf{K},t)\right\rangle, \quad (7.63)$$

and taking into account the definitions we consider in a discrete mode description for the electric field and vector potential operators

$$\hat{E}(t) = -if(t)\sum_{q=1}^{q_c}\text{g}(\omega_q)\left(\hat{a}_q^\dagger e^{i\omega_q t} - \hat{a}_q e^{-i\omega_q t}\right), \quad (7.64)$$

$$\hat{A}(t) = -\int \text{d}t\hat{E}(t) = i\sum_{q=1}^{q_c}\text{g}(\omega_q)\left[F_q(t)\hat{a}_q^\dagger - F_q^*(t)\hat{a}_q\right], \quad (7.65)$$

wherein Eq. (7.65) we have defined $F(t) \equiv \int \text{d}t f_q(t)$ with $f_q(t) \equiv f(t)e^{i\omega_q t}$, we can rewrite Eq. (7.63) as

$$i\hbar\frac{\partial}{\partial t}\left|\Phi_{\text{i,hom}}(\mathbf{K},t)\right\rangle \\ = i\sum_{q=1}^{q_c}g(\omega_q)\Big[\Big(-M_{\text{i,i}}^{(\text{ter})}(\mathbf{K},t)f_q(t) + M_{\text{i,i}}^{(\text{tra})}(\mathbf{K},t)F_q(t)\Big)\hat{a}_q^\dagger \\ - \Big(-M_{\text{i,i}}^{(\text{ter})}(\mathbf{K},t)f_q^*(t) + M_{\text{i,i}}^{(\text{tra})}(\mathbf{K},t)F_q^*(t)\Big)\hat{a}_q\Big]\left|\Phi_{\text{i,hom}}(\mathbf{K},t)\right\rangle. \quad (7.66)$$

As observed, the homogeneous part of this equation is given by a linear combination of creation and annihilation operators acting on the harmonic modes. This equation bears a resemblance to those derived in the analysis of HHG processes in atomic [11–13, 15] and molecular [21] systems. From these analysis, we learned that their solution can be expressed as

$$\left|\Phi_{\text{i,hom}}(\mathbf{K},t)\right\rangle = \hat{\mathcal{D}}\big(\chi_{\text{i}}(\mathbf{K},t,t_0)\big)\left|\Phi_{\text{i,hom}}(\mathbf{K},t_0)\right\rangle, \quad (7.67)$$

where in this expression we have defined

$$\hat{\mathcal{D}}(\chi_{\text{i}}(\mathbf{K},t,t_0)) \equiv \prod_{q=1}^{q_c}\left[e^{i\varphi_q(\mathbf{K},t,t_0)}\hat{D}_q(\chi_{\text{i}}^{(q)}(\mathbf{K},t,t_0))\right], \quad (7.68)$$

with $\varphi_q(\mathbf{K},t,t_0)$ a phase factor arising from the use of the Baker–Campbell–Hausdorff formula [12,13,56] in order to write the solution above, and where $\chi_{\mathrm{i}}^{(q)}(\mathbf{K},t,t_0)$ is given by

$$\chi_{\mathrm{i}}^{(q)}(\mathbf{K},t,t_0) = \frac{1}{\hbar} g(\omega_q) \int_{t_0}^{t} \mathrm{d}\tau \Big[ - M_{\mathrm{i,i}}^{(\mathrm{ter})}(\mathbf{K},\tau) f_q(\tau) + M_{\mathrm{i,i}}^{(\mathrm{tra})}(\mathbf{K},\tau) F_q(\tau) \Big]. \tag{7.69}$$

Thus, by adding to this solution a particular solution to the inhomogeneous equations, we find the following recursive relation for the solutions of Eqs. (7.62) and (7.61),

$$|\Phi_v(\mathbf{K},t)\rangle = \hat{\mathcal{D}}\big(\boldsymbol{\chi}_v(\mathbf{K},t,t_0)\big) |\Phi_v(\mathbf{K},t_0)\rangle - \frac{i}{\hbar} \int_{t_0}^{t} \mathrm{d}t' \hat{\mathcal{D}}\big(\boldsymbol{\chi}_v(\mathbf{K},t,t')\big) \hat{\mathcal{M}}_{v,c}(\mathbf{K},t') |\Phi_c(\mathbf{K},t')\rangle, \tag{7.70}$$

$$|\Phi_c(\mathbf{K},t)\rangle = \hat{\mathcal{D}}\big(\boldsymbol{\chi}_c(\mathbf{K},t,t_0)\big) |\Phi_c(\mathbf{K},t_0)\rangle - \frac{i}{\hbar} \int_{t_0}^{t} \mathrm{d}t' \hat{\mathcal{D}}\big(\boldsymbol{\chi}_c(\mathbf{K},t,t')\big) \hat{\mathcal{M}}_{c,v}(\mathbf{K},t') |\Phi_v(\mathbf{K},t')\rangle, \tag{7.71}$$

where in this expression we have defined for the sake of clarity

$$\hat{\mathcal{M}}_{\mathrm{i,j}}(\mathbf{K},t') \equiv M_{\mathrm{i,j}}^{(\mathrm{ter})}(\mathbf{K},t) \hat{E}(t) + M_{\mathrm{i,j}}^{(\mathrm{tra})} \hat{A}(t). \tag{7.72}$$

By combining these equations, while keeping terms up to first order in $g(\omega_q)$, we get after introducing the initial conditions

$$|\Phi_v(\mathbf{K},t)\rangle = \delta(\mathbf{K}-\mathbf{K}_0) \hat{\mathcal{D}}\big(\boldsymbol{\chi}_v(\mathbf{K},t,t_0)\big) \bigotimes_{q=1}^{q_c} |0_q\rangle \tag{7.73}$$

$$|\Phi_c(\mathbf{K},t)\rangle = -\delta(\mathbf{K}-\mathbf{K}_0) \frac{i}{\hbar} \int_{t_0}^{t} \mathrm{d}t' \hat{\mathcal{D}}\big(\boldsymbol{\chi}_c(\mathbf{K},t,t')\big) \hat{\mathcal{M}}_{c,v}(\mathbf{K},t') \bigotimes_{q=1}^{q_c} |0_q\rangle. \tag{7.74}$$

Likewise, for the joint state of the system we obtain, up to a normalization factor,

$$\begin{aligned} |\tilde{\Psi}'(t)\rangle &= \sum_{m=v,c} \int \mathrm{d}\mathbf{K}\, |\mathbf{K},m\rangle \otimes |\Phi_m(\mathbf{K},t)\rangle \\ &= \hat{\mathcal{D}}\big(\boldsymbol{\chi}_v(\mathbf{K}_0,t,t_0)\big) |\mathbf{K}_0,v\rangle \bigotimes_{q=1}^{q_c} |0_q\rangle \\ &\quad - \frac{i}{\hbar} \int_{t_0}^{t} \mathrm{d}t' \hat{\mathcal{D}}\big(\boldsymbol{\chi}_c(\mathbf{K}_0,t,t')\big) \hat{\mathcal{M}}_{c,v}(\mathbf{K}_0,t') |\mathbf{K}_0,c\rangle \bigotimes_{q=1}^{q_c} |0_q\rangle. \end{aligned} \tag{7.75}$$

It is worth noting that, similar to the quantum optical description of strong-field processes in atomic systems [11–13, 15], for solid-state systems each electromagnetic field mode experiences a displacement by an amount $\chi_{\mathrm{i}}^{(q)}(\mathbf{K}, t, t_0)$. However, given that the electron dynamics in solids are different to those occurring in atomic systems, this quantity gets influenced both by interband and intraband dynamics. In the remainder of this subsection, we study how these two impact the final quantum optical state. Nevertheless, before doing so, we return to the original frame of reference with respect to the electron dynamics, i.e.,

$$|\tilde{\Psi}(t)\rangle = e^{ie(A_{\mathrm{cl}}(t)+\hat{A}(t))\hat{r}_{i,\mathrm{tra}}/\hbar}\hat{U}_{\mathrm{sc}}(t, t_0)|\tilde{\Psi}'(t_0)\rangle, \tag{7.76}$$

which in the asymptotic limit $t \to \infty$ leads to

$$|\tilde{\Psi}(t)\rangle = \hat{U}_{\mathrm{sc}}(t, t_0)|\tilde{\Psi}'(t_0)\rangle. \tag{7.77}$$

At this point, it is crucial to note that in Eq. (7.76), the unitary transformation $e^{ie\hat{A}(t)\hat{r}_{i,\mathrm{tra}}/\hbar}$ inherently introduces entanglement between the canonical crystal momentum and the field modes. This arises as $\hat{r}_{i,\mathrm{tra}}$ is not diagonal with respect to $\mathbf{K}$. However, in the discrete mode description employed here, where we introduce the envelope function $f(t)$ to capture the field's envelope, it happens that $\lim_{t\to\infty} f(t) = 0$, leading to $\lim_{t\to\infty} \hat{A}(t) = 0$. Therefore, within the scope of our current framework, this additional operator does not entangle the field and the electron at the final time. Nevertheless, it does introduce entanglement at the intermediate steps of the dynamics, which are essentially caused by the coupling of the intraband dynamics to the different field modes.

The analysis we aim to perform mainly focuses on HHG processes, where the electron ends up in the valence band of the solid-state system. Here, we impose the electron to be found in the valence band by applying the projector operator $\hat{P}_v \equiv \int \mathrm{d}\mathbf{K}\, |\mathbf{K}, v\rangle\!\langle \mathbf{K}, v|$, and trace out the electronic degrees of freedom. By doing this, while assuming that the probability of finding in the valence band electrons initially born in the conduction band is small, we arrive at

$$|\Phi_v(\mathbf{K}, t)\rangle \approx \hat{\mathcal{D}}\big(\chi_v(\mathbf{K}_0, t, t_0)\big) \bigotimes_{q=1}^{q_c} |0_q\rangle, \tag{7.78}$$

similar to what is found for atomic systems. However, it is important to note here that, in this analysis, we have neglected contributions in the form $\langle \mathbf{K}, v|\, \hat{U}_{\mathrm{sc}}(t, t_0)\, |\mathbf{K}', c\rangle$, which are considerably smaller than the diagonal ones with respect to the valence band as they scale with $\mathrm{g}(\omega_q)$. The

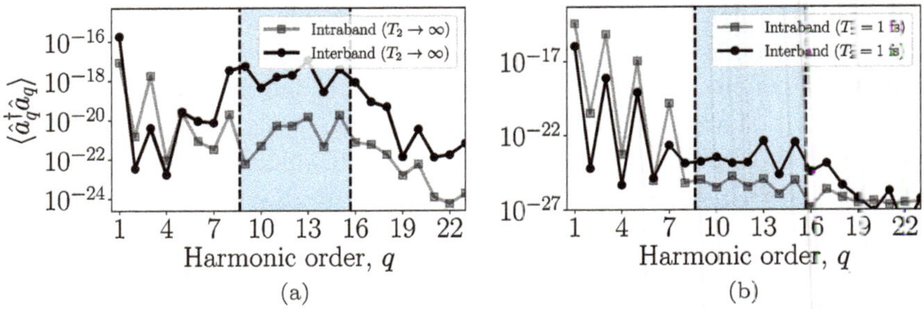


Fig. 7.5 Mean photon numbers calculated for various harmonics using Eq. (7.78). The analysis pertains to the ZnO material under the influence of a linearly polarized laser pulse aligned along the $\Gamma - A$ crystal direction. The laser settings include a peak field strength of $0.5\,\text{V/Å}$, a central wavelength of $\lambda_L = 3.25\,\mu\text{m}$, and a duration of $\Delta t \approx 96\,\text{fs}$ ($\approx$9 optical cycles). In panel (a), an ideal solid scenario is depicted, where dephasing effects are absent ($T_2 \to \infty$). In contrast, panel (b) accounts for scenarios where dephasing effects are included. The dashed blue region denotes the minimum and maximum energy band gaps in the specified solid direction.

introduction of these terms would indeed establish entanglement between the electron and electromagnetic field states. However, our subsequent exploration using a Wannier-based analysis — a more convenient representation for light-matter entanglement analyses — reveals that even in the scenario considered here, where ZnO is analyzed, the amount of light-matter entanglement remains marginal even in the many-electron limit.

To validate the outcomes derived from Eq. (7.78), we proceed to evaluate the mean photon number of the various harmonic modes at time $t \to \infty$. To some extent, this observable should mirror the harmonic spectra observed following HHG processes [77]. Figure 7.5 illustrates the outcomes of this computation for both (a) the no-dephasing scenario ($T_2 \to \infty$) and (b) $T_2 = 1\,\text{fs}$. Notably, the obtained quantity mimics key aspects of the HHG spectra, displaying a plateau region extending from the lowest to the maximum energy band gap, followed by a sharp cutoff. In both cases, the interband component prevails over the intraband one in the non-perturbative (plateau) region. Decreased dephasing values result in better resolved peaks and reduced overall harmonic intensities.

### 7.3.2.1 *An Extension to the Many-Electron Regime*

It is worth noting that our analysis revolves around the single-electron level. Nonetheless, in a semiconductor's ground state, the valence band is typically fully occupied, forming what is known as a Fermi sea. Utilizing the semiconductor Bloch equations holds a significant advantage in this scenario, as it inherently incorporates the typically weak many-electron interactions through effective couplings, such as, for instance, dephasing. Therefore, within this framework, each electron can be treated independently, possessing a distinct initial crystal momentum **K** spanning the entire Brillouin zone. Consequently, extending Eq. (7.78) to a many-body formulation, we have

$$\begin{aligned} |\Phi_v(t)\rangle &\approx \hat{\mathcal{D}}\Big(N_z \int \mathrm{d}\mathbf{K}\boldsymbol{\chi}_v(\mathbf{K}, t, t_0)\Big) \bigotimes_{q=1}^{q_c} |0_q\rangle, \\ &= \hat{\mathcal{D}}(\bar{\boldsymbol{\chi}}_v(t, t_0)) \bigotimes_{q=1}^{q_c} |0_q\rangle, \end{aligned} \tag{7.79}$$

where $N_z$ denotes the number of Brillouin zones excited by the laser field. In our numerical analysis, we employed $N_z$ within the range $10^6 - 10^7$, a value subject to variation based on distinct experimental conditions. These conditions involve parameters like the laser beam width, alignment of the crystal lattice concerning the field's polarization, and, in cases where the HHG process occurs during transmission, the sample's width. Such variations can significantly impact the specific value of $N_z$.

### 7.3.2.2 *Conditioning on HHG and Its Consequences*

The state presented in Eq. (7.79) corresponds to a classical state. For instance, if one computes the Wigner function of this state — regardless of the chosen harmonic mode — would yield a Gaussian distribution. However, additional operations can be applied to this state to generate non-classical states. In experimental setups [11, 78] extended to solid-state systems [79], the creation of non-classical states of light involves correlation measurements between the different harmonic modes and part of the input fundamental mode. Mathematically, this operation focuses on events where at least one photon is generated in some harmonic mode $q \neq 1$, while considering the correlations that emerge with the fundamental mode. This can be formulated as the projective operation [18, 19]

$$\hat{P}_{\text{cond}} = \mathbb{1} - \bigotimes_{q=1}^{q_c} |0_q\rangle\langle 0_q|, \tag{7.80}$$

which we shall refer to hereafter as the *conditioning on HHG* operation. Applying this operator to the state in Eq. (7.79) leads, up to a normalization factor, to

$$\left|\bar{\Phi}_v(t)\right\rangle = \bigotimes_{q=1}^{q_c} |\bar{\chi}_v^{(q)}(t,t_0)\rangle - \xi_{\text{IR}}\xi_{\text{UV}} \bigotimes_{q=1}^{q_c} |0_q\rangle, \tag{7.81}$$

where $\xi_{\text{IR}} \equiv \langle 0_q|\bar{\chi}_v^{(q=1)}(t,t_0)\rangle$ and $\xi_{\text{UV}} \equiv \prod_{q=2}^{q_c}\langle 0_q|\bar{\chi}_v^{(q)}(t,t_0)\rangle$. This state corresponds to a massively entangled state, where all the harmonic modes that get excited during the HHG process become entangled [19]. However, thus far, the existing experimental implementations focus only on the infrared mode by measuring the intensity of the generated harmonics [11–13,78]. In mathematical terms, this operation corresponds to projecting the conditioned state in Eq. (7.81) with respect to $\bigotimes_{q=2}^{q_c}|\bar{\chi}_v^{(q)}(t,t_0)\rangle$. This actually leads to

$$\left|\bar{\Phi}_{v,\text{IR}}(t)\right\rangle = |\bar{\chi}_v^{(q=1)}(t,t_0)\rangle - \xi_{\text{IR}}|\xi_{\text{UV}}|^2\,|0_{q=1}\rangle, \tag{7.82}$$

which is a superposition between two different coherent states.

In Fig. 7.6, we compute the Wigner function of the state in Eq. (7.82) for the case where ZnO is excited with a linearly polarized laser field along

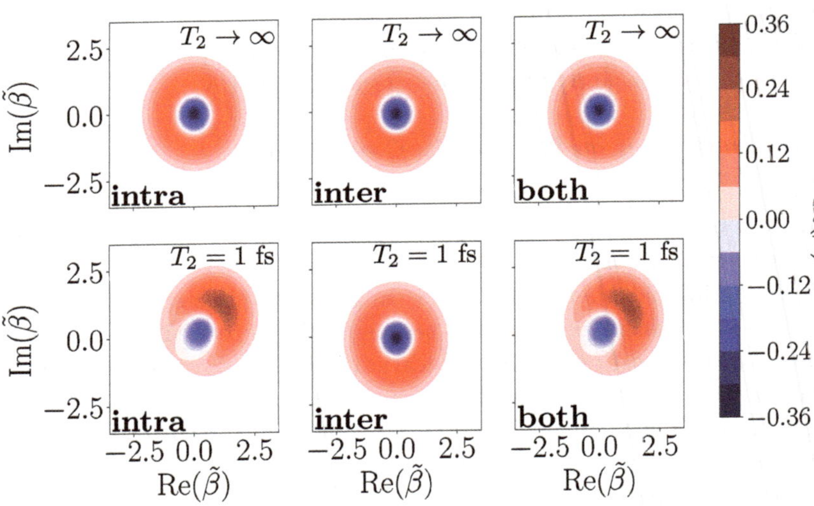


Fig. 7.6 Wigner functions of the state in Eq. (7.82) under various conditions. Each row represents different dephasing times, with the first two columns highlighting the separate contributions of intraband and interband effects (by intentionally disabling the other). The last column displays the combined contribution of both effects. The material analyzed is ZnO, excited with a linearly polarized laser field along the $\Gamma - A$ direction, exciting a total of $N_z = 1.6 \times 10^7$ Brillouin zones. The applied laser field has $\lambda_L = 3.25\,\mu$m, a field strength of 0.5 V/Å, and a duration of $\Delta t \approx 96$ fs (9 optical cycles). Adapted with permission from [32].

the $\Gamma - A$ direction which excites a total of $N_z = 1.6 \times 10^7$ Brillouin zones. We consider the applied laser field to have $\lambda_L = 3.25\,\mu$m, a field strength of 0.5 V/Å and a duration of $\Delta t \approx 96$ fs, corresponding to 9 optical cycles. In the first row, we show an ideal scenario without the introduction of dephasing, while in the second row a dephasing time $T_2 = 1$ fs has been set. On the other hand, each column corresponds to different contributions: intraband, interband and the combination of both, from left to right respectively. As observed, these results show that: (1) reduced values of $T_2$ lead to more unbalanced coherent state superpositions, yielding less homogeneous Wigner functions; and (2) that intraband effects seem to be dominant in dictating the shape of the final Wigner function. This can be understood by considering that a smaller dephasing time leads the electron to spend more time in the valence band, consequently increasing the contribution from intraband effects compared to interband effects.

Contrasting behaviors with respect to those found for the Wigner function are observed for the amount of entanglement between the infrared mode and the other harmonics, as shown in Fig. 7.7. The degree of entanglement in these scenarios is contingent upon the radiation generated in the harmonic modes, where the perturbative (plateau) region is primarily influenced by interband effects, as depicted in panel (a). In this case, interband effects dominate over intraband ones. Conversely, as shown in panel (b), the presence of dephasing effects significantly impacts the overall entanglement. Specifically, when considering $T_2 = 1$ fs, instead of no

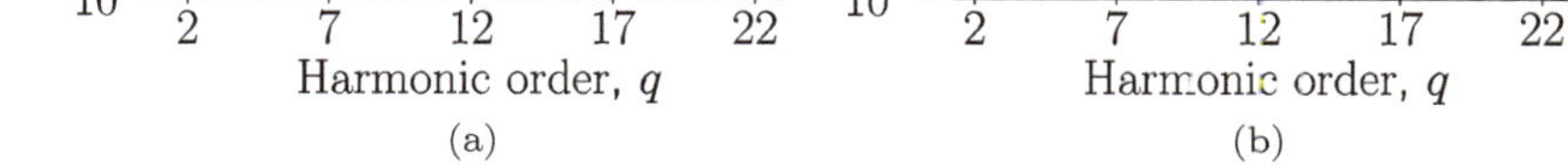


Fig. 7.7 The degree of entanglement between the fundamental mode and the $q$th harmonic mode is plotted against the harmonic order. Panel (a) demonstrates the distinct impacts of interband and intraband effects individually, along with their combined effect. Panel (b) showcases the effect of dephasing on this measure of entanglement. Identical excitation conditions to those in Fig. 7.6 have been applied.

dephasing effects, we observe a reduction of approximately nine orders of magnitude in these entanglement values.

### 7.3.3 *Wannier-Bloch Picture*

The use of a discrete basis set, such as the Wannier states, in the quantum optical analysis of HHG processes in solids provides a more manageable framework for exploring the entanglement between light and matter. Consequently, in this section, we aim to address the solution to the Schrödinger equation presented in Eq. (7.9) by employing a hybrid Wannier-Bloch basis representation akin to the methodology outlined in Section 7.2.2. To achieve this, we formulate the Schrödinger equation as

$$i\hbar \frac{\partial \left|\Psi(t)\right\rangle}{\partial t} = \left[\hat{H}_{\text{cr}} + \mathsf{e}\hat{r}\hat{E}(t) + \hat{H}_{\text{field}}\right], \tag{7.83}$$

where, unlike Eq. (7.9), here we do not explicitly differentiate between the intraband and interband components of the position operator. For the sake of simplicity, adhering to a single-active electron description similar to that of Section 7.2.2, we represent their initial state as

$$\left|\Psi(t_0)\right\rangle = \left|w_{j_0,v}\right\rangle \otimes \left|\alpha_L\right\rangle \bigotimes_{q=2}^{q_c} \left|0_q\right\rangle, \tag{7.84}$$

where the single-active electron initially occupies the $j_0$th Wannier site.

By applying similar unitary transformations as conducted in Section 7.3.2, excluding $e^{i\mathsf{e}(\hat{A}_{\text{cl}}(t)+\hat{A}(t))\hat{r}_{i,\text{tra}}/\hbar}$, we can rewrite Eq. (7.83) as follows

$$i\hbar \frac{\partial |\tilde{\Psi}'(t)\rangle}{\partial t} = \mathsf{e}\hat{r}(t)\hat{E}(t)|\tilde{\Psi}'(t)\rangle, \tag{7.85}$$

where $\hat{r}(t) \equiv \hat{U}_{\text{sc}}^{\dagger}(t)\hat{r}\hat{U}_{\text{sc}}(t)$. This is the pivotal point where the main distinction between the two approaches becomes evident. Here, the identity in the Wannier-Bloch approximation is introduced, namely

$$\mathbb{1} = \sum_j |w_{j,v}\rangle\langle w_{j,v}| + \int \mathrm{d}\mathbf{k}\, |\phi_{\mathbf{k},c}\rangle\langle\phi_{\mathbf{k},c}|, \tag{7.86}$$

which inputted in Eq. (7.85) leads to

$$\begin{aligned} i\hbar \frac{\partial |\tilde{\Psi}'(t)\rangle}{\partial t} &= \mathsf{e}\sum_j \hat{r}(t)\hat{E}(t)\, \langle w_{j,v}|\tilde{\Psi}'(t)\rangle\, |w_{j,v}\rangle \\ &\quad + \mathsf{e}\int \mathrm{d}\mathbf{k}\ \hat{r}(t)\hat{E}(t)\, \langle \phi_{\mathbf{k},c}|\tilde{\Psi}'(t)\rangle\, |\phi_{\mathbf{k},c}\rangle. \end{aligned} \tag{7.87}$$

In the following, we perform a similar assumption to that at the conclusion of the Bloch basis analysis in Section 7.3.2. This assumption involves considering the probability of residing in the conduction band to be extremely minimal at all times. We anticipate this approximation to be valid within the realm of weak depletion of the valence band, a scenario aligning with the frequencies and field strengths employed here [38]. Consequently, we approximate Eq. (7.54) as

$$i\hbar\frac{\partial|\tilde{\Psi}'(t)\rangle}{\partial t} \approx \mathsf{e}\sum_j \hat{r}(t)\hat{E}(t)\,\langle w_{j,v}|\tilde{\Psi}'(t)\rangle\,|w_{j,v}\rangle, \tag{7.88}$$

and by projecting with respect to the set of Wannier states, we arrive at the following system of differential equations

$$i\hbar\frac{\partial\,|\Phi_i(t)\rangle}{\partial t} = \sum_j M_{i,j}(t)\hat{E}(t)\,|\Phi_j(t)\rangle, \quad \forall j \in \mathbb{W}, \tag{7.89}$$

where we have defined $|\Phi_i(t)\rangle \equiv \langle w_{i,v}|\tilde{\Psi}(t)\rangle$ and $M_{i,j}(t) \equiv \mathsf{e}\,\langle w_{i,v}|\,\hat{r}(t)\,|w_{j,v}\rangle$. It is worth noting that $M_{i,j}(t)$ is a non-diagonal matrix since an electron initially on site $i$ can end up in site $j$ either through an intra-band transition, described by off-diagonal elements, or via an excitation-recombination mechanism, mainly described by the diagonal elements.

In the following, we investigate HHG processes under the regime of weak electron delocalization, that is, $|M_{i,j}(t)| \ll |M_{i,i}(t)|$ with $i \neq j$. For simplicity, we solely consider nearest neighbor contributions, restricting our focus to the non-diagonal elements of $M_{i,j}(t)$ for which $j = i \pm 1$. Treating these as a perturbation parameter, we expand our state $|\Phi_i(t)\rangle$ up to first order as

$$|\Phi_i(t)\rangle \approx |\Phi_i^{(0)}(t)\rangle + |\Phi_i^{(1)}(t)\rangle, \tag{7.90}$$

such that the zeroth-order term satisfies

$$i\hbar\frac{\mathrm{d}|\Phi_i^{(0)}(t)\rangle}{\mathrm{d}t} = M_{i,i}(t)\hat{E}(t)|\Phi_i^{(0)}(t)\rangle, \tag{7.91}$$

and whose solution can be written as

$$\begin{aligned}|\Phi_i^{(0)}(t)\rangle &= \hat{\mathcal{D}}\big(\boldsymbol{\chi}_{i,i}(t,t_0)\big)|\Phi_i^{(0)}(t_0)\rangle\\ &\equiv \prod_q e^{i\varphi_{i,i}^{(q)}(t,t_0)}\hat{D}_q\big(\chi_{i,i}^{(q)}(t,t_0)\big)|\Phi_i^{(0)}(t_0)\rangle,\end{aligned} \tag{7.92}$$

where $\hat{D}_q(\alpha) = \exp\left[\alpha\hat{a}_q^\dagger - \alpha^*\hat{a}_q\right]$ is the displacement operator acting on mode $q$, $\varphi_{i,i}^{(q)}(t)$ is a prefactor arising from applying the Baker–Campbell–Hausdorff relation, and $\chi_{i,i}^{(q)}(t,t_0)$ is a coherent state amplitude given by

$$\chi_{i,i}^{(q)}(t,t_0) = -\frac{1}{\hbar}\int_{t_0}^{t} \mathrm{d}\tau e^{i\omega_q\tau}\mathrm{g}(\omega_q)M_{i,i}(\tau). \tag{7.93}$$

It is important to note that, while we have explicitly depicted the dependence of $\chi_{i,i}^{(q)}(t,t_0)$ on the Wannier site, specifically the $i$th site, the dynamics across all sites are equivalent. Consequently, they result in identical HHG spectra, enabling us to disregard the $i$ index in $\chi_{i,i}^{(q)}(t,t_0)$. Nevertheless, for the sake of completeness, we maintain this notation in the subsequent discussion.

Regarding the first-order term we get the following differential equation

$$\begin{aligned} i\hbar\frac{\mathrm{d}|\Phi_i^{(1)}(t)\rangle}{\mathrm{d}t} &= M_{i,i}(t)\cdot\hat{E}(t)|\Phi_i^{(1)}(t)\rangle \\ &\quad + \sum_{j\in\{i\pm1\}} M_{i,j}(t)\hat{E}(t)|\Phi_j^{(0)}(t)\rangle, \end{aligned} \tag{7.94}$$

whose solution can be generally written as

$$\begin{aligned} |\Phi_i^{(1)}(t)\rangle &= \hat{\mathcal{D}}\big(\boldsymbol{\chi}_{i,i}(t,t_0)\big)|\Phi_i^{(1)}(t_0)\rangle \\ &\quad - \frac{i}{\hbar}\sum_{j\in\{i\pm1\}}\int_{t_0}^{t}\mathrm{d}t'\hat{\mathcal{D}}\big(\boldsymbol{\chi}_{i,i}(t,t')\big)M_{i,j}(t')\hat{E}(t') \\ &\quad \times\hat{\mathcal{D}}\big(\boldsymbol{\chi}_{j,j}(t',t_0)\big)|\Phi_j^{(0)}(t_0)\rangle. \end{aligned} \tag{7.95}$$

Introducing the initial conditions at this point, we observe two distinct contributions based on the final site, $j$, where the electron ends up:

- If the electron concludes its dynamics in the initial Wannier site, $j_0$, then the final state is given by

$$|\Phi_{j_0}(t)\rangle = \hat{\mathcal{D}}\big(\boldsymbol{\chi}_{i,i}(t,t_0)\big)\bigotimes_q|0_q\rangle; \tag{7.96}$$

- Conversely, if the electron concludes its dynamics in a different Wannier site from where it started, $j$, the final state becomes

$$\begin{aligned} |\Phi_j(t)\rangle &= -\frac{i}{\hbar}\int_{t_0}^{t}\mathrm{d}t'\hat{\mathcal{D}}\big(\boldsymbol{\chi}_{j,j}(t,t')\big)M_{j,j_0}(t')\cdot\hat{E}(t') \\ &\quad \times\hat{\mathcal{D}}\big(\boldsymbol{\chi}_{j_0,j_0}(t',t_0)\big)\bigotimes_q|0_q\rangle. \end{aligned} \tag{7.97}$$

It is worth noting that, by taking into account that $\chi_{i,i}^{(q)}(t) = \chi_{j,j}^{(q)}(t) = \chi_q(t)$, we can rewrite Eq. (7.97) as

$$\begin{aligned} |\Phi_j(t)\rangle = &-\frac{i}{\hbar}e^{i\varphi(t)}\hat{\mathcal{D}}(\boldsymbol{\chi}(t,t_0)) \\ &\times \int_{t_0}^{t} \mathrm{d}t' e^{-i\theta(t')}\hat{\mathcal{D}}^\dagger(\boldsymbol{\chi}(t',t_0)) \\ &\times \big(M_{j,j_0}(t')\hat{E}(t')\big)\hat{\mathcal{D}}(\boldsymbol{\chi}(t',t_0)) \bigotimes_{q=1} |0_q\rangle, \end{aligned} \tag{7.98}$$

where we have defined $\theta(t') \equiv \sum_q \mathrm{Im}\{\chi_q(t,t_0)\chi_q^*(t',t_0)\}$. Furthermore, from numerical analysis, we find that the integral above differ in the fifth significant decimal when $j = \pm 1$. Hence, in the following we approximate $|\Phi_{+1}(t)\rangle \approx |\Phi_{-1}(t)\rangle \equiv |\Phi_{\text{NN}}(t)\rangle$.

For a one-dimensional lattice, the joint state of the system can be written as

$$|\tilde{\Psi}'(t)\rangle = |w_{j_0,v}\rangle \otimes |\Phi_{j_0}(t)\rangle + \sqrt{2}\,|w_{\text{NN},v}\rangle \otimes |\Phi_{\text{NN}}(t)\rangle, \tag{7.99}$$

where we defined $|w_{\text{NN},v}\rangle = (1/\sqrt{2})(|w_{j_0+1,v}\rangle + |w_{j_0-1,v}\rangle)$. If we denote the coordinate number of the lattice by $f_\text{c}$ (i.e., $f_\text{c} = 2d$ in hypercubic lattices in $d$-dimensions, $f_\text{c} = 3$ for honeycomb lattice in 2D, $f_\text{c} = 6$ for triangular lattice in 2D), we can provide a generalization of our state to arbitrary dimensions

$$|\tilde{\Psi}'(t)\rangle = |w_{j_0,v}\rangle \otimes |\Phi_{j_0}(t)\rangle + \sqrt{f_\text{c}}\,|w_{\text{NN},v}\rangle \otimes |\Phi_{\text{NN}}(t)\rangle, \tag{7.100}$$

where now $|w_{\text{NN},v}\rangle = (1/\sqrt{f_\text{c}})(\sum_{j\in\text{NN}} |w_{j_0+j,v}\rangle)$.

#### 7.3.3.1 *An Extension to the Many-Electron Regime*

Although until now we have worked under the single-active electron approximation, as mentioned in the Bloch-Bloch analysis of Sec. 7.3.2, the initial state of the system corresponds to a completely filled Fermi sea, i.e., all Wannier sites are occupied. Given that electron-electron correlations are weak in typical semiconductor materials, we can consequently treat the interaction of all electrons with the laser field as independent. Thus, by introducing the same unitary transformations as for the single-electron analysis, and neglecting the contributions of the conduction band, the Schrödinger equation reads

$$i\hbar\frac{\mathrm{d}|\tilde{\Psi}'(t)\rangle}{\mathrm{d}t} \approx \mathsf{e}\sum_{\mathbf{m}} \hat{\tilde{r}}(t)\hat{E}(t)\langle w_{\mathbf{m},v}|\tilde{\Psi}'(t)\rangle\,|w_{\mathbf{m},v}\rangle, \tag{7.101}$$

where $|w_{\mathbf{m},v}\rangle \equiv |w_{m_0,v}\rangle \otimes |w_{m_1,v}\rangle \otimes \cdots \otimes |w_{m_{N_e},v}\rangle$; and $\hat{\tilde{R}} = \sum_{j=0}^{N_e-1} \hat{r}_{i,j}$, with $N_e$ being the number of electrons contributing in a phase-matched way to the HHG process, and $r_{i,j}$ the position operator along direction $i$ acting on the $j$th electron.

Similarly to before, we assume the regime of weak delocalization, ensuring the single-electron wavepacket does not significantly spread across different Wannier sites. Consequently, we separate the contribution of $\tilde{r}(t)$ into diagonal and off-diagonal elements, allowing us to treat the latter using perturbation theory. Thus, by projecting Eq. (7.101) with respect to $|w_{\mathbf{i},v}\rangle$, we obtain

$$i\hbar\frac{\mathrm{d}\,|\Phi_{\mathbf{i}}(t)\rangle}{\mathrm{d}t} = \mathsf{e}\sum_{\mathbf{m}} \langle w_{\mathbf{i},v}|\tilde{r}(t)|w_{\mathbf{m},v}\rangle\, \hat{E}(t)\,|\Phi_{\mathbf{m}}(t)\rangle, \tag{7.102}$$

where we have defined $|\Phi_{\mathbf{i}}(t)\rangle = \langle w_{\mathbf{i},v}|\tilde{\psi}(t)\rangle$, and where the matrix elements of $\tilde{r}(t)$ are given by

$$\langle w_{\mathbf{n},v}|\tilde{r}(t)|w_{\mathbf{m},v}\rangle = \sum_{j=0}^{N_e-1} \langle w_{n_j,v}|\hat{r}_{i,j}|w_{m_j,v}\rangle \prod_{l\in\mathcal{O}_j} \delta_{n_l,m_l}, \tag{7.103}$$

wherein the previous expression $\mathcal{O}_j := \{l : l \in \mathbb{N} - \{j\}\}$.

Up to the first order in perturbation theory, the equations we get are identical to the ones found in the single-electron regime. Thus, taking into account the initial conditions, we find two kinds of contributions depending on the final sites $j$ where the electrons end up in:

- If the electrons end up in the initial Wannier sites where they were initially located, then the final state reads

$$|\Phi_{\mathbf{i}}(t)\rangle = \hat{\mathcal{D}}\big(N_e\boldsymbol{\chi}(t,t_0)\big)\bigotimes_q |0_q\rangle; \tag{7.104}$$

- If one of the electrons ends up in a different Wannier site $j$ from which it initially started the dynamics, then the final state reads

$$\begin{aligned}|\Phi_{\mathbf{i}}(t)\rangle = &-\frac{i}{\hbar}\hat{\mathcal{D}}\big(N_e\boldsymbol{\chi}(t,t_0)\big)\\ &\times\int_{t_0}^{t}\mathrm{d}t' e^{-i\theta(t,t',t_0)}M_{i_n,i_{n,0}}(t')\\ &\times\hat{\mathcal{D}}^\dagger\big(N_e\boldsymbol{\chi}(t',t_0)\big)\hat{E}(t')\hat{\mathcal{D}}\big(N_e\boldsymbol{\chi}(t',t_0)\big)\\ &\bigotimes_q|0_q\rangle,\end{aligned} \tag{7.105}$$

where we have considered that one of the electrons has transitioned from its initial site $i_{n,0}$ to site $i_n$. It is worth noting that the first-order perturbation theory term describes events where only one of the electrons ends up in a different Wannier site from which it initially started the dynamics. In order to consider events where we find $n$ transitions, we then have to perform a perturbation theory expansion up to the $n$th order term. However, in the many-electron scenario, we restrict ourselves to solids for which the HHG process is highly localized, as it happens for instance with solid argon [80], or for specific crystal directions of certain solids like ZnO (see Fig. 7.2).

With all this, we can extend Eq. (7.100) to the many-electron scenario as

$$|\tilde{\Psi}'(t)\rangle = |w_{\boldsymbol{j_0},v}\rangle \otimes |\Phi_{\boldsymbol{j_0}}(t)\rangle + \sqrt{f_c N_{\mathrm{e}}}|w_{\overline{\mathrm{NN}},v}\rangle \otimes |\Phi_{\overline{\mathrm{NN}}}(t)\rangle, \tag{7.106}$$

where it is worth noting that, like in Eq. (7.75), we get an entangled state between light and matter. However, the main advantage in this case is that we have characterized the electronic states through a discrete basis set. This allows us to perform a simple light-matter entanglement characterization, assessing whether the approximations performed in the many-body analysis, which essentially neglect the entanglement between light and matter, hold valid for the considered material.

#### 7.3.3.2 *Conditioning on HHG and Its Consequences*

To compare the implications of the light-matter entanglement outlined in Eq. (7.106) against the findings in Section 7.3.2, we consider here the use of the conditioning on HHG operation presented in Eq. (7.80). By applying this operation to Eq. (7.106) and subsequently projecting it onto the coherent state $|\gamma\rangle \equiv \bigotimes_{q\geq 2} |\chi_q(t,t_0)\rangle$, representing the measured harmonics, we obtain

$$|\Psi_{\mathrm{cond}}(t)\rangle = \langle\gamma|\Psi(t)\rangle - \xi(\gamma,t)\,|\alpha_L\rangle, \tag{7.107}$$

with $\xi(\gamma,t) = \langle\alpha, 0_{\mathrm{HH}}|\Psi(t)\rangle\,\langle\gamma|0_{\mathrm{HH}}\rangle$.

We begin the comparison by first computing the Wigner function of the infrared field mode. However, unlike the Bloch-basis analysis, in this scenario we include in our state the possibility of the electrons ending up in a valence band state different from the initial one. Therefore, assuming a lack of knowledge about the final state with which the electron has recombined, we can express the quantum optical state of the system in terms of the following density matrix

$$\hat{\rho}_{\text{field}}(t) = \text{tr}_{\text{elec}}(|\Psi_{\text{cond}}(t)\rangle\langle\Psi_{\text{cond}}(t)|), \tag{7.108}$$

where we have essentially traced out the electronic degrees of freedom.

When exploring the interaction between ZnO and a linearly polarized electric field along the $\Gamma - A$ crystal direction, featuring a peak field strength $E_0 \approx 0.2\,\text{V/Å}$, $\lambda_L = 3.25\,\mu\text{m}$ and duration $\Delta t \approx 40\,\text{fs}$, we arrive at the Wigner functions shown in Fig. 7.8. Each panel corresponds to a different number of electrons $N_e$, specifically $N_e = 1\times10^7, 1\times10^8, 5\times10^8$ and $1\times10^9$ from panels (a) to (d), respectively. Notably, these results do not account for dephasing effects, yet exhibit a strong resemblance to those illustrated in Fig. 7.6 (c) from the Bloch-based analysis, where $N_z = 1.6\times10^7$.

Furthermore, as can be seen here, explicitly increasing the number of electrons contributing in a phase-matched way to the HHG process leads to a Wigner distribution that shifts from a displaced Fock state

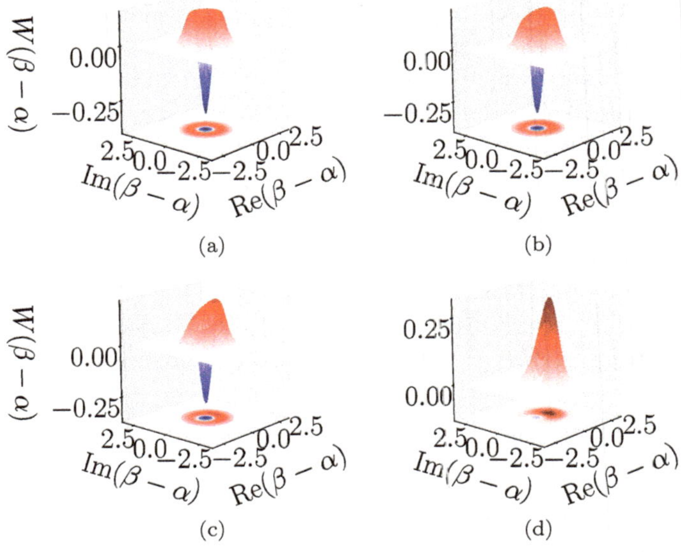


Fig. 7.8 Wigner function of the state in Eq. (7.108) when considering different numbers of electrons independently contributing to the HHG process. Specifically, from (a) to (d), the values $N_e = 1\times10^7, 1\times10^8, 5\times10^8$ and $1\times10^9$ have been used. Here, we consider the interaction of ZnO with a linearly polarized electric field along the $\Gamma - A$ crystal direction, with peak field strength $E_0 \approx 0.2\,\text{V/Å}$, $\lambda_L = 3.25\,\mu\text{m}$ and duration $\Delta t \approx 40\,\text{fs}$ (≈4 optical cycles). Adapted with permission from https://arxiv.org/abs/2211.00033.

(Fig. 7.8 (a)), to an unbalanced superposition between two close coherent states (Fig. 7.8 (b) and (c)) until it reaches a regime where one of the coherent states in the superposition dominates (Fig. 7.8 (d)). This behavior bears resemblance to what is found for atomic HHG [11–13, 15], which is a consequence of the natural localization of the HHG process in atomic and molecular targets. However, and as mentioned earlier, in this case the electron dynamics are different to the case of atoms, and they lead to non-trivial consequences on the generated coherent state superpositions.

As was already stated, a main advantage of introducing a Wannier-based analysis lies in the discreteness of the electronic basis set. This discretization significantly simplifies the characterization of entanglement compared to continuous-variable basis sets, as observed for the Bloch-based approach outlined in Section 7.3.2. Moreover, working within the regime of weak delocalization allows us to focus on nearest-neighbor interactions, further streamlining the entanglement analysis. This particular simplification stems from treating the electronic degrees of freedom in Eq. (7.106) as qubits. Consequently, the entanglement entropy in Eq. (7.43) serves as a suitable entanglement measure. Additionally, examining the fidelity of the state with respect to the $|w_{\overline{\text{NN}},v}\rangle$, offers insights into the likelihood of finding electrons in their nearest neighbors. For a generic density matrix $\hat{\sigma}$, this quantity can be computed as follows

$$\mathcal{F}_{\text{del}}(\hat{\sigma}) = \text{tr}_{\text{field}}\Big(\langle w_{\overline{\text{NN}},v}|\hat{\sigma}|w_{\overline{\text{NN}},v}\rangle\Big). \tag{7.109}$$

In Fig. 7.9, we illustrate these two quantities for different lattice geometries, and their dependency on the number of electrons $N_{\text{e}}$. More specifically, panel (a) demonstrates the fidelity, while panel (b) showcases the entanglement entropy. The dashed curves represent the state without conditioning, i.e., when considering Eq. (7.106). On the other hand, the solid curves depict the case with Eq. (7.107), where the conditioning on HHG operation has been applied. Throughout both cases, the observed values of these measures generally remain within the order of $10^{-6}$ at maximum. This observation suggests that, under the considered excitation conditions, the entanglement features are nearly negligible. This underscores the validity of the approximations utilized in Section 7.3.2. Nevertheless, there is a noticeable trend: the amount of entanglement tends to increase with the number of electrons involved in the interaction region. However, for larger values of $N_{\text{e}}$, introducing more than nearest neighbors might become necessary.

Despite the notably reduced values observed for the degree of entanglement and the fidelity in Fig. 7.9, several noteworthy features emerge.

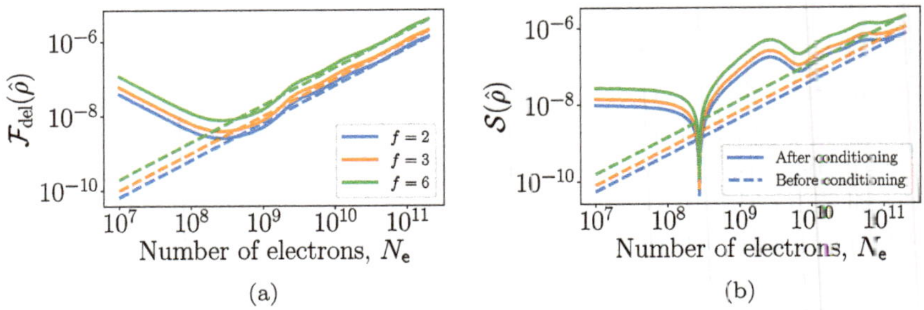


Fig. 7.9 In (a), the fidelity measure outlined in Eq. (7.109) is plotted against the number of electrons contributing independently to the HHG process. In (b), the entanglement entropy is depicted instead. Dashed lines represent scenarios without conditioning on HHG operation, while continuous curves signify cases where this conditioning has been applied. Each color corresponds to a different coordinate number. The laser-solid interaction conditions mirror those of Fig. 7.8. Adapted with permission from https://arxiv.org/abs/2211.00033.

Firstly, for $N_e \lesssim 2 \times 10^8$, both the fidelity and the entanglement after the conditioning (solid curves) exhibit higher values compared to those before the conditioning (dashed curves). This enhancement is primarily attributed to $|\boldsymbol{\chi}(t, t_0)|$, which remains a small quantity ($\sim 10^{-2}$), allowing the conditioning to accentuate the contributions from electron recombination in the nearest neighbors. Consequently, this increase in fidelity is accompanied by a growth in the value of $\mathcal{S}(\hat{\rho})$. Secondly, an intriguing observation occurs around $N_e \simeq 2.5 \times 10^8$: an abrupt drop in $\mathcal{S}(\hat{\rho})$ after the conditioning, which is absent for $\mathcal{F}_{\rm del}(\hat{\rho})$. This phenomenon stems from the cancellation of quantum optical contributions arising from the $|w_{\overline{\rm NN},v}\rangle$ term in Eq. (7.107), i.e., $\langle\gamma|\Phi_{\overline{\rm NN}}(t)\rangle - \xi(\gamma, t)\,|\alpha_L\rangle$. As $N_e$ increases, $\xi(\gamma, t)$ diminishes with the increment of $|\boldsymbol{\chi}(t, t_0)|$, resulting in a minimal overlap with the initial state. Consequently, in this regime where the conditioning on HHG operation exhibits this minimal effect, $\mathcal{F}_{\rm del}(\hat{\rho})$ and $\mathcal{S}(\hat{\rho})$ for the state before and after conditioning coincide. Furthermore, it is important to note that these two quantities demonstrate a marginal increment with the coordinate number $f_c$, despite the modifications being far from noticeable.

## 7.4 Outlook

In this chapter, we have provided an in-depth study of strongly driven laser-solid interactions from a quantum optical perspective. This analysis, focused on solids exhibiting minimal electron-electron correlations, produces outcomes akin to those observed in atomic systems [15] and molecular setups [21]. Rather than disheartening, these aligned findings serve as a catalyst for transitioning from prevailing atomic-based experimental platforms to solid-state counterparts. Notably, considering the diminutive sizes of typical semiconductor samples [34, 47] — integral components in existing technologies — the similarity in observed results following strong laser-solid interactions suggests the feasibility of conducting these experiments on a smaller scale. This marks the initial stride from fundamental physics inquiries toward practical technological advancements.

## Acknowledgments

ICFO group acknowledges support from: ERC AdG NOQIA; MICIN/AEI (PGC2018-0910.13039/501100011033, CEX2019-000910-S/10.13039/501100011033, Plan National FIDEUA PID2019-106901GB-I00, FPI; MICIIN with funding from European Union NextGenerationEU (PRTR-C17.I1): QUANTERA MAQS PCI2019-111828-2); MCIN/AEI/10.13039/501100011033 and by the "European Union NextGeneration EU/PRTR" QUANTERA DYNAMITE PCI2022-132919 within the QuantERA II Programme that has received funding from the European Union's Horizon 2020 research and innovation programme under Grant Agreement No 101017733Proyectos de I+D+I "Retos Colaboración" QUSPIN RTC2019-007196-7); Fundació Cellex; Fundació Mir-Puig; Generalitat de Catalunya (European Social Fund FEDER and CERCA program, AGAUR Grant No. 2021 SGR 01452, QuantumCAT U16-011424, co-funded by ERDF Operational Program of Catalonia 2014-2020); Barcelona Supercomputing Center MareNostrum (FI-2023-1-0013); EU (PASQuanS2.1, 101113690); EU Horizon 2020 FET-OPEN OPTOlogic (Grant No 899794); EU Horizon Europe Program (Grant Agreement 101080086 — NeQST), National Science Centre, Poland (Symfonia Grant No. 2016/20/W/ST4/00314); ICFO Internal "QuantumGaudi" project; European Union's Horizon 2020 research and innovation program under the Marie-Skłodowska-Curie grant agreement No 101029393 (STREDCH) and No 847648 ("La Caixa" Junior Leaders fellowships ID100010434: LCF/BQ/PI19/11690013, LCF/BQ/PI20/11760031, LCF/BQ/PR20/11770012, LCF/BQ/PR21/11840013). Views and opinions expressed are, however, those of the author(s) only and do not necessarily reflect those of the European Union, European Commission, European Climate, Infrastructure and Environment Executive Agency (CINEA), nor any other granting authority. Neither the European Union nor any granting authority can be held responsible for them.

P. Tzallas group at FORTH acknowledges support from: LASER-LABEUROPE V (H2020-EU.1.4.1.2 grant no.871124), the H2020 project IMPULSE (GA 871161) and ELI–ALPS. ELI–ALPS is supported by the European Union and co-financed by the European Regional Development Fund (GINOP Grant No. 2.3.6-15-2015-00001).

J. R-D. acknowledges funding from the Government of Spain (Severo Ochoa CEX2019-000910-S and TRANQI), Fundació Cellex, Fundació Mir-Puig, Generalitat de Catalunya (CERCA program) and the ERC AdG CERQUTE.

P. S. acknowledges funding from the European Union's Horizon 2020 research and innovation programme under the Marie Skłodowska-Curie grant agreement No 847517.

A. S. M. acknowledges funding support from the European Union's Horizon 2020 research and innovation programme under the Marie Skłodowska-Curie grant agreement, SSFI No. 887153.

E. P. acknowledges the Royal Society for University Research Fellowship funding under URF\R1\211390.

M. F. C. acknowledges support by the National Key Research and Development Program of China (Grant No. 2023YFA1407100), Guangdong Province Science and Technology Major Project (Future functional materials under extreme conditions - 2021B0301030005) and the Guangdong Natural Science Foundation (General Program project No. 2023A1515010871).

## References

[1] A. Gilchrist, K. Nemoto, W. J. Munro, T. C. Ralph, S. Glancy, S. L. Braunstein and G. J. Milburn, Schrödinger cats and their power for quantum information processing, *Journal of Optics B: Quantum and Semiclassical Optics* **6**, 8, p. S828 (2004).

[2] N. Gisin and R. Thew, Quantum communication, *Nature Photonics* **1**, 3, pp. 165–171 (2007).

[3] J. L. O'Brien, A. Furusawa and J. Vučković, Photonic quantum technologies, *Nature Photonics* **3**, 12, pp. 687–695 (2009).

[4] D. V. Strekalov and G. Leuchs, Nonlinear Interactions and Non-classical Light, in R. W. Boyd, S. G. Lukishova and V. N. Zadkov (Eds.), *Quantum Photonics: Pioneering Advances and Emerging Applications*, Springer, pp. 51–101 (2019).

[5] P. G. Kwiat, K. Mattle, H. Weinfurter, A. Zeilinger, A. V. Sergienko and Y. Shih, New high-intensity source of polarization-entangled photon pairs, *Physical Review Letters* **75**, 24, pp. 4337–4341 (1995).

[6] A. McPherson, G. Gibson, H. Jara, U. Johann, T. S. Luk, I. A. McIntyre, K. Boyer and C. K. Rhodes, Studies of multiphoton production of vacuum-ultraviolet radiation in the rare gases, *JOSA B* **4**, 4, pp. 595–601 (1987).

[7] M. Ferray, A. L'Huillier, X. F. Li, L. A. Lompre, G. Mainfray and C. Manus, Multiple-harmonic conversion of 1064 nm radiation in rare gases, *Journal of Physics B: Atomic, Molecular and Optical Physics* **21**, 3, p. L31 (1988).

[8] T. Popmintchev, M.-C. Chen, D. Popmintchev, P. Arpin, S. Brown, S. Ališauskas, G. Andriukaitis, T. Balčiunas, O. D. Mücke, A. Pugzlys, A. Baltuška, B. Shim, S. E. Schrauth, A. Gaeta, C. Hernández-García, L. Plaja, A. Becker, A. Jaron-Becker, M. M. Murnane and H. C. Kapteyn, Bright coherent ultrahigh harmonics in the keV X-ray regime from mid-infrared femtosecond lasers, *Science* **336**, 6086, pp. 1287–1291 (2012).

[9] J. Rivera-Dean, *Quantum-optical analysis of high-order harmonic generation*, Master thesis, Universitat Politècnica de Catalunya (2019), https://upcommons.upc.edu/handle/2117/168580.

[10] A. Gorlach, O. Neufeld, N. Rivera, O. Cohen and I. Kaminer, The quantum-optical nature of high harmonic generation, *Nature Communications* **11**, 1, p. 4598 (2020).

[11] M. Lewenstein, M. F. Ciappina, E. Pisanty, J. Rivera-Dean, P. Stammer, T. Lamprou and P. Tzallas, Generation of optical Schrödinger cat states in intense laser–matter interactions, *Nature Physics* **17**, 10, pp. 1104–1108 (2021).

[12] J. Rivera-Dean, T. Lamprou, E. Pisanty, P. Stammer, A. F. Ordóñez, A. S. Maxwell, M. F. Ciappina, M. Lewenstein and P. Tzallas, Strong laser fields and their power to generate controllable high-photon-number coherent-state superpositions, *Physical Review A* **105**, 3, p. 033714 (2022).

[13] P. Stammer, J. Rivera-Dean, A. Maxwell, T. Lamprou, A. Ordóñez, M. F. Ciappina, P. Tzallas and M. Lewenstein, Quantum electrodynamics of intense laser-matter interactions: A tool for quantum state engineering, *PRX Quantum* **4**, 1, p. 010201 (2023).

[14] M. Lewenstein, N. Baldelli, U. Bhattacharya, J. Biegert, M. F. Ciappina, U. Elu, T. Grass, P. T. Grochowski, A. Johnson, T. Lamprou, A. S. Maxwell, A. Ordóñez, E. Pisanty, J. Rivera-Dean, P. Stammer, I. Tyulnev and P. Tzallas, Attosecond physics and quantum information science, arXiv:2208.14769 [physics, physics:quant-ph] (2022).

[15] U. Bhattacharya, T. Lamprou, A. S. Maxwell, A. Ordóñez, E. Pisanty, J. Rivera-Dean, P. Stammer, M. F. Ciappina, M. Lewenstein and P. Tzallas, Strong–laser–field physics, non–classical light states and quantum information science, *Reports on Progress in Physics* **86**, 9, p. 094401 (2023).

[16] A. Pizzi, A. Gorlach, N. Rivera, A. Nunnenkamp and I. Kaminer, Light emission from strongly driven many-body systems, *Nature Physics* **19**, pp. 551–561 (2023).

[17] T. Lamprou, J. Rivera-Dean, P. Stammer, M. Lewenstein and P. Tzallas, Nonlinear optics using intense optical Schrödinger "cat" states, arXiv:2306.14480 [physics, physics:quant-ph] (2023).

[18] P. Stammer, J. Rivera-Dean, T. Lamprou, E. Pisanty, M. F. Ciappina, P. Tzallas and M. Lewenstein, High photon number entangled states and coherent state superposition from the extreme ultraviolet to the far infrared, *Physical Review Letters* **128**, 12, p. 123603 (2022).
[19] P. Stammer, Theory of entanglement and measurement in high-order harmonic generation, *Physical Review A* **106**, 5, p. L050402 (2022).
[20] J. Rivera-Dean, P. Stammer, A. S. Maxwell, T. Lamprou, P. Tzallas, M. Lewenstein and M. F. Ciappina, Light-matter entanglement after above-threshold ionization processes in atoms, *Physical Review A* **106**, 6, p. 063705 (2022).
[21] J. Rivera-Dean, P. Stammer, A. S. Maxwell, Th. Lamprou, E. Pisanty, P. Tzallas, M. Lewenstein, and M. F. Ciappina, Quantum-optical analysis of high-order harmonic generation in $H_2^+$ molecules, *Physical Review A* **109**, 3, p. 033706 (2024).
[22] P. Stammer and M. Lewenstein, Quantum optics as applied quantum electrodynamics is back in town, *Acta Physica Polonica A* **143**, 6, pp. S42–S42 (2023).
[23] P. Stammer, J. Rivera-Dean, A. S. Maxwell, T. Lamprou, J. Argüello-Luengo, P. Tzallas, M. F. Ciappina and M. Lewenstein, Entanglement and squeezing of the optical field modes in high harmonic generation, *Physical Review Letters* **132**, 14, p. 143603 (2024).
[24] M. Even Tzur, M. Birk, A. Gorlach, M. Krüger, I. Kaminer and O. Cohen, Photon-statistics force in ultrafast electron dynamics, *Nature Photonics* **17**, 6, pp. 501–509 (2023).
[25] Y. Fang, F.-X. Sun, Q. He and Y. Liu, Strong-field ionization of hydrogen atoms with quantum light, *Physical Review Letters* **130**, 25, p. 253201 (2023).
[26] M. E. Tzur and O. Cohen, Motion of charged particles in bright squeezed vacuum, *Light: Science & Applications* **13**, p. 41 (2024).
[27] A. Gorlach, M. E. Tzur, M. Birk, M. Krüger, N. Rivera, O. Cohen and I. Kaminer, High-harmonic generation driven by quantum light, *Nature Physics* **19**, pp. 1689–1696 (2023).
[28] P. Stammer, On the limitations of the semi-classical picture in high harmonic generation, arXiv:2308.15087 [physics, physics:quant-ph] (2023).
[29] P. Stammer, On the role of the optical phase and coherence in high harmonic generation, arXiv:2309.05010 [quant-ph] (2023).
[30] I. Gonoskov, R. Sondenheimer, C. Hünecke, D. Kartashov, U. Peschel and S. Gräfe, Nonclassical light generation and control from laser-driven semiconductor intraband excitations, *Physical Review B* **109**, 12, p. 125110 (2024).
[31] J. Rivera-Dean, P. Stammer, A. S. Maxwell, T. Lamprou, A. F. Ordóñez, E. Pisanty, P. Tzallas, M. Lewenstein and M. F. Ciappina, Entanglement and non-classical states of light in a strong-laser driven solid-state system, arXiv:2211.00033 [physics, physics:quant-ph] (2023).
[32] J. Rivera-Dean, P. Stammer, A. S. Maxwell, Th. Lamprou, A. F. Ordóñez, E. Pisanty, P. Tzallas, M. Lewenstein, and M. F. Ciappina, Nonclassical

states of light after high-harmonic generation in semiconductors: A Bloch-based perspective, *Physical Review B* **109**, 3, p. 035203 (2024).

[33] A. H. Chin, O. G. Calderón and J. Kono, Extreme midinfrared nonlinear optics in semiconductors, *Physical Review Letters* **86**, 15, pp. 3292–3295 (2001).

[34] S. Ghimire, A. D. DiChiara, E. Sistrunk, P. Agostini, L. F. DiMauro and D. A. Reis, Observation of high-order harmonic generation in a bulk crystal, *Nature Physics* **7**, 2, pp. 138–141 (2011).

[35] H. Haug and S. W. Koch, Periodic Lattice of Atoms, in *Quantum Theory of the Optical and Electronic Properties of Semiconductors* 4th Edition, World Scientific, pp. 41–44 (2004).

[36] G. Vampa, C. McDonald, G. Orlando, D. Klug, P. Corkum and T. Brabec, Theoretical analysis of high-harmonic generation in solids, *Physical Review Letters* **113**, 7, p. 073901 (2014).

[37] G. Vampa, C. R. McDonald, G. Orlando, P. B. Corkum and T. Brabec, Semiclassical analysis of high harmonic generation in bulk crystals, *Physical Review B* **91**, 6, p. 064302 (2015).

[38] E. N. Osika, A. Chacón, L. Ortmann, N. Suárez, J. A. Pérez-Hernández, B. Szafran, M. F. Ciappina, F. Sols, A. S. Landsman and M. Lewenstein, Wannier-Bloch approach to localization in high-harmonics generation in solids, *Physical Review X* **7**, 2, p. 021017 (2017).

[39] L. Li, P. Lan, X. Zhu and P. Lu, High harmonic generation in solids: particle and wave perspectives, *Reports on Progress in Physics* **86**, 11, p. 116401 (2023).

[40] L. Yue and M. B. Gaarde, Introduction to theory of high-harmonic generation in solids: tutorial, *JOSA B* **39**, 2, pp. 535–555 (2022).

[41] S. Jiang, J. Chen, H. Wei, C. Yu, R. Lu and C. Lin, Role of the transition dipole amplitude and phase on the generation of odd and even high-order harmonics in crystals, *Physical Review Letters* **120**, 25, p. 253201 (2018).

[42] J. L. Krause, K. J. Schafer and K. C. Kulander, High-order harmonic generation from atoms and ions in the high intensity regime, *Physical Review Letters* **68**, 24, pp. 3535–3538 (1992).

[43] P. B. Corkum, Plasma perspective on strong field multiphoton ionization, *Physical Review Letters* **71**, 13, pp. 1994–1997 (1993).

[44] K. C. Kulander, K. J. Schafer and J. L. Krause, Dynamics of Short-Pulse Excitation, Ionization and Harmonic Conversion, in B. Piraux, A. L'Huillier and K. Rzążewski (Eds.), Super-Intense Laser — Atom Physics. Springer, pp. 95–110 (1993).

[45] M. Lewenstein, P. Balcou, M. Y. Ivanov, A. L'Huillier and P. B. Corkum, Theory of high-harmonic generation by low-frequency laser fields, *Physical Review A* **49**, 3, pp. 2117–2132 (1994).

[46] F. Bloch, Über die Quantenmechanik der Elektronen in Kristallgittern, *Zeitschrift für Physik* **52**, 7, pp. 555–600 (1929).

[47] E. Goulielmakis and T. Brabec, High harmonic generation in condensed matter, *Nature Photonics* **16**, 6, pp. 411–421 (2022).

[48] P. C. Becker, H. L. Fragnito, C. H. B. Cruz, R. L. Fork, J. E. Cunningham, J. E. Henry and C. V. Shank, Femtosecond photon echoes from band-to-band transitions in GaAs, *Physical Review Letters* **61**, 14, pp. 1647–1649 (1988).
[49] M. T. Portella, J. Bigot, R. W. Schoenlein, J. E. Cunningham and C. V. Shank, k-space carrier dynamics in GaAs, *Applied Physics Letters* **60**, 17, pp. 2123–2125 (1992).
[50] I. Kilen, M. Kolesik, J. Hader, J. V. Moloney, U. Huttner, M. K. Hagen and S. W. Koch, Propagation induced dephasing in semiconductor high-harmonic generation, *Physical Review Letters* **125**, 8, p. 083901 (2020).
[51] M. Kolesik and J. V. Moloney, Numerical discreteness and dephasing in high-harmonic calculations in solids, *Physical Review B* **108**, 11, p. 115433 (2023).
[52] K. Amini, J. Biegert, F. Calegari, A. Chacón, M. F. Ciappina, A. Dauphin, D. K. Efimov, C. F. d. M. Faria, K. Giergiel, P. Gniewek, A. S. Landsman, M. Lesiuk, M. Mandrysz, A. S. Maxwell, R. Moszyński, L. Ortmann, J. A. Pérez-Hernández, A. Picón, E. Pisanty, J. Prauzner-Bechcicki, K. Sacha, N. Suárez, A. Zaïr, J. Zakrzewski and M. Lewenstein, Symphony on strong field approximation, *Reports on Progress in Physics* **82**, 11, p. 116001 (2019).
[53] W. Kohn, Analytic Properties of Bloch waves and Wannier functions, *Physical Review* **115**, 4, pp. 809–821 (1959).
[54] G. G. Brown, A. Jiménez-Galán, R. E. F. Silva and M. Ivanov, A real-space perspective on dephasing in solid-state high harmonic generation, arXiv:2210.16889 [cond-mat, physics:physics] (2022).
[55] M. O. Scully and M. S. Zubairy, Quantum Theory of Radiation, in *Quantum Optics.* Cambridge University Press, pp. 1–45 (2001).
[56] C. Gerry and P. Knight, Field Quantization, in *Introductory Quantum Optics.* Cambridge University Press, pp. 10–42 (2005).
[57] R. Loudon, Quantization of the Radiation Field, in *The Quantum Theory of Light.* Oxford University Press, pp. 125–179 (1973).
[58] W. Vogel and D.-G. Welsch, Elements of Quantum Electrodynamics, in *Quantum Optics.* John Wiley & Sons, pp. 15–72 (2006).
[59] R. J. Glauber, Photon correlations, *Physical Review Letters* **10**, 3, pp. 84–86 (1963).
[60] R. J. Glauber, Coherent and incoherent states of the radiation field, *Physical Review* **131**, 6, pp. 2766–2788 (1963).
[61] R. J. Glauber, The quantum theory of optical coherence, *Physical Review* **130**, 6, pp. 2529–2539 (1963).
[62] E. Wigner, On the quantum correction for thermodynamic equilibrium, *Physical Review* **40**, 5, pp. 749–759 (1932).
[63] W. P. Schleich, Wigner Function, in *Quantum Optics in Phase Space.* John Wiley & Sons, pp. 67–98 (2001).
[64] R. L. Hudson, When is the Wigner quasi-probability density non-negative? *Reports on Mathematical Physics* **6**, 2, pp. 249–252 (1974).
[65] D. T. Smithey, M. Beck, M. G. Raymer and A. Faridani, Measurement of the Wigner distribution and the density matrix of a light mode using optical

homodyne tomography: Application to squeezed states and the vacuum, *Physical Review Letters* **70**, 9, pp. 1244–1247 (1993).
[66] A. Royer, Wigner function as the expectation value of a parity operator, *Physical Review A* **15**, 2, pp. 449–450 (1977).
[67] M. A. Nielsen and I. L. Chuang, Introduction to Quantum Mechanics, in *Quantum Computation and Quantum Information: 10th Anniversary Edition*. Cambridge University Press, pp. 60–119 (2010).
[68] L. Gurvits, Classical complexity and quantum entanglement, *Journal of Computer and System Sciences* **69**, 3, pp. 448–484 (2004).
[69] M. A. Nielsen and I. L. Chuang, Entropy and Information, in *Quantum Computation and Quantum Information: 10th Anniversary Edition*. Cambridge University Press, pp. 500–527 (2010).
[70] M. B. Plenio and S. Virmani, An introduction to entanglement measures, *Quantum Information & Computation* **7**, 1, pp. 1–51 (2007).
[71] C. H. Bennett and G. Brassard, Quantum cryptography: Public key distribution and coin tossing, *Theoretical Computer Science* **560**, pp. 7–11 (2014).
[72] A. K. Ekert, Quantum cryptography based on Bell's theorem, *Physical Review Letters* **67**, 6, pp. 661–663 (1991).
[73] L. K. Grover, A fast quantum mechanical algorithm for database search, in *Proceedings of the Twenty-Eighth Annual ACM Symposium on Theory of Computing*, STOC '96. Association for Computing Machinery, pp. 212–219 (1996).
[74] P. W. Shor, Polynomial-time algorithms for prime factorization and discrete logarithms on a quantum computer, *SIAM Review* **41**, 2, pp. 303–332 (1999).
[75] M. D. Reid, P. D. Drummond, W. P. Bowen, E. G. Cavalcanti, P. K. Lam, H. A. Bachor, U. L. Andersen and G. Leuchs, Colloquium: The Einstein-Podolsky-Rosen paradox: From concepts to applications, *Reviews of Modern Physics* **81**, 4, pp. 1727–1751 (2009).
[76] G. Wang and T.-Y. Du, Quantum decoherence in high-order harmonic generation from solids, *Physical Review A* **103**, 6, p. 063109 (2021).
[77] B. Sundaram and P. W. Milonni, High-order harmonic generation: Simplified model and relevance of single-atom theories to experiment, *Physical Review A* **41**, 11, pp. 6571–6573 (1990).
[78] N. Tsatrafyllis, I. K. Kominis, I. A. Gonoskov and P. Tzallas, High-order harmonics measured by the photon statistics of the infrared driving-field exiting the atomic medium, *Nature Communications* **8**, 1, p. 15170 (2017).
[79] N. Tsatrafyllis, S. Kühn, M. Dumergue, P. Foldi, S. Kahaly, E. Cormier, I. Gonoskov, B. Kiss, K. Varju, S. Varro and P. Tzallas, Quantum optical signatures in a strong laser pulse after interaction with semiconductors, *Physical Review Letters* **122**, 19, p. 193602 (2019).
[80] G. Ndabashimiye, S. Ghimire, M. Wu, D. A. Browne, K. J. Schafer, M. B. Gaarde and D. A. Reis, Solid-state harmonics beyond the atomic limit, *Nature* **534**, 7608, pp. 520–523 (2016).

https://doi.org/10.1142/9789811279560_0008

# Chapter 8

# Theoretical Explorations of High-Order Harmonic Generation in Solids: From 1D to 3D

Xiao-Shuang Kong[1,*], Xiao-Yuan Wu[1,*], Jian-Zhao Jin[1], Wan-Dong Yu[1] and Liang-You Peng[1,2,3]

[1]*State Key Laboratory for Mesoscopic Physics and Frontiers Science Center for Nano-optoelectronics, School of Physics, Peking University, Beijing 100871, China*
[2]*Collaborative Innovation Center of Extreme Optics, Shanxi University, Taiyuan, Shanxi 030006, China*
[3]*Peking University Yangtze Delta Institute of Optoelectronics, Nantong, Jiangsu 226010, China*
*These authors equally contributed to this work.*

## 8.1 Introduction

Ultrashort and ultraintense laser pulses have emerged as essential tools for manipulating and detecting quantum states in materials. Fundamental to the development of attosecond science, research into high-order harmonic generation (HHG) is constantly evolving. Today, HHG can be induced by focusing intense lasers on gases [1–4], solids [5–9], and even liquids [10, 11]. The detection of HHG from bulk crystals in 2010 marked a significant milestone [5], heralding a new era in attosecond physics and strong field physics. Since then, HHG in solids has been extensively employed in cutting-edge experiments, such as lattice structure reconstruction [12] and ultrafast electron dynamics detection [13]. The HHG-related technique holds a promising potential for creating novel, compact extreme ultraviolet light sources.

HHG mechanisms in solids are notably more complex than the gas-phase counterparts. Typically, the HHG mechanism in solids is attributed to intricate intraband and interband processes [14]. Echoing the conventional three-step model in atomic gases, the mechanism of HHG in solids can also be characterized in three stages: initial electronic excitation from the valence band (VB) to the conduction band (CB) under the laser field, subsequent intraband acceleration of excited electrons in CB and residual holes in VB, and finally the electron-hole recombination instigating the time-dependent interband current. Both intraband and interband contributions are essential to HHG. Compared to the gas-phase HHG, studies have demonstrated that the HHG and related ultrafast electronic dynamics in solids contain unique and fundamentally different aspects. For instance, the cutoff law of HHG in solids differs from that in atoms or gases. In general, for solid targets, the cutoff energy of intraband harmonics linearly correlates with the laser wavelength and field strength, while the cutoff energy of interband harmonics depends on the maximum energy difference between the energy bands.

HHG in solids has been experimentally measured in numerous two-dimensional (2D) [15, 16] and three-dimensional (3D) [5, 7, 17] materials. Concurrently, various theoretical methods undergo a rapid development, including widely utilized methods like solving the time-dependent Schrödinger equation (TDSE), the semiconductor Bloch equations (SBE), and the time-dependent density functional theory (TDDFT). These methods represent a subset of the numerous theoretical approaches available for exploring HHG in solids. In these simulations, the calculation of HHG is based on the principle that materials are periodically described, in compliance with Bloch's theorem in solids. There are marked differences between these methods, stemming from different levels of approximations. For instance, while TDSE focuses on solving using a model potential, it faces challenges when aiming for an accurate description of HHG in realistic materials, primarily due to the increase of many-electron correlations. This gap provides great opportunities for SBE and TDDFT methods, which facilitate simulations of HHG in realistic materials. Encouragingly, both the two methods can be coupled with Maxwell equations to investigate the propagation effect in dense bulk materials [18, 19].

As the complexity increases, including factors like spatial dimensionality or the size of the primitive cell, the exploration of HHG in solids reveals more surprises and introduces additional challenges. Reliable numerical simulations have been crucial in understanding the mechanisms of HHG in solids. In this chapter, we will delve into some of the most influential models of HHG in solids and examine typical simulations in multi-dimensional systems. In Section 8.2, we begin by introducing the primary objectives of commonly used theoretical models and numerical methods for the HHG in solids, including TDSE, SBE, TDDFT, and multiscale numerical simulations. In Section 8.3, we review some typical simulations of HHG in solids, covering one-dimensional (1D) to 3D materials. We also discuss novel HHG mechanisms, such as the contributions of Floquet-Bloch states in 1D systems, laser-manipulated HHG in 2D materials, and propagation effects in 3D bulk materials. Finally, in Section 8.4, we provide a brief summary and outlook.

## 8.2 Theoretical Models for HHG in Solids

As HHG experiments in solids rapidly advance, there is a corresponding swift development in the associated theoretical research. The primary numerical methods commonly employed in the study of HHG include TDSE, SBE, TDDFT, among others. In this section, we will briefly introduce these methods. In studying HHG, a central aim of these methods is to effectively simulate electronic dynamics by solving specific equations. The HHG spectra can be extracted by calculating the modulus square of the Fourier transform of the electric current.

### 8.2.1 *Time-Dependent Schrödinger Equation*

The quasi-electron description of the solid uses a single-particle wavefunction $\psi(\mathbf{r}, t)$, and the interacting system of the solid with a laser pulse is governed by the TDSE,

$$i\hbar\frac{\partial}{\partial t}|\psi(t)\rangle = \left(\frac{(\mathbf{p}+e\mathbf{A})^2}{2m} + V(\mathbf{r})\right)|\psi(t)\rangle, \tag{8.1}$$

where the arbitrary single-particle wavefunction $\psi(\mathbf{r}, t)$ can be expanded by the augmented plane-wave basis, the Bloch state basis, the Houston state basis, etc. The $V(\mathbf{r})$ is the periodic model potential. One can first solve the Schrödinger equation to obtain the initial wavefunction, and then propagate the wavefunction via the TDSE. The electric current at time $t$ is given by,

$$\mathbf{J}(t) = -\frac{e}{m}[\langle\psi(t)|\mathbf{p}|\psi(t)\rangle]. \tag{8.2}$$

If the wavefunction is expanded with the Bloch basis, one can get the total electric current. When it is expanded with the Houston basis, the electronic dynamics can be naturally separated into intraband and interband contributions. Formally, the Houston states $\widetilde{\phi}_{n,\mathbf{k}}$ are related to the Bloch states $\phi_{n,\mathbf{k}}$ with a lattice momentum $|\widetilde{\phi}_{n,\mathbf{k}_0}(t)\rangle = e^{-ie\mathbf{A}\hat{\mathbf{x}}/\hbar}|\phi_{n,\mathbf{k}_0}\rangle$, where $\hat{\mathbf{x}}$ is the position operator. The time-dependent wavefunction with an initial lattice momentum $\hbar\mathbf{k}_0$ is expressed in Houston states:

$$|\psi(t)\rangle = \sum_n a_{n,\mathbf{k}_0}(t)\left|\widetilde{\phi}_{n,\mathbf{k}_0}(t)\right\rangle. \tag{8.3}$$

Finally, the total electric current can be calculated as:

$$\mathbf{J}_{\text{tot}} = -\frac{e}{m}\text{Re}\left[\sum_{nn'} a^*_{n,\mathbf{k}_0}a_{n',\mathbf{k}_0}\left\langle\phi_{n,\mathbf{k}(t)}\left|\hat{\mathbf{p}}\right|\phi_{n',\mathbf{k}(t)}\right\rangle\right], \tag{8.4}$$

which is composed of the intraband and interband contributions:

$$\mathbf{J}_{\text{intra}} = -\frac{e}{m}\sum_n \left|a_{n,\mathbf{k}_0}\right|^2\left\langle\phi_{n,\mathbf{k}(t)}\left|\hat{\mathbf{p}}\right|\phi_{n,\mathbf{k}(t)}\right\rangle, \tag{8.5}$$

$$\mathbf{J}_{\text{inter}} = -\frac{e}{m}\text{Re}\left[\sum_{nn'(n\neq n')} a^*_{n,\mathbf{k}_0}a_{n',\mathbf{k}_0}\left\langle\phi_{n,\mathbf{k}(t)}\left|\hat{\mathbf{p}}\right|\phi_{n',\mathbf{k}(t)}\right\rangle\right]. \tag{8.6}$$

The TDSE simulations can be performed in either momentum space or real space. Although the results from both methods are fundamentally consistent, solving the TDSE in momentum space tends to be more convenient, especially for studying HHG in solids.

### 8.2.2 *Semiconductor Bloch Equations*

In this subsection, we introduce the methods for studying HHG in solids with consideration of the band structures of realistic systems. Based on the realistic electronic structures of semiconductors [14, 20], the time evolution of Bloch electrons in semiconductors interacting with a strong laser can be described by the SBE [21], which can be derived from the single-electron TDSE. The SBE is extensively used to investigate the solid-state HHG [14]. It can separate the intraband and interband mechanisms, and discuss the dephasing effect.

The TDSE in the velocity gauge is expressed as (in atomic units) [22]

$$\begin{aligned}[H_0 + H_{\text{int}}(t)]\psi_{m,\mathbf{k}}(\mathbf{r},t) &= \left\{\frac{[\mathbf{p}+\mathbf{A}(t)]^2}{2} + V_{lat}(\mathbf{r})\right\}\psi_{m,\mathbf{k}}(\mathbf{r},t)\\ &= i\frac{\partial\psi_{m,\mathbf{k}}(\mathbf{r},t)}{\partial t},\end{aligned} \tag{8.7}$$

where the dipole approximation is applied for the laser-matter interaction. $H_0 = \frac{1}{2}\mathbf{p}^2 + V_{lat}(\mathbf{r})$ and $H_{\text{int}} = \mathbf{p}\cdot\mathbf{A}(t)$ denote the field-free Hamiltonian and the interaction Hamiltonian between the laser and the crystal, respectively. $\mathbf{A}(t)$ is the laser vector potential and $V_{lat}(\mathbf{r})$ is the periodic potential of the crystal. By expanding the Bloch wavefunction $\psi_{m,\mathbf{k}}(\mathbf{r},t)$ under the Houston state basis and substituting it into Eq. (8.7), one can get

$$\frac{\partial a_{m,\mathbf{k}}(t)}{\partial t} = -i\varepsilon_m^{\mathbf{k}+\mathbf{A}(t)} a_{m,\mathbf{k}}(t) - i\mathbf{E}(t)\cdot\sum_l \mathbf{d}_{ml}^{\mathbf{k}+\mathbf{A}(t)} a_{l,\mathbf{k}}(t), \tag{8.8}$$

where $a_{m,\mathbf{k}}(t)$ is the expansion coefficient of the wavefunction, $\varepsilon_m^{\mathbf{k}+\mathbf{A}(t)}$ is the band dispersion, $\mathbf{d}_{ml}^{\mathbf{k}}(t)$ is the dipole-matrix element between the $m$-th and the $l$-th energy band at $\mathbf{k}$, and the laser electric field $\mathbf{E}(t) = -\partial\mathbf{A}/\partial t$. For better describing the time evolution of polarization and carrier occupation of solids radiated by the intense field, one can introduce the density matrix elements $\rho_{mn}^{\mathbf{k}}(t) = a_{m,\mathbf{k}}^*(t)a_{n,\mathbf{k}}(t)$. Equation (8.8) will be further simplified into the so-called SBEs:

$$\begin{aligned}\frac{\partial \rho_{mn}^{\mathbf{k}}(t)}{\partial t} &= i[\varepsilon_m^{\mathbf{k}+\mathbf{A}(t)} - \varepsilon_n^{\mathbf{k}+\mathbf{A}(t)}]\rho_{mn}^{\mathbf{k}}(t) \\ &\quad + i\mathbf{E}(t)\cdot\sum_l [\mathbf{d}_{lm}^{\mathbf{k}+\mathbf{A}(t)}\rho_{ln}^{\mathbf{k}}(t) - \mathbf{d}_{nl}^{\mathbf{k}+\mathbf{A}(t)}\rho_{ml}^{\mathbf{k}}(t)] \\ &\quad - \frac{[\rho_{mn}^{\mathbf{k}}(t) - f_m^0(\mathbf{k})]\delta_{mn}}{T_1} - \frac{(1-\delta_{mn})\rho_{mn}^{\mathbf{k}}(t)}{T_2}.\end{aligned} \tag{8.9}$$

Here the last two relaxation terms are added phenomenologically, where $T_1$ is the relaxation time for the electron and hole distributions, $T_2$ is the dephasing time for the interband polarization, and $f_m^0(\mathbf{k})$ denotes the initial electron distribution of the $m$-th band [23]. The relationship between the dephasing time satisfies $T_2 \ll T_1$. Therefore, the $T_1$ term can be ignored in the calculation. However, the value of $T_2$ is usually difficult to be given exactly and is still a subject of ongoing debate [14].

By means of the density matrix elements, the laser-induced electric current within solids can be divided into the intraband current $\mathbf{J}_{\text{intra}}(t)$ and the interband current $\mathbf{J}_{\text{inter}}(t)$:

$$\mathbf{J}_{\text{intra}}(t) = \frac{1}{\Omega}\sum_m \int_{BZ} \mathbf{p}_{mm}^{\mathbf{k}+\mathbf{A}(t)}\rho_{mm}^{\mathbf{k}}(t)d\mathbf{k}, \tag{8.10}$$

$$\mathbf{J}_{\text{inter}}(t) = \frac{1}{\Omega}\sum_{m\neq n} \int_{BZ} \mathbf{p}_{mn}^{\mathbf{k}+\mathbf{A}(t)}\rho_{mn}^{\mathbf{k}}(t)d\mathbf{k}, \tag{8.11}$$

where the $\Omega$ is the volume of the unit cell in real space, and the total current $\mathbf{J}_{\text{tot}}(t) = \mathbf{J}_{\text{intra}}(t) + \mathbf{J}_{\text{inter}}(t)$. The $\mathbf{J}_{\text{intra}}(t)$ arises from the electric current in individual bands, while the $\mathbf{J}_{\text{inter}}(t)$ arises from the electron-hole polarization between the CBs and VBs [24].

In principle, after obtaining the band structure, the transition dipole-matrix elements, and the momentum matrix elements of the material, we can solve the SBE for any real material to analyze the corresponding HHG spectrum. But how to obtain the accurate phase of the transition matrix elements is still a challenge for the present SBE [25].

### 8.2.3 *Time-Dependent Density Functional Theory*

The TDSE and SBE methods are usually based on the strong field approximation and several empirical hypotheses. The empirical hypotheses primarily encompass various approximations utilized during the solution process, which include the single active electron approximation, the omission of electron-electron interactions, and the focus on specific energy band structures, among others. However, the many-body effect usually plays an important role in the interaction between the laser and the condensed matter systems. The TDDFT method [26, 27] incorporates the electron-electron interactions and encompasses a broad range of electronic band structures, which is a widely used method to investigate the laser-matter interaction.

In the TDDFT, the time evolution of electronic dynamics is usually described by the time-dependent Kohn-Sham (KS) equation (in atomic units):

$$i\frac{\partial}{\partial t}\psi_{n,\mathbf{k}}(\mathbf{r},t) = \hat{H}_{\text{KS}}(\mathbf{r},t)\psi_{n,\mathbf{k}}(\mathbf{r},t), \tag{8.12}$$

where $\psi_{n,\mathbf{k}}$ is the KS orbital, $n$ is a band index, and $\mathbf{k}$ is a point in the first Brillouin zone (BZ). The time-dependent KS Hamiltonian, $\hat{H}_{\text{KS}}$, is given by

$$\begin{aligned}\hat{H}_{\text{KS}}(\mathbf{r},t) = \frac{1}{2}\left[-i\nabla + \frac{\mathbf{A}(t)}{c}\right]^2 + v_{ion}(\mathbf{r},t) \\ + v_H(\mathbf{r},t) + v_{xc}(\mathbf{r},t).\end{aligned} \tag{8.13}$$

where $\mathbf{A}(t)$ is the vector potential, which is related to the electric field by $\mathbf{A}(t) = -c\int_{-\infty}^{t}\mathbf{E}(t')dt'$. $v_{ion}$ is the ionic potential. $v_H$ is the classical Hartree potential, which describes the electron-electron Coulomb interaction. The exchange-correlation potential $v_{xc}$ contains non-trivial interactions between the electrons, which need to be approximated in practice. After solving the time-dependent KS equation, the average time-dependent electric current density in a unit cell is evaluated by

$$\mathbf{J}(t) = \frac{1}{\Omega} \int_{\Omega} d^3\mathbf{r}\mathbf{j}(\mathbf{r}, t), \tag{8.14}$$

where $\mathbf{j}(\mathbf{r}, t)$ is the microscopic electric current density,

$$\mathbf{j}(\mathbf{r}, t) = -\sum_{n,\mathbf{k}} \mathrm{Re}\left[\psi^*_{n,\mathbf{k}}(\mathbf{r}, t)\left(-i\nabla + \frac{\mathbf{A}}{c}(t)\right)\psi_{n,\mathbf{k}}(\mathbf{r}, t)\right]. \tag{8.15}$$

Then, the resulting high-harmonic spectrum can be calculated by

$$\mathrm{HHG}(\omega) = \left|\mathrm{FT}\left[\frac{\partial}{\partial t}\int_{\Omega} d^3\mathbf{r}\mathbf{j}(\mathbf{r}, t)\right]\right|^2, \tag{8.16}$$

where FT denotes the Fourier transform, and $\Omega$ is the volume of the simulation box.

#### 8.2.3.1 *Multiscale Numerical Simulation*

A laser is a coherent electromagnetic wave and its macroscopic propagation obeys the fundamental laws of electromagnetism, namely the Maxwell equations. To explore the HHG in solids, the laser propagation effect may play an important role. Thus, an approach about coupling the macroscopic Maxwell equations with the microscopic electronic dynamics has been proposed to investigate this issue. The relevant studies have been reported based on the SBE and TDDFT methods [18, 19].

In this section, we account for the propagation effect by combining the microscopic SBE solution as described in Sec. 8.2.2 with the macroscopic Maxwell equations. In general, the Maxwell equations contain four partial differential equations. For convenience in numerical simulations, under the Coulomb gauge where the scalar potential is ignored, the four Maxwell equations can be converted into only one Helmholtz wave equation in the vector potential form,

$$\nabla \times \nabla \times \mathbf{A}(t) + \frac{1}{c^2}\frac{\partial^2 \mathbf{A}(t)}{\partial t^2} = \frac{4\pi}{c}\mathbf{J}(t), \tag{8.17}$$

where $c$ is the speed of light. The laser field $\mathbf{E}(t)$ and the magnetic induction $\mathbf{B}(t)$ can be derived from the vector potential,

$$\mathbf{E}(t) = -\frac{\partial \mathbf{A}(t)}{\partial t}, \quad \mathbf{B}(t) = \nabla \times \mathbf{A}(t). \tag{8.18}$$

If the laser field propagates along the $\hat{\mathbf{z}}$ direction ($\hat{\mathbf{z}}$ is the unit vector), forward $\mathbf{E}^+(t)$ and backward electric fields $\mathbf{E}^-(t)$ can be defined as,

$$\mathbf{E}^+(t) = \frac{\mathbf{E}(t) + c\mathbf{B}(t) \times \hat{\mathbf{z}}}{2}, \tag{8.19}$$

$$\mathbf{E}^-(t) = \frac{\mathbf{E}(t) - c\mathbf{B}(t) \times \hat{\mathbf{z}}}{2}. \tag{8.20}$$

From Eqs. (8.19) and (8.20), one can distinguish among the incident wave, the reflected wave, and the transmitted wave. It means that the propagation of strong laser field in bulk solids can be investigated by solving Eqs. (8.17)–(8.20).

Figure 8.1 shows the schematic illustration of multiscale calculations with a light propagation through a material thin film. In this multiscale numerical model, the laser pulse propagates along the $Z$ axis and the polarization plane of laser field is perpendicular to the $Z$ axis. Meanwhile, the spatial dependence of the laser field in the polarization plane tends to be ignored. In this case, the generalized Helmholtz wave equation in Eq. (8.17) can be simplified as

$$-\frac{\partial^2 \mathbf{A}_Z(t)}{\partial Z^2} + \frac{1}{c^2}\frac{\partial^2 \mathbf{A}_Z(t)}{\partial t^2} = \frac{4\pi}{c}\mathbf{J}_Z(t), \tag{8.21}$$

where $Z$ denotes the macroscopic grid coordinate along the propagation direction as shown in Fig. 8.1. The incident laser field is taken as,

$$\mathbf{A}_Z^{inc}(t) = \hat{\mathbf{n}} A_0 f(t - Z/c) \cos\left[\omega_0(t - Z/c) + \phi\right], \tag{8.22}$$

where $\hat{\mathbf{n}}$ is the unit vector in the polarization direction, $A_0$ the maximum value of the vector potential, and $\omega_0$ the carrier central frequency of the laser pulse. $f(t)$ is the envelope of the laser pulse, and $\phi$ is the carrier envelope phase. At the same time, due to the limited size of the vacuum region on both sides of the material thin film (see Fig. 8.1), the following absorbing boundary condition is added for solving the Maxwell equation in Eq. (8.21), i.e.,

$$\mathbf{A}_{Z_M}(t + \Delta t) = \mathbf{A}_{Z_{M-1}}(t) - \frac{\Delta Z - c\Delta t}{\Delta Z + c\Delta t}\left[\mathbf{A}_{Z_M}(t) - \mathbf{A}_{Z_{M-1}}(t + \Delta t)\right], \tag{8.23}$$

where $Z_M$ denotes the macroscopic grid coordinate at the boundary.

At the time step $t$, the macroscopic electromagnetic field can be obtained by solving Eq. (8.21). Feeding the solved macroscopic electromagnetic field into the SBEs, the current density $\mathbf{J}(t)$ can be obtained by solving the

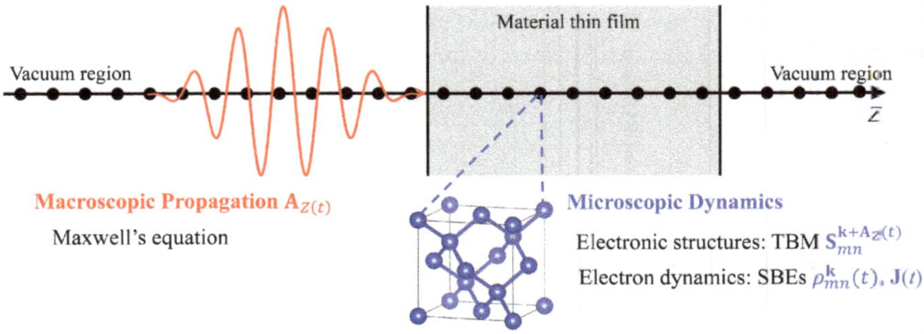


Fig. 8.1 Schematic illustration of multiscale calculations with light propagation through a thin film material. The macroscopic spatial coordinate is expressed by $Z$ and the black dots represent the macroscopic grid for solving the Maxwell equation. Every macroscopic grid point in the material thin film contains a unit cell of crystal, as shown by the blue dashed lines. Reproduced with permission from [22].

SBEs at every macroscopic point of the material thin film, as shown in Fig. 8.1. Inserting the solved current density into Eq. (8.21), the macroscopic electromagnetic field can be computed for the next time step. By continuing in this way, the propagation of macroscopic electromagnetic field will be coupled with the microscopic electronic dynamics. The grid size of propagation of electromagnetic field (about 10 nm) is 100 times larger than that of the microscopic electronic dynamics (about 1 Å). In this way, one can realize the multiscale coupling between the Maxwell equations and the SBEs.

## 8.3 Novel Phenomena of HHG in Solids

### 8.3.1 *Michelson Interferometry and Contribution of Floquet-Bloch States*

#### 8.3.1.1 *Michelson Interferometry of HHG based on the Simplest 1D Model*

In this subsection, we establish a Michelson interference model of electrons in the energy band to analyze the interference phenomenon in the harmonic spectrum based on the TDSE method [28]. By varying the laser peak intensity, we identify an unusual overall oscillation in the high-energy region of the harmonic spectra. This oscillation can be regarded as a result of the Michelson interferometry of the Bloch electrons when they are driven to the BZ boundary [29]. The interference fringes can be nicely reproduced using an analytical model based on the principle of the Michelson interferometer.

For the demonstration of the proposed Michelson interferometer of the Bloch electrons in solids, it is sufficient to adopt the simplest 1D model. In this model, the periodic potential, expressed in atomic units (a.u.), is defined as

$$V(x) = V_0\left(1 + \cos\frac{2\pi x}{a}\right), \tag{8.24}$$

where $V_0 = -0.37$ and the lattice constant $a = 8$. As shown in Fig. 8.2(a), $q$ is the crystal momentum, and the lowest four bands contain two valence bands (V1, V2) and two conduction bands (C1, C2). We initialize the state

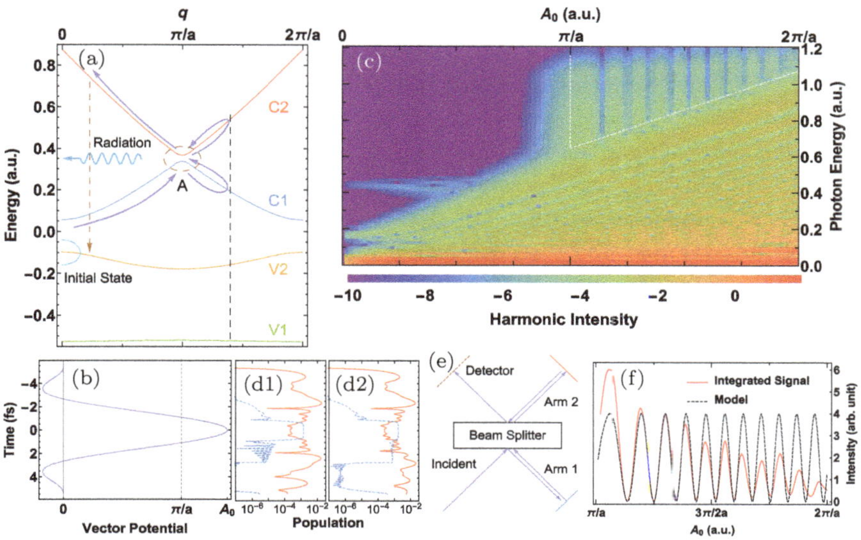


Fig. 8.2 The realization of the Michelson interferometer of Bloch electrons in the HHG spectra. For the model with the first four bands shown in (a), the harmonic spectra are shown in (c) as a function of $A_0$ of the pulse, as shown in (b). The Michelson interferometry is present at the top-right corner of (c), enclosed by the white dotted lines. The Bloch electron Michelson interferometer is in the same principle with the usual optical one, as sketched in (e). The interference minimum and maximum can be verified by checking the populations of C1 (solid red line) and C2 (dashed blue line) as a function of time, as respectively shown in (d1) for $A_{01} = 1.14\pi/a$ and (d2) for $A_{02} = 1.18\pi/a$. The interference patterns can be successfully be explained by the analytical model, as shown in (f). Reproduced with permission from [28].

at the $q = 0$ point of the band V2, as indicated in Fig. 8.2(a), where the transition probability to the band C1 is the highest.

In order to show interference fringes of the Michelson interferometer clearly, we first use a single-cycle laser pulse with a $\cos^2$ envelope which is plotted in Fig. 8.2(b). Figure 8.2(c) shows the harmonic radiation spectrum as a function of $A_0$ in a wide range of $[0, 2\pi/a]$. It can be seen from the figure that at all $A_0$ values, the maximum probability of photon emission occurs in the range of photon energy below 0.1 a.u., which is caused by the intraband motion of electrons on V2 [30]. The interband polarizations between the electrons excited to C1 and the holes in the V2 may contribute to the interband harmonic emission. Its cutoff energy grows linearly with the increase of $A_0$. The obvious diagonal interference patterns come from the two-path contributions for the same photon energy on the rising and falling part of the vector potential [Fig. 8.2(b)]. The most important feature in Fig. 8.2(c) is the distinct vertical stripes in the top-right corner when $A_0 > \pi/a$, which is enclosed by white dotted lines. These interference patterns originate from the Michelson interferometry of the Bloch electrons.

Figure 8.2(c) shows that the interference structures in the region of $A_0 > \pi/a$ only rely on the value of $A_0$, independent of the photoemission energy. To interpret this phenomenon, we show the populations of C1 and C2 as a function of time for $A_{01} = 1.14\pi/a$ and $A_{02} = 1.18\pi/a$ in Figs. 8.2(d1) and (d2), respectively. They are corresponding to the interference minimum and maximum. As can be seen, when the vector potential increases to $\pi/a$, a considerable population is transferred from C1 to C2, and it will jump back when the vector potential decreases down to $\pi/a$ again. However, for the case of interference maximum, as shown in Fig. 8.2(d2), only a small amount of populations can transfer back to C1. It means that the vertical interference patterns in the radiation spectrum are contributed by the oscillation of population remaining in C2.

The electrons are initially located at the top of the V2. Under the interaction of the laser field, they are excited to C1. With the increase of the vector potential, the electrons can reach the BZ boundary. Landau-Zener tunneling occurs when they move to the avoid crossing point "A". The excited electrons will be split into two parts: one keeps moving along the band C1, while the other follows the path of the band C2. Due to the energy difference between the two bands, when the two parts of electrons recombine, their wavefunctions will produce the difference of phase accumulation. It will affect the population remaining in C2 and then determine the intensity of harmonic radiation that can be measured. The above scenario is exactly equivalent to the usual optical Michelson interferometer, as depicted in Fig. 8.2(e). We also calculate the harmonic radiation analytically. The analytical result is compared with that calculated by the TDSE in Fig. 8.2(f). It is clear that the oscillation behaviour of the two curves agrees well with each other, except for the region of $A_0$ close to $2\pi/a$. This is because we do not consider the transition to the higher band C3 in the model.

Besides the single-cycle pulse with $\cos^2$ envelope, we demonstrate that the phenomenon of Michelson interferometry in HHG also exist under the laser field with other parameters. For example, Fig. 8.3 shows the HHG spectra under a longer pulse with eight optical cycles and different laser envelopes. For the case of $\cos^2$ envelope in Fig. 8.3(a), it can be seen that the vertical patterns are still clear but much more complicated because of the different maximal vector potentials in each half-cycle. Because the interference in the first half period has affected the electron population in

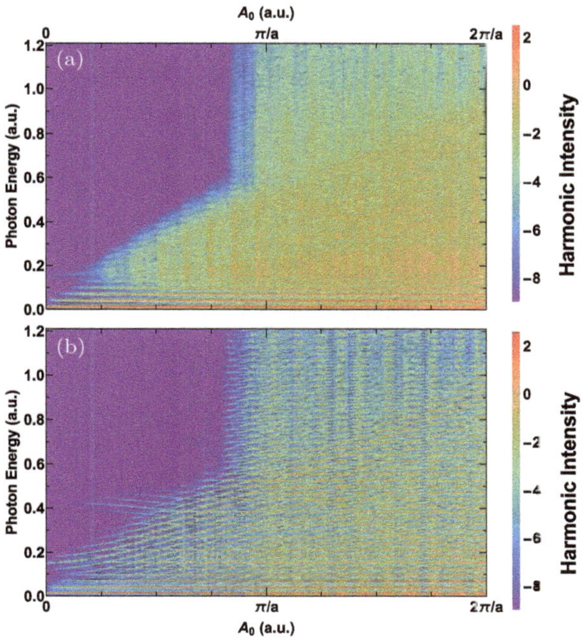


Fig. 8.3 Harmonics spectra as a function of peak vector potential $A_0$ for a longer pulse of an 8-cycle at the wavelength of 3200 nm for: (a) a cosine square, and (b) a trapezoid envelope. Reproduced with permission from [28].

the second half period, the vertical stripes extend downward to low-order harmonics. The phenomenon has been shown in some figures of previous publications [30–32], but without any discussions about it.

Theoretically, the Michelson interferometry should also exist for other forms of the pulse envelope. Figure 8.3(b) shows the result driven by a trapezoid-envelope pulse which guarantees the same maximum vector potentials in each half-cycle. The sharper vertical stripes are observed and the intervals are similar to that in Fig. 8.2(c). It is the product of multiple interferences. However, for $A_0 < \pi/a$, we can also observe sharp parallel signals, which may be attributed to the effects of the Floquet states for the multiple-cycle pulse [33]. In this region, there are resonance-like vertical lines formed at the crossing points of the dressed-state curves and the low-order harmonics. The structures are different from the interference fringes formed by the Michelson interferometer.

#### 8.3.1.2 *Study on the Contribution of Floquet-Bloch States to HHG under a 1D Model*

By numerically solving the TDSE of a 1D model, we study the contribution of Floquet-Bloch states to HHG in solids [34]. The model potential parameters are the same as those in subsection 8.3.1.1. Two VBs and one CB are used, as shown in the left part of Fig. 8.4(a). Firstly, we take the single point at $k = 0$ on V2 as the initial state. When the model is driven by a laser with a trapezoidal envelope, eight cycles, $\omega_0 = 0.014$ a.u., the calculated HHG spectrum is presented in Fig. 8.4(c). Obviously, with the increase of the laser peak intensity, a series of resonance peaks is observed, due to the resonance enhancement between different Floquet-Bloch states. The spectrum signal can be divided into the horizontal odd harmonic part and the slanted fringe part. The intensity-dependent slanted fringe signal can be easily associated with the channel-closing effect caused by $U_p$ shift in the atomic ionization [35].

In order to study the influence of Floquet-Bloch states on the HHG, the time-averaged energy difference $\overline{\Delta E}$ within one optical cycle as a function of $A_0$ is plotted in Fig. 8.4(c) with black dashed line. Indeed, it is in good agreement with one of the slanted fringes. However, the energy difference between bands C1 and V2 (i.e., $\Delta E$) only qualitatively gives a rough prediction of the cutoff when $A_0$ increases, as illustrated by the white dashed line in Fig. 8.4(c). We demonstrate that slanted fringes are caused by the transitions between various Floquet-Bloch states with different parities. In addition, to further examine the signal enhancement near the crossing

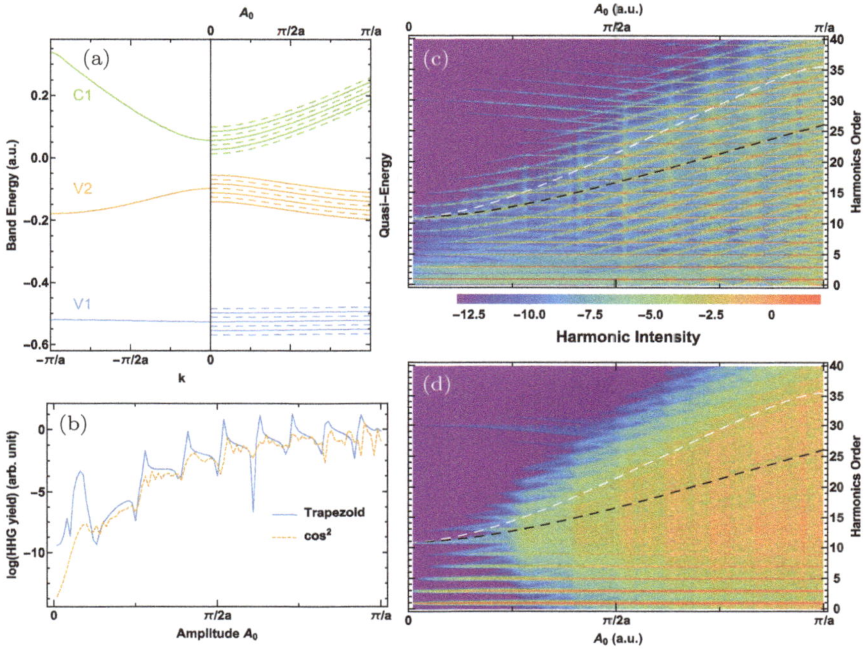


Fig. 8.4 (a) Left panel: the first three energy bands of the 1D model. Right panel: corresponding quasi-energies as a function of the amplitude of the vector potential. The solid and dashed lines represent states with different parities. (b) The yield of the 11th order harmonic induced by a trapezoidal (blue solid line) and cosine square (yellow dashed line) envelope at various $A_0$. The full HHG spectrum is respectively shown in (c) for the trapezoidal envelope and (d) for the cosine square envelope. The white dashed line is the energy difference between the C1 and V2, the black one denotes the time-averaged energy difference within one optical cycle. Reproduced with permission from [34].

points of the horizontal and slanted signals in Fig. 8.4(c), we plot the yield of the 11th harmonic as a function of the $A_0$ in Fig. 8.4(b). It shows that the HHG yield increases non-monotonically with the increasing $A_0$ and presents asymmetric Fano-type resonance peaks, which illustrates the existence of resonance behavior of the discrete and continuous state. Meanwhile, we change the laser envelope into $\cos^2$ shape to calculate the HHG spectrum. Figure 8.4(d) shows that the slanted structures also appear, but are slightly blurred. Importantly, the integrated yield for the 11th harmonic also exhibits step-like resonance structures, as indicated by the dashed line in Fig. 8.4(b).

We utilize the Floquet-Bloch approach and the strong-field approximation (SFA) [14, 36] method to interpret the observed structures in the harmonic spectra. The bands of C1 and V2 are considered. Figure 8.5(c1) shows the HHG spectra as a function of $A_0$ for the trapezoidal pulse. The odd harmonics and the slanted fringes contributed by the Floquet-Bloch states are reproduced, which is consistent with that in Fig. 8.4(c). We compare the HHG spectra calculated by the TDSE and SFA at different laser intensities in Figs. 8.5(b1)–(b4). It shows that the spectra in the plateau region agree well with each other. However, the SFA does not consider the intraband current and the discrete summation of

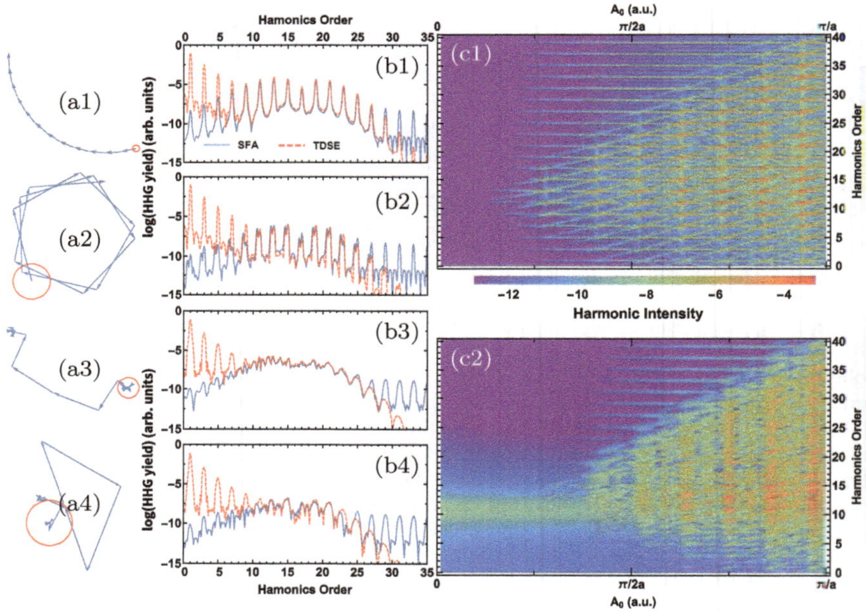


Fig. 8.5 The harmonic spectra calculated by SFA at different peak vector potential $A_0$ of laser pulses with a trapezoidal or cosine square envelope. (a1)–(a4) The amplitude of the conduction band for the laser pulse: (a1) a trapezoidal pulse with $A_0 = 0.52\ \pi/a$, (a2) a trapezoidal pulse with $A_0 = 0.50\ \pi/a$, (a3) a cosine square pulse with $A_0 = 0.52\ \pi/a$, and (a4) a cosine square pulse with $A_0 = 0.50\ \pi/a$. In these four subfigures, the red circle centered at the origin has the same absolute size to indicate the scale and the small arrows represent the tunneling amplitude in each saddle point. (b1)–(b4) The corresponding HHG spectra, calculated by TDSE (red dashed line) and by SFA (blue solid line). (c1),(c2) An overview of the HHG spectra calculated by SFA as a function of $A_0$ for (c1) a trapezoidal pulse and (c2) a cosine square pulse. Reproduced with permission from [34].

the saddle points would lead to the unphysical high-frequency Fourier components. Therefore, the SFA underestimates lower-order harmonic yields and overestimates higher-order harmonic yields. It is noted that when $A_0 = 0.52\pi/a$, odd harmonics are clearly shown in Fig. 8.5(b1), but when $A_0 = 0.5\pi/a$, each peak is split into two peaks shown in Fig. 8.5(b2). This is because the choice of $A_0$ could determine a near-resonance and an off-resonance case.

## 8.3.2 *Distinct Features of HHG in 2D Materials*

### 8.3.2.1 *Anomalous Ellipticity Dependence of HHG in Graphene*

In recent years, 2D materials have been widely used in the study of HHG because of their unique electronic and optical properties. Graphene is not only one of the simplest 2D materials, but it has also emerged as a popular material for laser-matter interaction studies in recent years. The structure of graphene is relatively simple and it is suitable for benchmark calculations. In the following, we focus on the HHG in monolayer graphene driven by elliptically polarized pulses based on the SBE method.

The crystal structure of monolayer graphene is shown in Fig. 8.6(a) and the unit cell is framed by magenta lines. As one can see, the structure of graphene is hexagonally symmetric and there are two carbon atoms in the unit cell with C-C bond length of 1.42 Å. The band structure is obtained using the tight-binding model (TBM), which takes into account the first and second nearest-neighbors coupling. Figures 8.6(c) and (d) show the graphene energy band along the path Γ-K-M-Γ and the corresponding energy dispersion surface, respectively. The calculated energy bands are in good agreement with those in previous publications [37, 38].

In 2017, Yoshikawa *et al.* published a groundbreaking paper on the anomalous ellipticity dependence of HHG from graphene [16]. We reproduce the main findings of this paper using our own numerical tool, considering the following laser vector potential (in atomic units),

$$\mathbf{A}(t) = \frac{\sqrt{I_0}}{\omega_0} \sin^2\left(\frac{\omega_0 t}{2n_c}\right) \left[\frac{1}{\sqrt{1+\epsilon^2}} \cos(\omega_0 t)\hat{\mathbf{e}}_x \frac{\epsilon}{\sqrt{1+\epsilon^2}} \sin(\omega_0 t)\hat{\mathbf{e}}_y\right], \tag{8.25}$$

where the peak intensity $I_0 = 1.7 \times 10^{10}\,\mathrm{W/cm^2}$, the carrier frequency $\omega_0 = 0.0097\,\mathrm{a.u.}$ (corresponding to the wavelength $\lambda = 4700\,\mathrm{nm}$), and the pulse cycles $n_c = 10$. The ellipticity $\epsilon$ ranges from 0.0 to 1.0 in the calculations.

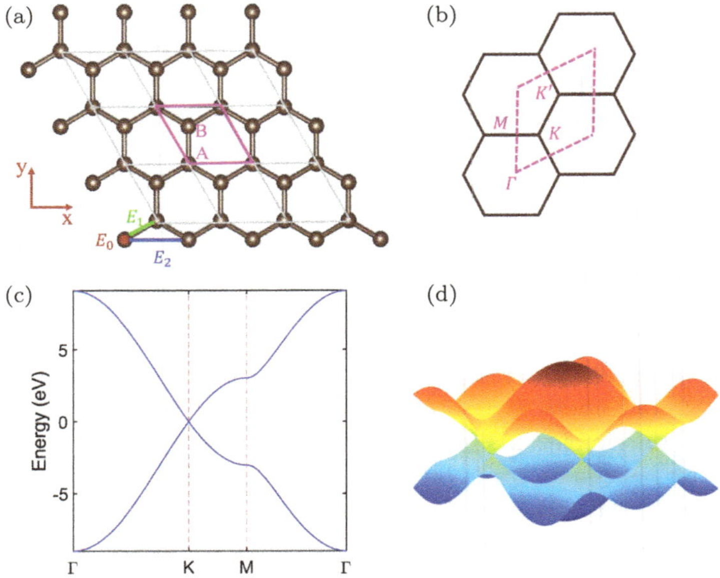


Fig. 8.6 The electronic structures of graphene obtained from our tool. (a) Top view of monolayer graphene. The parallelogram in magenta lines is the unit cell and there are two different types of carbon atom (namely A and B) in the unit cell. $E_0$ is the on-site energy of the carbon atom, $E_1$ the nearest-neighbor hopping energy and $E_2$ the second-neighbor hopping energy. (b) The reciprocal unit cell of graphene (the parallelogram in dashed magenta lines). (c) The energy band along the path Γ-K-M-Γ. (d) The energy band in the plane of the monolayer. Reproduced with permission from [22].

The $\hat{\mathbf{e}}_x$ and $\hat{\mathbf{e}}_y$ are unit vectors in the $x$ and $y$ directions, respectively, as illustrated in Fig. 8.6(a). Due to the atomic thickness structure of monolayer graphene, the propagation effect can be ignored when we calculate the HHG from graphene. The HHG yield can be calculated from the time-dependent current density as mentioned in Sec. 8.2.2. In the SBE simulations, we employ a dense $k$-space grid of $1000 \times 1000$ and use a time interval of $\Delta t = 0.1$ a.u. for the time-dependent propagation. The dephasing time is 0.3 $T_0$, where $T_0 = 2\pi/\omega_0$ is the period of laser pulse.

Figure 8.7(a) compares the ellipticity dependence of the 7th harmonic between our results and experiments [16]. It can be seen that our numerical results agree well with the experimental measurements. As discussed in Ref. [16], the yield of the 7th harmonic along the $y$ direction takes a maximum value around ellipticity $\epsilon = 0.25$, which is different from that

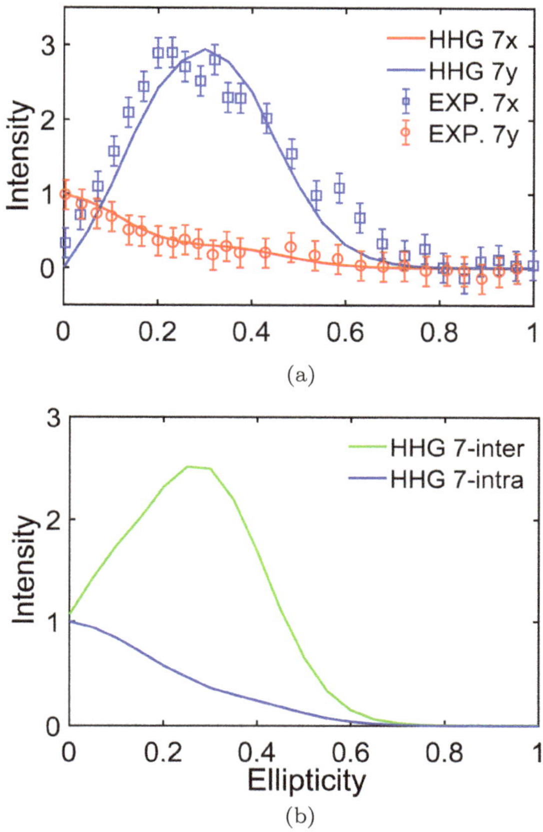


Fig. 8.7 (a) Comparisons of the ellipticity dependence of the 7th harmonic between the results from our numerical calculation and those from the experiment in Ref. [16], where $x$ denotes the horizontal direction of Fig. 8.6(a) and $y$ the vertical direction. The circles and squares are the experimental data in Ref. [16] and solid lines are the numerical data in our calculations. For comparison, the numerical harmonic yield is normalized to the maximum value of the experimental data. (b) Ellipticity dependence of the 7th harmonic from the contributions of the intraband (blue curve) and the interband (green curve) currents. The harmonic yield is normalized to the intraband harmonic yield at the ellipticity $\epsilon = 0$. Reproduced with permission from [22].

in the atomic HHG. For further analyzing the mechanism of the HHG, we show the ellipticity dependence of the 7th harmonic from the intraband (the blue curve) and the interband (the green curve) currents in Fig. 8.7(b). We find that the anomalous ellipticity dependence is mainly attributed to the contribution of interband current.

#### 8.3.2.2 *Symmetry Analyses of HHG in Monolayer Hexagonal Boron Nitride*

Symmetry usually leads to conservation laws and selection rules, which is one of the most generic and useful concepts in physics [40]. In the following, we will analyze the symmetry of the laser-matter interacting system to understand the symmetric characteristics of the harmonic spectra [39]. We verify that the TDDFT results are consistent with the selection rules. In general, the Hamiltonian ($\hat{\mathrm{H}}_{\mathrm{HG}}$) of the interacting system is given by (in atomic units):

$$\hat{\mathrm{H}}_{\mathrm{HG}}(\mathrm{t}) = -\frac{1}{2}\sum_{\mathrm{j}} \nabla_{\mathrm{j}}^2 + \frac{1}{2}\sum_{\mathrm{i}\neq\mathrm{j}} \frac{1}{|\mathbf{r}_{\mathrm{i}} - \mathbf{r}_{\mathrm{j}}|} + \sum_{\mathrm{j}} \mathrm{V}(\mathbf{r}_{\mathrm{j}}) + \sum_{\mathrm{j}} \mathbf{E}(\mathrm{t}) \cdot \mathbf{r}_{\mathrm{j}}, \quad (8.26)$$

where $\nabla_j^2$ is the Laplacian operator with respect to the coordinates of the $j$th electron, and $V(\mathbf{r}_j)$ is the electron-ion potential. Without the nuclear motion, the first three terms are time-independent. The symmetry of the time-dependent term in $\hat{\mathrm{H}}_{\mathrm{HG}}$, i.e., the laser-matter interaction $\mathbf{E}(t) \cdot \mathbf{r}_j$, determines the symmetry of harmonics. In other words, the HHG must adhere to the symmetry rules determined by the target and the input field [40, 41].

Taking the monolayer hexagonal boron nitride (hBN) as a typical example, its geometric structure in real space and the first BZ in reciprocal space are shown in Fig. 8.8. For the laser-matter interaction, we take the near-infrared fundamental incident laser field with $\sin^2$ envelope and wavelength of 1600 nm. As shown in Fig. 8.9, when the incident pulse is linearly polarized, the HHG spectrum emitted from the monolayer hBN shows apparent anisotropic properties under different laser polarization

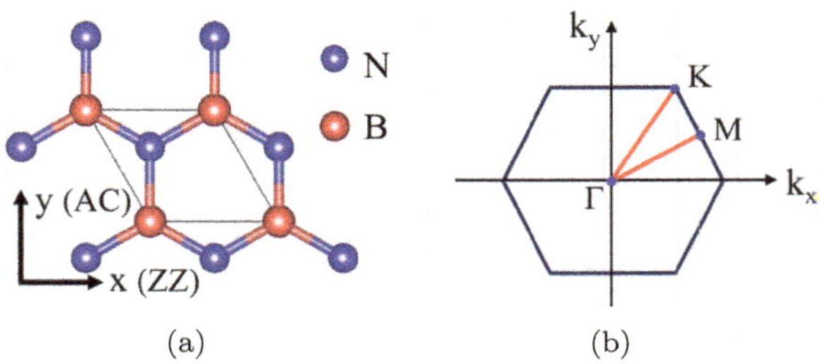


Fig. 8.8 (a) The geometric structure of monolayer hBN crystal in real space, where the blue balls denote the nitrogen atoms and the red balls denote the boron atoms. (b) First BZ of hBN crystal in reciprocal space with high symmetric points Γ, M, and K. Reproduced with permission from [39].

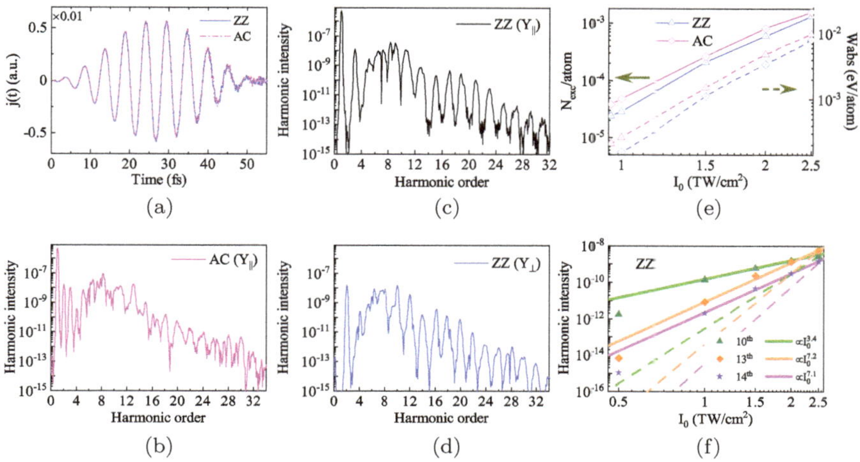


Fig. 8.9 (a) The induced electric current in a monolayer hBN when the linear pulse is polarized along ZZ (blue solid line) or AC (magenta dash-dotted line) direction with a laser wavelength of 1600 nm, a pulse duration of 53 fs, and a peak intensity of $10^{12}$ W/cm$^2$. (b)–(d) Computed HHG spectra from monolayer hBN parallel $Y_{\parallel}$ or perpendicular $Y_{\perp}$ to the laser polarizations. (e) Number of electrons excited to the CBs (solid lines and hollow symbols) and the energy absorption (dashed lines and hollow symbols) after the laser pulse. (f) Pump power dependence of the intensity of the 10th (triangles), the 13th (diamonds), and the 14th (stars) harmonics. Solid lines are fits of the calculated data to power laws, yielding exponents of 3.4, 7.2 and 7.1 for the the 10th, the 13th, and the 14th harmonics, respectively. Reproduced with permission from [39].

directions. The monolayer hBN lattice has the reflection symmetry in the $x$ axis and none in the $y$ axis. The linearly polarized laser field has the time-translation symmetry and the reflection symmetry about a specific axis after transmitting for half an optical cycle. Then, according to Eq. (8.26), when the laser is polarized parallel to the B-N bond axis (armchair direction, AC), both odd and even harmonics appear along the laser-polarized direction. Meanwhile there is no harmonic emission in the perpendicular direction. However, when the laser field is polarized perpendicular to the B-N bond direction, i.e., the zigzag direction (ZZ), pure odd (even) harmonics are only generated in the ZZ (AC) direction. The difference in the AC and ZZ directions may be attributed to the anisotropic band dispersion along the Γ-M and Γ-K.

When the target is driven by a fundamental single circularly polarized pulse, according to the selection rules [42], the harmonic order $n_H$ of the material with $N$-fold rotational symmetry satisfies $n_H = \sigma + Nj$ $(j \in \mathbf{N})$, where $\sigma = \pm 1$ indicates the helicity of the harmonic is the same or opposite to that of the single circularly polarized pulse, respectively. Here, the monolayer hBN and the circularly polarized pulse both have the three-fold rotational symmetry in the $xy$ plane, and thus the interaction system is three-fold symmetric. Then, as shown in Fig. 8.10, the harmonic orders with $n_H = 3m$ are forbidden in the HHG spectrum. It shows the harmonic spectrum radiated in the AC (magenta dash-dotted line) and the ZZ (blue solid line) directions. It also demonstrates that the harmonics with order $n_H = 3m - 1$ and $n_H = 3m + 1$ are circularly polarized, and they have the opposite helicities. Our calculated results are consistent with the selection rules.

Next, we investigate HHG in the monolayer hBN crystal driven by a circularly polarized bichromatic field. The bichromatic field is combined a circularly polarized fundamental field of frequency $\omega_1$ (1600 nm) with its counter-rotating or co-rotating second harmonic $\omega_2$ (800 nm) field. The intensities of the two lasers are set to be $I_1 = I_2 = 5 \times 10^{11}$ W/cm$^2$ and the total pulse fluence and durations are the same as the case of single linearly polarized pulse. Figure 8.11 shows that the $3m$-th harmonics disappear in the case of the counter-rotating circularly polarized field and both the odd and even harmonics exist in the case of the co-rotating circularly polarized field.

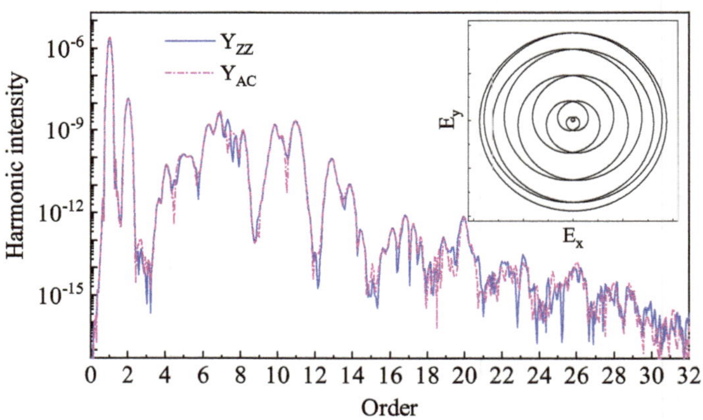


Fig. 8.10 HHG spectra in hBN driven by a single circularly polarized laser field with $I_1 = 10^{12}$ W/cm$^2$, $\omega_1 = 0.0285$ a.u. The emitted harmonic component along the ZZ and AC directions are plotted with the blue solid line and the magenta dash-dotted line, respectively. The inset is the Lissajous curves of the applied circularly polarized laser field. Reproduced with permission from [39].

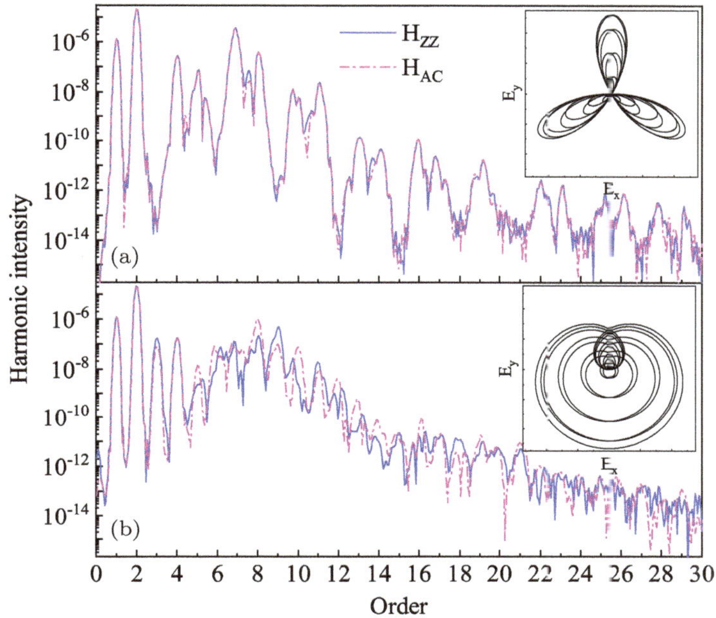


Fig. 8.11 HHG spectra in the monolayer hBN crystal driven by a counter-rotating (a) and co-rotating (b) bichromatic circularly polarized laser field. The emitted harmonic along the ZZ and AC directions are plotted with the blue solid line and the magenta dash-dotted line. The inset is the Lissajous curve of the applied circularly polarized laser field. Reproduced with permission from [39].

The counter-rotating bichromatic circularly polarized laser field with a three-fold rosette shape has the same three-fold rotational symmetry as the monolayer hBN lattice. Then, the dynamical symmetry gives rise to the harmonics with orders $n = 3m$ missing over the entire HHG spectra. It proves that the generated harmonics are circularly polarized. However, for the interaction between the co-rotating circularly polarized driving field and the monolayer hBN, the symmetry of the interacting system is broken, giving rise to the existence of both the odd and even harmonics. Also, the harmonics are not circularly polarized. The harmonic yields obtained from the two cases are comparable, which further indicates that the wavefunctions of the electrons in both VB and CB are delocalized and it is unnecessary for the electrons to recombine with its original site. Once again, our results show that the symmetry of the interacting system plays essential roles in HHG.

#### 8.3.2.3 *Manipulation of HHG in Monolayer Hexagonal Boron Nitride by Two-Color Laser Field*

In atomic systems, a two-color laser field with a parallel or orthogonal polarization has been used to optimize or control the harmonic generation [44–49] by modifying the electron trajectories during the three-step process [50]. Compared with the atoms, the solid material has various lattice structures and complex band structures. Controlling and tailoring the HHG radiation properties can be realized by changing the interband and intraband dynamics [12, 14, 51, 52]. In Figs. 8.12 and 8.13, we show the manipulation of HHG in the monolayer hBN by a two-color linearly polarized laser field.

By keeping the incident pulse energy fixed, we investigate the total HHG spectra of the monolayer hBN crystal using the TDDFT method in three different laser configurations. The laser pulses plotted in Fig. 8.12(a) are for the fundamental pulse only (black line), and the two-color pulses whose fundamental and the second-harmonic pulses have the same maximum electric-field amplitude but with a relative phase of $\Delta\phi = 0$ (red line) and $\Delta\phi = \frac{\pi}{2}$ (blue line). It is apparently that the two-color laser field can drive HHG more effectively than the monochromatic laser field. In general, the HHG spectrum in solids is insensitive to the carrier-envelope phase for a monochromatic long pulse. While for the two-color field, it can be seen that the pulse shape is sensitive to the relative phase between two components, leading to a significant change of the harmonic spectra. Under the two-color laser field, the symmetry of the electronic dynamics within half a cycle is

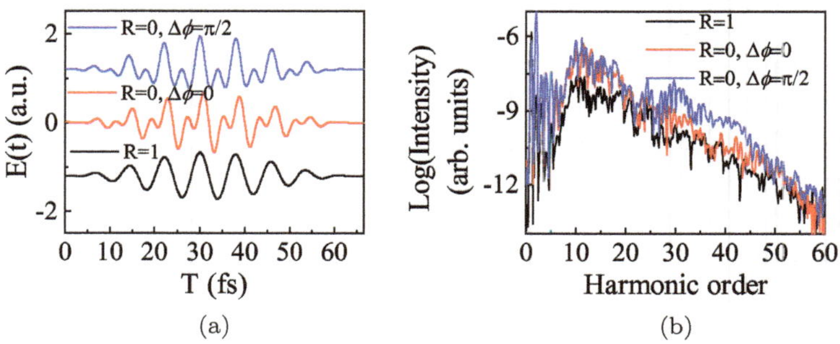


Fig. 8.12 (a) The two-color pulses with different relative phases ($\Delta\phi = \frac{\pi}{2}$, blue; $\Delta\phi = 0$, red) and the fundamental only pulse (black). The curves are shifted downward or upward for a better clarity. (b) Corresponding HHG spectra. Reproduced with permission from [43].

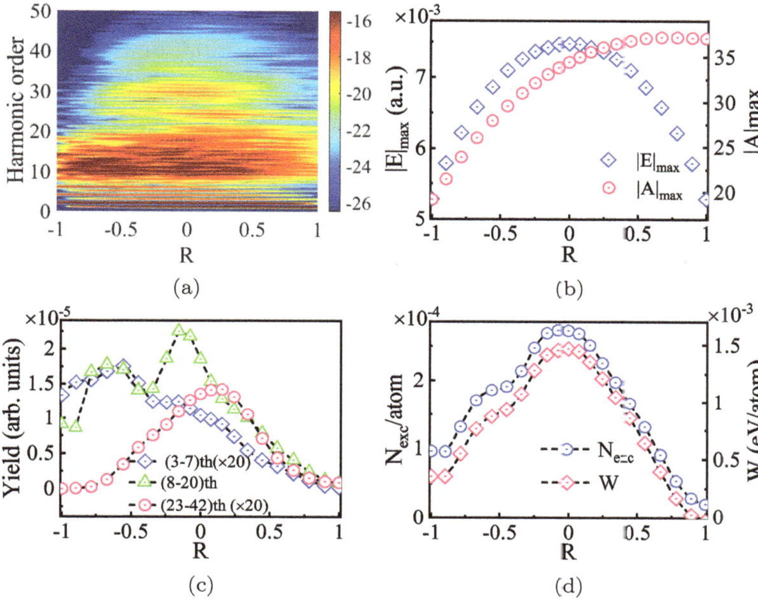


Fig. 8.13 (a) The calculated harmonic spectra as a function of the amplitude ratio $R$ between the fundamental and second-harmonic pulses. (b) The maximum value of the two-color electric field $|E|_{max}$ and vector potential $|A|_{max}$ as functions of $R$. (c) The harmonic yield in the regions of the below band gap, the first plateau, and the higher plateaus in the HHG spectra. (d) The number of excited electrons ($N_{exc}$, solid lines and blue circle symbols) and the energy transfer ($W$, solid lines and red diamond symbols) from the laser field to the monolayer hBN after the end of the pulses, as functions of $R$. Reproduced with permission from [43].

broken, then both even and odd harmonics can be generated in both the $x$ and the $y$ directions. The results indicate that the two-color linearly polarized laser field can greatly enhance the yield of the HHG spectrum in the plateaus and efficiently control the HHG by tuning $\Delta\phi$.

In fact, the laser amplitude of the two-color field is varied with the relative phase. The amplitudes of the electric field and the vector potential play important roles in HHG. Figure 8.13 shows the impact of the amplitude difference between the fundamental and the second-harmonic pulse, $R = (E_{01} - E_{02})/\sqrt{E_{01}^2 + E_{02}^2}$, on the HHG spectrum. It is obvious that the participation of the frequency-doubled pulse results in the enhancement of harmonic yield. The maximum harmonic yield occurs around $R = 0$, where the fundamental and second-harmonic pulses have the same electric-field amplitude. As shown in Fig. 8.13(d), the calculated $N_{exc}$ and absorbed energy ($W$) have the same trend of variation as that of the maximum value of the electric field ($|E|_{max}$), where $\mathrm{W}(t) = \Omega \int_{-\infty}^{t} dt' \mathbf{E}(t') \cdot \mathbf{J}(t')$. The harmonic yields in the first and higher plateaus are dominantly contributed by the interband polarization, which is also modulated by the $|E|_{max}$ and the $|A|_{max}$ simultaneously. A larger $|E|_{max}$ makes more electrons transit from the VB to CB, and a larger $|A|_{max}$ has the potential to drive a part of the excited electrons to the top of the first CB and transfer to the second CB. The electrons pumped to the higher CBs contribute to the higher HHG plateaus. It indicates that the HHG is affected by the electron population, which is influenced by both the electric field and the vector potential.

#### 8.3.2.4 *Enhancement of HHG in 2D Materials by Plasmonic Fields*

In this subsection, we investigate the HHG enhancement in monolayer and bilayer graphene driven by out-of-plane polarized plasmonic fields. The plasmonic field can be generated with the aid of nanoscale Au dimers, which is simulated by the finite-difference time-domain (FDTD) method using SALMON code [54, 55]. Figure 8.14(a) shows the schematic illustration of the composite system composed of 2D materials and nanoscale Au dimers for HHG. The graphene spreads in the $xy$ plane and is placed at the middle of the gap of Au dimers. The Au dimers are composed of a trapezoid cylinder and a circular cylinder. The incident laser pulse with a $\sin^2$ envelope is polarized along the $z$ direction. Its maximal electric field strength is 2.3 V/nm, the wavelength is 1600 nm, and the pulse duration is 32 fs.

In our simulations, the wavelength-dependent dielectric constant of the Au dimers is described by the generalized Lorentz-Drude model [56]. Figure 8.14(b) shows the plasmonic field enhancement in the $xz$ plane obtained from the FDTD simulation. The red line denotes the position profile of plasmonic field strength along the axial line of the dimers. The time profiles between the incident pulse (black dashed line) and the induced plasmonic field (red solid line) at the gap's center of the Au dimers, as shown in Fig. 8.14(c), have a similar shape but with a time delay. The size of the perpendicular dimension (the $y$ axis) of Au dimers is hundreds of times larger than the size of the unit cell of graphene. It means that at the $xy$ plane of $z = 0$, we can approximately regard the plasmonic field as spatially homogeneous on a scale of tens of nanometers, as shown in Fig. 8.14(d).

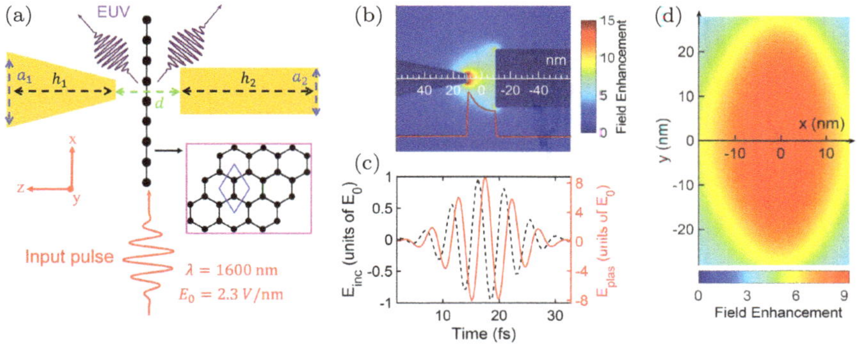


Fig. 8.14 (a) Schematic illustration of the composite system composed of 2D materials and nanoscale Au dimers for HHG. The incident laser pulse is polarized along the z-axis and the monolayer or bilayer graphene spread in $xy$ plane are placed at the middle of the gap of Au dimers. The inset on the bottom right (the magenta box) shows the structure of graphene and its unit cell used in our simulation (the blue diamond). $E_0$ is the maximal incident electric field strength. (b) Plasmonic field enhancement in the $xz$ plane obtained from FDTD simulation. The red line is the position profile of plasmonic field strength along the axial line of the dimers. (c) The time profiles for the incident pulse (black dashed line) and the induced plasmonic field (red solid line) at the crossing point of the axial line of the dimers and the plane of the graphene ($z = 0$). (d) Plasmonic field enhancement in the plane of $z = 0$ (the graphene layer) obtained from FDTD simulation. Reproduced with permission from [53].

We mainly focus on the out-of-plane HHG from graphene and simulate the HHG under plasmonic field enhancement by the TDDFT method. In the TDDFT simulation, to ensure the maximal electron excursion along the out-of-plane direction, a simulation box of 200 Å along the $z$ direction is used, which includes an absorbing region of 20 Å on each side of the graphene. The HHG spectrum will be obtained from the total time-dependent electric current $\mathbf{j}(\mathbf{r}, t)$. Figure 8.15(a) presents the HHG spectra from monolayer graphene driven by a spatial homogeneous field (black line) and a spatial inhomogeneous field (blue line). Both spectra show the atomic-like HHG cutoff. By comparing the two cases, we find that the HHG spectrum obtained with the spatial inhomogeneous plasmonic fields shows significant enhancement in the high-energy region of 60 eV–80 eV. The continuous harmonic spectra in this region (between red arrows in Fig. 8.15(a)) support an ultrashort pulse with a time width of 280 fs, as shown in Fig. 8.15(b). Therefore, the HHG in 2D materials by plasmonic fields may provide an effective way to produce isolated attosecond pulses. Compared with atomic

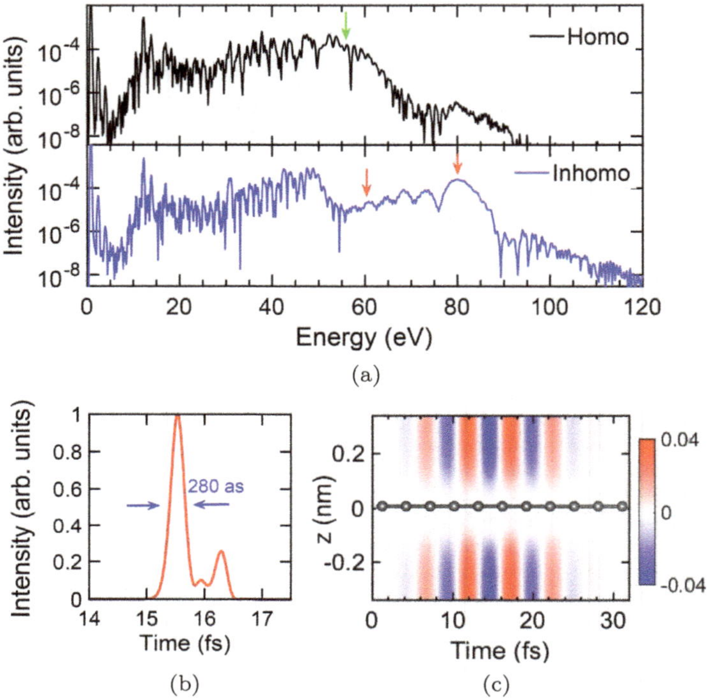


Fig. 8.15 (a) HHG spectra from monolayer graphene ignoring the spatial dependence of plasmonic field (black line) and considering the spatial inhomogeneity (blue line). The green arrow shows the atomic-like HHG cutoff position. (b) The time profile of the generated ultrashort pulse obtained by performing inverse Fourier transformation of the region between red arrows in (a). (c) Total electric field distribution along the out-of-plane direction (z-axis) near the gap's center. The monolayer graphene is placed at $z = 0$. Reproduced with permission from [53].

or molecular gas targets, for the out-of-plane polarization in 2D materials, the "local field effects" [57,58] will be induced and play an important role for the HHG process. The total electric field distribution along the $z$ direction near the center of the dimers' gap is shown in Fig. 8.15(c).

Time-frequency analysis [59] provides a tool to distinguish the harmonics emission at different moments. To obtain physical insights of the enhancement of harmonics observed in Fig. 8.15(a), Figs. 8.16(a) and (b) present the time-frequency analysis of HHG spectra from monolayer graphene driven by the field without or with the consideration of the spatial

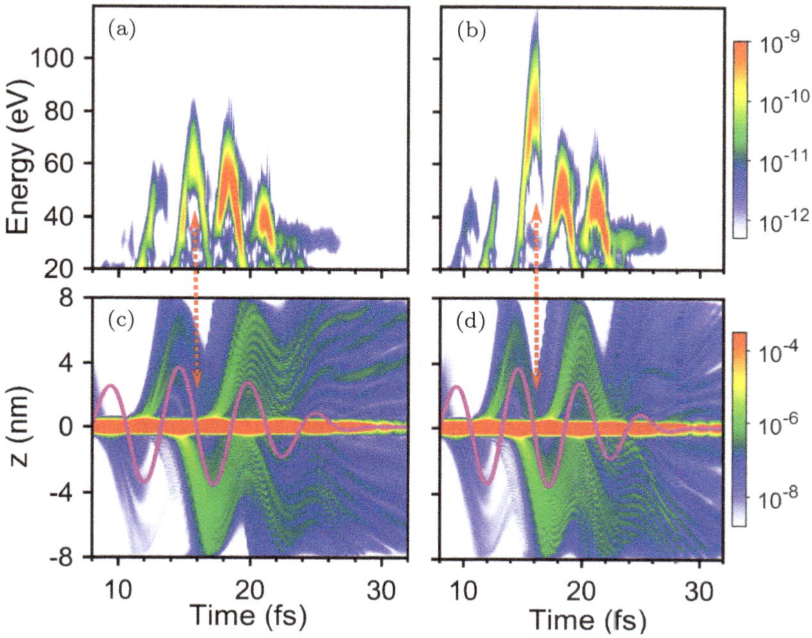


Fig. 8.16 The time-frequency analyses of HHG spectra from monolayer graphene (a) ignoring the spatial dependence of plasmonic field (corresponding to the black line in Fig. 8.15(a)) and (b) considering the spatial inhomogeneity (corresponding to the blue line in Fig. 8.15(a)). The evolution of the induced electronic density, integrated in the plane of 2D material (c) ignoring the spatial dependence of plasmonic field (corresponding to (a)) and (d) considering the spatial inhomogeneity (corresponding to (b)). The magenta curves in (c) and (d) represent the time profiles of the driving plasmonic field. Reproduced with permission from [53].

inhomogeneity, respectively. Like in gases, the well-defined electron trajectories that give rise to the different harmonics are observed. We find that the enhancement of HHG yield in the spatial inhomogeneous field mainly comes from the contributions near a special time $t = 16$ fs. We further analyze the corresponding evolution of the induced electronic density along the out-of-plane direction in Figs. 8.16(c) and (d). It is clear that a large amount of the excited electrons collectively return to the monolayer near $t = 16$ fs when the instantaneous spatial inhomogeneous plasmonic field strength is close to zero, which would contribute an ultrashort isolated pulse as shown in Fig. 8.15(b). It suggests that the spatial dependence of plasmonic field is not ignorable but of vital importance for the HHG enhancement.

#### 8.3.2.5 *Strain Effects on HHG in Monolayer Hexagonal Boron Nitride*

Laser control and the band structure modulation are powerful ways to manipulate HHG in solids. Due to the atomic thickness, the electronic and optical properties are sensitive to external perturbations. In this subsection, the stretched strain effects on the symmetry and HHG of the monolayer hBN crystal will be discussed.

The structure of strain-free monolayer hBN is the same as that in Fig. 8.8(a). The minimum direct band gap is located at K point, which is 4.5 eV. Stretching the crystal lattice in the $x$ $[\varepsilon_{uni.(x)}]$ or $y$ $[\varepsilon_{uni.(y)}]$ direction leads to the uniaxial strains, while the biaxial deformation is stretching the lattice in both the $x$ and $y$ directions ($\varepsilon_{bi.}$). Under different strain intensities, the symmetry of the reciprocal space will be changed. One can find that the band structure depends largely on the strength and direction of the applied strains, and the band energies change when the strain is applied. Here, the stretching percentage is limited to the range of 0–6%.

Compared with the strain-free case, the increasing tensile strain always results in the decrease of the minimum band gap. Thus, the harmonic emission may be altered. Figure 8.17(a) shows the total HHG spectrum in monolayer hBN under different stretching strains with $\varepsilon = 0$ and $\varepsilon = 6\%$. It shows that both uniaxial and biaxial strains lead to a slight enhancement in the HHG yield. The strain-dependent $N_{exc}$ of monolayer hBN after the

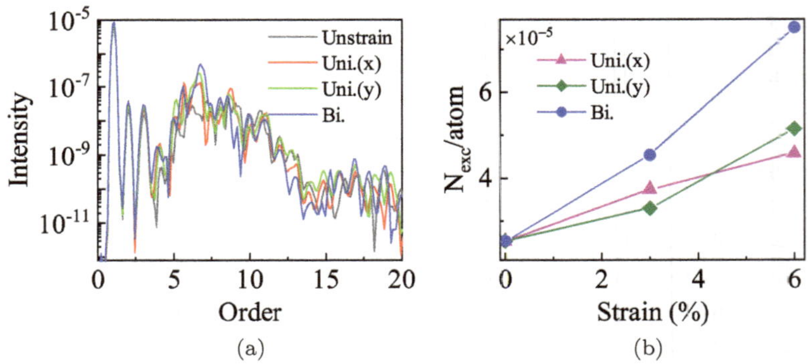


Fig. 8.17 (a) The total HHG spectra in the monolayer hBN with the stretching percent of $\varepsilon = 6\%$ under the linear pulse polarized along the $x$ direction. The laser wavelength is 1600 nm with a peak intensity of $10^{12}$ W/cm$^2$. The gray solid line is for the HHG spectra in the strain-free case. (b) The strain-dependent excited electron number after the laser ends for different strain strengths. Reproduced with permission from [60].

laser field action is shown in Fig. 8.17(b). The $N_{exc}$ under biaxial stretching is larger. It contributes slightly higher HHG intensities than that in the uniaxial strain cases. This is because reducing the band gap makes it relatively easier for electrons to transfer from VB to CB. The finding is similar to the strain-controlled HHG mechanism in silicene, in which both the intraband and interband HHG can be enhanced due to the increase of electronic population in the CB [61].

Obviously, the uniaxial strain will change the rotational symmetry of monolayer hBN. The HHG spectrum can be used to reflect the corresponding symmetry of the deformed structure. Figure 8.18 shows the HHG spectra in monolayer hBN under $\varepsilon_{uni.(y)}$ stretching driven by a bichromatic counter-rotating circularly polarized laser pulse. Such a laser pulse combines a circularly polarized fundamental field of frequency $\omega_1$ (1600 nm) with its counter-rotating second harmonic $\omega_2$ (800 nm) field, and their intensities are set to be $I_1 = I_2 = 5 \times 10^{11}$ W/cm$^2$. The total pulse exhibits a three-fold rosette pattern as shown in the insert of Fig. 8.18. For the strain-free case, governed by the selection rule, it is obvious that the $3m$-th harmonics are forbidden, as described in Sec. 8.3.2.2. However, as the uniaxial stretching percent increases, the three-fold rotation symmetry is broken, and the forbidden $3m$-th harmonics will gradually restore. Comparing with the results of $\varepsilon_{uni.(y)} = 3\%$ and $\varepsilon_{uni.(y)} = 6\%$,

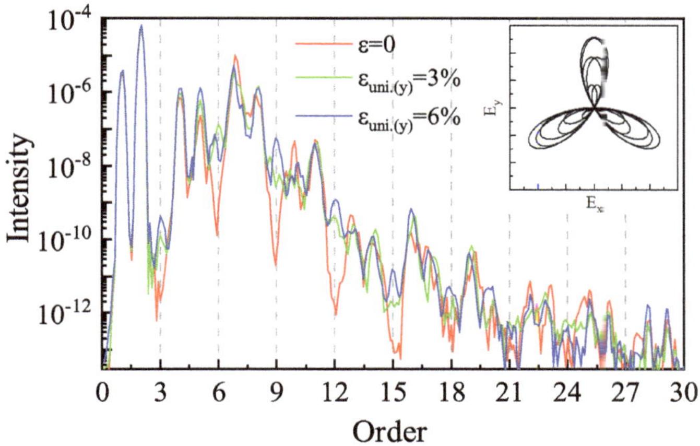


Fig. 8.18 The HHG in the stretching and strain-free monolayer hBN driven by a two-color counter-rotated circularly polarized laser pulse. The insert is the Lissajous curve of the applied laser field. Reproduced with permission from [60].

it can be found that a larger tensile strength gives rise to a stronger $3m$-th harmonic signal, which suggests that applying a stronger external perturbation may lead to a larger deviation. This also shows that the recovery of the $3m$-th harmonic signal can be used to detect the lattice deformation.

### 8.3.3 *Propagation Effects in 3D Crystals*

#### 8.3.3.1 *Nonlinear Propagation and Energy Transfer of Short Pulses in a Film Silicon*

For the process of radiating high-order harmonics from bulk crystal materials, the nonlinear propagation effect might impact the macroscopic characteristics of harmonics. In this subsection, we investigate the interaction between the strong laser field and crystalline silicon. We simulate the nonlinear propagation of the ultrashort laser pulse and the energy transfer in a film silicon with the method introduced in Sec. 8.2.3.1 [22].

As shown in Fig. 8.19, crystalline silicon is an indirect band gap semiconductor with the face-centered cubic structure. Its Wigner-Seitz cell in the reciprocal space is a truncated octahedron. In the simulation, we first get the TBM Hamiltonian in real space from DFT calculation.

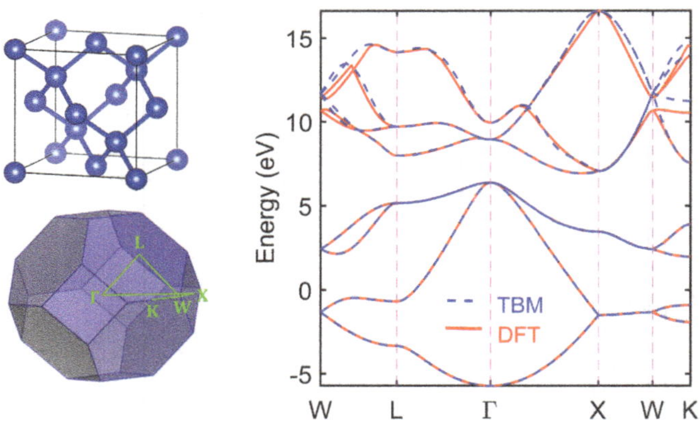


(a) (b)

Fig. 8.19 (a) The face-centered cubic geometry structure of the crystalline silicon in real space (the upper subfigure) and its Wigner-Seitz cell in the reciprocal space (the bottom subfigure). (b) The energy band structures of silicon along the path W-L-Γ-X-W-K from the TBM calculation (the blue dashed lines) and the DFT calculation (the red solid lines). Reproduced with permission from [22].

After the DFT calculation, we use the $sp^3$ hybrid orbitals of silicon atom to calculate the matrix elements of the TBM Hamiltonian by the Wannier90 package [62]. As the result shows in Fig. 8.19(b), the energy bands from the TBM calculation (blue dashed lines) and that from the DFT calculation (red solid lines) agree well with each other. It means that the TBM Hamiltonian we obtained is accurate.

For the multiscale time-dependent calculation, a silicon thin film with a thickness of 100 nm is considered and the macroscopic gird size is $\Delta Z = 5\,\text{nm}$ for the FDTD calculation. We take a linearly polarized laser field with a wavelength of 800 nm, a peak intensity of $1 \times 10^{12}\,\text{W/cm}^2$, and a pulse duration of 15 fs to irradiate the thin film along the $z$ direction (see Fig. 8.1). In the SBE simulation, the reciprocal space is discretized into $60 \times 60 \times 60$ grid points. Both the time intervals for the SBE and the FDTD calculations are $\Delta t = 0.1$ a.u. We do not consider the dephasing time in the numerical calculations.

Figure 8.20 shows typical results of the microscopic Maxwell-SBE. The incident pulse is split into the reflected wave and the transmitted wave after the interaction between the electric field and the silicon thin film, as shown in Fig. 8.20(a). We can retrieve the HHG spectra under the multiscale calculation by the Fourier transform of the reflected wave and the transmitted wave as follows,

$$\text{HHG}_{\hat{\mathbf{n}}}^{\text{ref}}(\omega) = |\text{FT}(\mathbf{E}_{\text{ref}}(t) \cdot \hat{\mathbf{n}})|^2, \tag{8.27}$$

$$\text{HHG}_{\hat{\mathbf{n}}}^{\text{tra}}(\omega) = |\text{FT}(\mathbf{E}_{\text{tra}}(t) \cdot \hat{\mathbf{n}})|^2, \tag{8.28}$$

where $\hat{\mathbf{n}}$ is the unit vector of the polarization direction, $\mathbf{E}_{\text{ref}}(t)$ the reflected electric field and $\mathbf{E}_{\text{tra}}(t)$ the transmitted electric field. The corresponding HHG spectra from the contributions of the transmitted field (the blue line) and the reflected field (the black line) are shown in Fig. 8.20(c). It is apparent that the HHG intensity from the transmitted pulse is higher than that from the reflected pulse.

Figures 8.20(b) and 8.20(d) show the time evolution of the electromagnetic field energy at the back surface macroscopic grid point of the silicon film and absorbing energy at the same macroscopic grid point, respectively. During the propagation, the electromagnetic field energy can be calculated as,

$$E_{\text{em}}(t) = \frac{\Omega}{2}[\mathbf{E}^2(t) + c\mathbf{B}^2(t)], \tag{8.29}$$

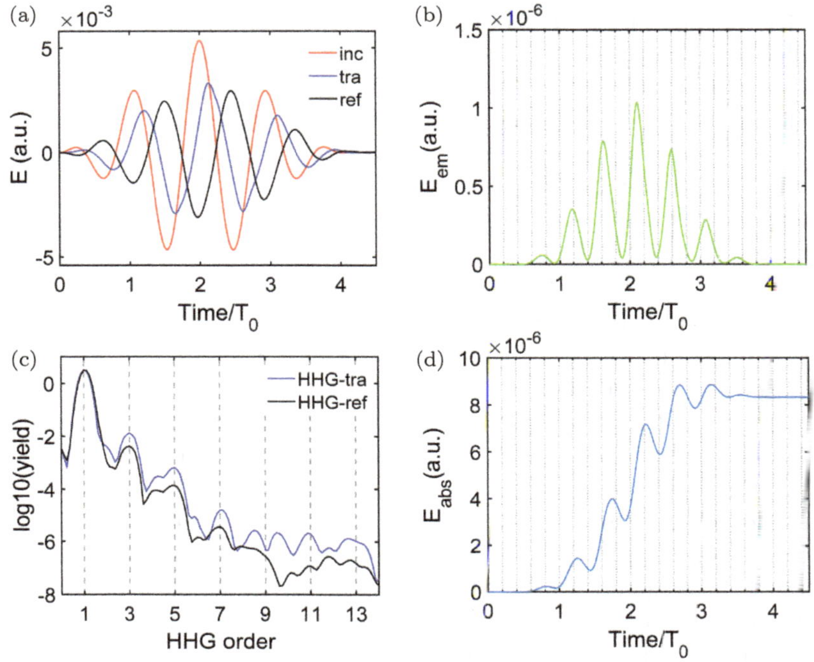


Fig. 8.20 (a) The time evolution of incident wave, reflected wave and transmitted wave (electric field) of silicon thin film. (b) The time evolution of the energy of electromagnetic field at the back surface macroscopic grid point of silicon film (see Fig. 8.1). (c) The HHG spectra from the contributions of transmitted wave (blue line) and reflected wave (black line). (d) The absorbing energy at the back surface macroscopic grid point of silicon film (see Fig. 8.1). Reproduced with permission from [22].

where $\Omega$ is the volume of unit cell in real space. From the two figures, it can be seen that with the increase of electromagnetic field energy, the absorbed energy increases in an oscillating manner.

#### 8.3.3.2 *HHG in $\alpha$-quartz Crystal*

The electronic structure of $\alpha$-quartz, such as the transition dipole moment and the Berry curvature, plays an important role in HHG. In 2018, Luu *et al.* retrieved the Berry curvature of $\alpha$-quartz by high-order harmonic emission [63]. Using our numerical tool, we also study the strong-field nonlinear responses of $\alpha$-quartz crystal.

Figure 8.21 shows the unit cell, the projection of the BZ along the $z$ direction, and the band structure of $\alpha$-quartz. The unit cell contains three

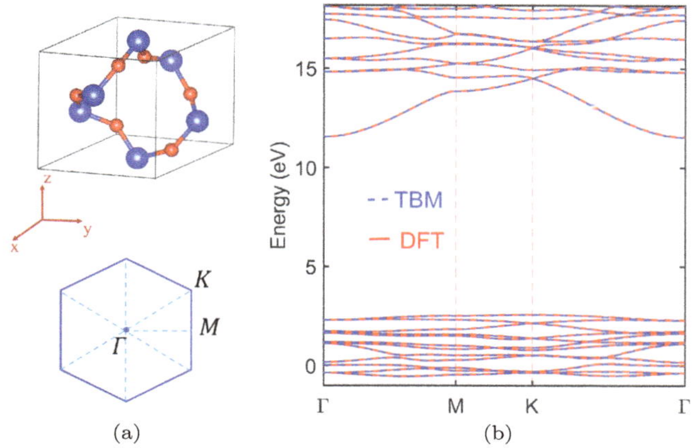


Fig. 8.21 (a) The unit cell of $\alpha$-quartz crystal in real space (the upper subfigure) and the projection of the BZ along $z$ direction for a $z$-cut $\alpha$-quartz crystal (the bottom subfigure). (b) The energy band structures of silicon along the path Γ-M-K-Γ from TBM calculation (blue dashed lines) and DFT calculation (red solid lines). Reproduced with permission from [22].

silicon atoms and six oxygen atoms with lattice constants of 4.913 Å in both the $x$ and $y$ directions, and 5.404 Å in the $z$ direction. We extract the TBM Hamiltonian of $\alpha$-quartz from DFT calculation. A $16 \times 16 \times 16$ Monkhorst-Pack grid is used to sample the BZ. After DFT calculation, the $sp^3$ hybrid orbitals of silicon atom and oxygen atom are used to calculate matrix elements of the TBM Hamiltonian with the Wannier90 package [62]. As shown in Fig. 8.21(b), the band structure from the TBM calculation (blue dashed lines) shows a good agreement with that from the DFT calculation (red solid lines), which ensures the correctness of our calculation.

In the nonlinear multiscale response regime, an $\alpha$-quartz film with a thickness of $3.2\,\mu$m is considered, and the macroscopic grid size for the FDTD calculation is $\Delta Z = 10\,$nm. A linear-polarized laser pulse is used with frequency of 1.55 eV, peak intensity of $3 \times 10^{12}\,$W/cm$^2$, and pulse duration of 30 fs. The direction of the pulse propagation is taken along the $z$ direction, as shown in Fig. 8.1. In the SBE simulation, a $100 \times 100 \times 100$ Monkhorst-Pack $k$-space grid is adopted in the reciprocal unit cell. We take two VBs and six CBs for the time-dependent calculation.

Because the $\alpha$-quartz is almost transparent at the wavelength $\lambda =$ 800 nm, we calculate the HHG spectrum by taking the Fourier transform of the transmitted wave in the subsection. In Fig. 8.22, we compare the

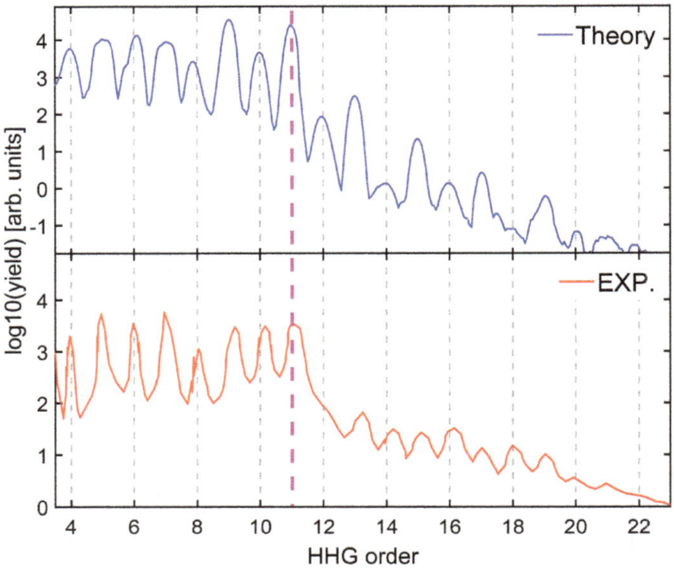


Fig. 8.22 The HHG spectra from our numerical simulation and the experimental results in Ref. [63]. The blue line at the top is our theoretical HHG spectrum while the red line is that presented in Ref. [63]. To compare with the experimental result, the first three order harmonics are not shown in the figure. The laser parameters in both theoretical and experimental results are consistent, that is, central carrier wavelength $\lambda = 800\,\text{nm}$, peak intensity $I_0 = 3 \times 10^{12}\,\text{W/cm}^2$ and pulse duration $\tau = 30\,\text{fs}$. The dephasing time is ignored in our numerical calculations. Reproduced with permission from [22].

calculated HHG with that in the experiment, where the experimental result is redrawn from Ref. [63]. It can be seen that our theoretical result is in good agreement with the experimental result. There are two plateaus in the spectra. The cutoffs of the first platform and the secondary platform are at 17 eV and 32 eV for both cases. It corresponds to around the 11th and 21st harmonics, respectively.

Figure 8.23 shows the polarization-dependent high-order harmonic spectra. When the incident laser field is polarized along the Γ-M direction, as shown in Fig. 8.23(a), the polarization of odd harmonics is perpendicular to that of the even harmonics. Meanwhile, the odd harmonics are polarized parallel to the incident driven pulse. On the contrary, when the incident laser field is polarized along the Γ-K direction, Fig. 8.23(b) shows that the polarization of odd harmonics is parallel to that of the even harmonics. Both the odd and even harmonics are polarized parallel to the incident

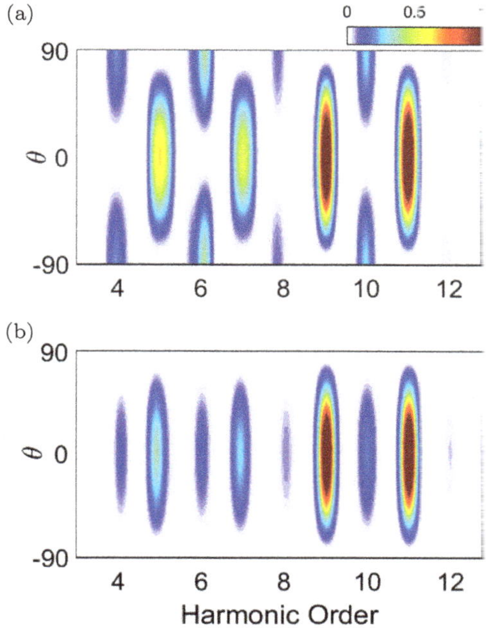


Fig. 8.23 The polarization dependence of HHG spectra. (a) The polarization-dependent HHG spectra with the incident wave polarized along Γ-M direction [see the bottom subfigure in Fig. 8.21(a)]. (b) The polarization-dependent HHG spectra with the incident wave polarized along Γ-K direction. In both (a) and (b), $\theta = 0$ corresponds to the polarization direction of incident wave. Reproduced with permission from [22].

electric field. The two manifestations of the symmetry breaking along these two directions in $\alpha$-quartz are observed, which agrees well with the experimental results in Ref. [63]. Further, for the case of laser polarized along the Γ-M direction, by scanning a series of laser peak intensities, one can measure the Berry curvature of $\alpha$-quartz by even harmonics.

## 8.4 Summary and Outlook

In this chapter, we provide an overview of several extensively employed theoretical models and numerical methods for simulating HHG in solids. We explore their advantages and limitations in dissecting the HHG mechanism. TDSE and SBE rely on the single-electron approximation, making them apt for systems where many-electron correlation effects are not critical. In contrast, the TDDFT method considers many-electron effects to a certain degree and requires more computing resources. By integrating Maxwell equations, we can investigate the propagation effect of the laser pulse and its interplay with thin materials.

We then summarize our recent theoretical studies on HHG in solids, spanning systems from 1D to 3D. We explain the interference phenomenon in the high-energy region of the HHG using the Michelson interference model, and delve into the impact of Floquet-Bloch states by employing the TDSE. Interestingly, we observe that the ultrafast electronic dynamics in 2D systems are more diverse. For instance, the HHG in monolayer graphene under the elliptically polarized laser field shows an unusual ellipticity dependence dominated by the interband current. By adjusting the laser field, we can significantly manipulate the HHG. Taking the monolayer hBN as an example, we find that the HHG depends on the symmetry of both the target and the incident laser field, and that the harmonic yield in the plateau region is markedly enhanced by the two-color laser field. We also examine the enhancement effects of the plasma field and the tensile strain on the HHG in graphene and hBN, respectively. Meanwhile, we introduce the influence of tensile strain on the symmetry of lattice structure. Lastly, we discuss the influence of the propagation effect on HHG in 3D bulk materials.

Based on the content summarized in this chapter, several comprehensive studies can be conducted, offering unique insights for future research. Currently, our focus is on the light-matter interaction at 0 K. An increase in temperature may significantly alter the electronic dynamics, potentially leading to pronounced effects on the HHG. While considering systems driven by intense laser fields, we have not yet accounted for the magnetic effects. Further exploration is required to understand the diverse effects produced by ferromagnetic materials or materials under the magnetic field. In addition, it would be promising to extend TDSE and SBE to more complicated multi-electron systems. This enhancement could boost the universality and accuracy of the methods, facilitating more thorough simulations of complex materials and a deeper understanding of HHG phenomena in solid-state systems.

## Acknowledgements

This work is supported by the National Natural Science Foundation of China (NSFC) under Grant Nos. 12234002, 92250303, and 12104019 and the National Key R&D Program of China under Grant Nos. 2022YFA1603200 and 2022YFA1603201.

## References


[1] M. Lewenstein, P. Balcou, M. Y. Ivanov, A. L'Huillier and P. B. Corkum, Theory of high-harmonic generation by low-frequency laser fields, *Phys. Rev. A* **49**, p. 2117 (1994).

[2] F. Krausz and M. Ivanov, Attosecond physics, *Rev. Mod. Phys.* **81**, p. 163 (2009).

[3] J. Cheng, Z. Xiao-Xin and Z. Song-Feng, Destructive and constructive interference of high-order harmonic generation in mixture of He and Ne gases, *Chinese Phys. Lett.* **27**, p. 033301 (2010).

[4] T. Joyce and A. Jaron-Becker, Macroscopic properties of high-order harmonic generation from molecular ions, *Opt. Lett.* **45**, p. 1954 (2020).

[5] S. Ghimire, A. D. Dichiara, E. Sistrunk, P. Agostini, L. F. Dimauro and D. A. Reis, Observation of high-order harmonic generation in a bulk crystal, *Nat. Phys.* **7**, p. 138 (2011).

[6] N. Tancogne-Dejean, O. D. Mücke, F. X. Kärtner and A. Rubio, Impact of the electronic band structure in high-harmonic generation spectra of solids, *Phys. Rev. Lett.* **118**, p. 087403 (2017).

[7] S. Ghimire and D. A. Reis, High-harmonic generation from solids, *Nat. Phys.* **15**, p. 10 (2019).

[8] O. Zurrón-Cifuentes, R. Boyero-García, C. Hernández-García and L. Plaja, High harmonic generation in armchair carbon nanotubes, *Opt. Express* **28**, p. 19760 (2020).

[9] Y.-T. Zhao, S.-C. Jiang, X. Zhao, J.-G. Chen and Y.-J. Yang, Effect of interband polarization on a solid's high-order-harmonic generation just below the band gap, *Opt. Lett.* **45**, p. 2874 (2020).

[10] T. T. Luu, Z. Yin, A. Jain, T. Gaumnitz, Y. Pertot, J. Ma and H. J. Wörner, Extreme-ultraviolet high-harmonic generation in liquids *Nat. Commun.* **9**, p. 3723 (2018).

[11] A.-W. Zeng and X.-B. Bian, Impact of statistical fluctuations on high harmonic generation in liquids, *Phys. Rev. Lett.* **124**, p. 203901 (2020).

[12] G. Vampa, T. J. Hammond, N. Thiré, B. E. Schmidt, F. Légaré, C. R. McDonald, T. Brabec, D. D. Klug and P. B. Corkum, All-optical reconstruction of crystal band structure, *Phys. Rev. Lett.* **115**, p. 193603 (2015).

[13] M. Schultze, K. Ramasesha, C. Pemmaraju, S. Sato, D. Whitmore, A. Gandman, J. S. Prell, L. J. Borja, D. Prendergast, K. Yabana, D. M. Neumark and S. R. Leone, Attosecond band-gap dynamics in silicon, *Science* **346**, p. 1348 (2014).

[14] G. Vampa, C. R. McDonald, G. Orlando, D. D. Klug, P. B. Corkum and T. Brabec, Theoretical analysis of high-harmonic generation in solids, *Phys. Rev. Lett.* **113**, p. 073901 (2014).

[15] H. Liu, Y. Li, Y. S. You, S. Ghimire, T. F. Heinz and D. A. Reis, High-harmonic generation from an atomically thin semiconductor, *Nat. Phys.* **13**, p. 262 (2016).

[16] N. Yoshikawa, T. Tamaya and K. Tanaka, High-harmonic generation in graphene enhanced by elliptically polarized light excitation, *Science* **356**, p. 736 (2017).
[17] J. Li, J. Lu, A. Chew, S. Han, J. Li, Y. Wu, H. Wang, S. Ghimire and Z. Chang, Attosecond science based on high harmonic generation from gases and solids, *Nat. Commun.* **11**, p. 2748 (2020).
[18] I. Floss, C. Lemell, G. Wachter, V. Smejkal, S. A. Sato, X.-M. Tong, K. Yabana and J. Burgdörfer, Ab initio multiscale simulation of high-order harmonic generation in solids, *Phys. Rev. A* **97**, p. 011401 (2018).
[19] S. Yamada, T. Otobe, D. Freeman, A. Kheifets and K. Yabana, Propagation effects in high-harmonic generation from dielectric thin films, *Phys. Rev. B* **107**, p. 035132 (2023).
[20] M. Hohenleutner, F. Langer, O. Schubert, M. Knorr, U. Huttner, S. W. Koch, M. Kira and R. Huber, Real-time observation of interfering crystal electrons in high-harmonic generation, *Nature* **523**, p. 572 (2015).
[21] H. Haug and S. W. Koch, *Quantum Theory of the Optical and Electronic Properties of Semiconductors.* World Scientific (2009).
[22] X.-Y. Wu, H. Liang, X.-S. Kong, Q. Gong and L.-Y. Peng, Multiscale numerical tool for studying nonlinear dynamics in solids induced by strong laser pulses, *Phys. Rev. E* **105**, p. 055306 (2022).
[23] C. Liu, Y. Zheng, Z. Zeng and R. Li, Polarization-resolved analysis of high-order harmonic generation in monolayer $MoS_2$, *New J. Phys.* **22**, p. 073046 (2020).
[24] D. Golde, T. Meier and S. W. Koch, High harmonics generated in semiconductor nanostructures by the coupled dynamics of optical inter- and intraband excitations, *Phys. Rev. B* **77**, p. 075330 (2008).
[25] C. Yu, S. Jiang and R. Lu, High order harmonic generation in solids: a review on recent numerical methods, *Adv. Phys. X* **4**, p. 1562982 (2019).
[26] E. Runge and E. K. Gross, Density-functional theory for time-dependent systems, *Phys. Rev. Lett.* **52**, p. 997 (1984).
[27] R. van Leeuwen, Mapping from densities to potentials in time-dependent density-functional theory, *Phys. Rev. Lett.* **82**, p. 3863 (1999).
[28] J.-Z. Jin, H. Liang, X.-R. Xiao, M.-X. Wang, S.-G. Chen, X.-Y. Wu, Q. Gong and L.-Y. Peng, Michelson interferometry of high-order harmonic generation in solids, *J. Phys. B* **51**, p. 16LT01 (2018).
[29] C. Zener, A theory of the electrical breakdown of solid dielectrics, *Proc. R. Soc. Lond. A* **145**, p. 523 (1934).
[30] M. Wu, S. Ghimire, D. A. Reis, K. J. Schafer and M. B. Gaarde, High-harmonic generation from Bloch electrons in solids, *Phys. Rev. A* **91**, p. 043839 (2015).
[31] T. Higuchi, M. I. Stockman and P. Hommelhoff, Strong-field perspective on high-harmonic radiation from bulk solids, *Phys. Rev. Lett.* **113**, p. 213901 (2014).
[32] J.-Z. Jin, X.-R. Xiao, H. Liang, M.-X. Wang, S.-G. Chen, Q. Gong and L.-Y. Peng, High-order harmonic generation from a two-dimensional band structure, *Phys. Rev. A* **97**, p. 043420 (2018).

[33] F. H. M. Faisal and J. Z. Kamiński, Floquet-Bloch theory of high-harmonic generation in periodic structures, *Phys. Rev. A* **56**, p. 748 (1997).

[34] J.-Z. Jin, H. Liang, X.-R. Xiao, M.-X. Wang, S.-G. Chen, X.-Y. Wu, Q. Gong and L.-Y. Peng, Contribution of Floquet-Bloch states to high-order harmonic generation in solids, *Phys. Rev. A* **100**, p. 013412 (2019).

[35] R. Kopold, W. Becker, M. Kleber and G. G. Paulus, Channel-closing effects in high-order above-threshold ionization and high-order harmonic generation, *J. Phys. B* **35**, p. 217 (2002).

[36] G. Vampa, C. R. McDonald, G. Orlando, P. B. Corkum and T. Brabec, Semiclassical analysis of high harmonic generation in bulk crystals, *Phys. Rev. B* **91**, p. 064302 (2015).

[37] S. Fang, S. Carr, M. A. Cazalilla and E. Kaxiras, Electronic structure theory of strained two-dimensional materials with hexagonal symmetry, *Phys. Rev. B* **98**, p. 075106 (2018).

[38] A. H. Castro Neto, F. Guinea, N. M. R. Peres, K. S. Novoselov and A. K. Geim, The electronic properties of graphene, *Rev. Mod. Phys.* **81**, p. 109 (2009).

[39] X.-S. Kong, H. Liang, X.-Y. Wu and L.-Y. Peng, Symmetry analyses of high-order harmonic generation in monolayer hexagonal boron nitride, *J. Phys. B* **54**, p. 124004 (2021).

[40] O. Neufeld, D. Podolsky and O. Cohen, Floquet group theory and its application to selection rules in harmonic generation, *Nat. Commun.* **10**, p. 405 (2019).

[41] O. E. Alon, V. Averbukh and N. Moiseyev, Selection rules for the high harmonic generation spectra, *Phys. Rev. Lett.* **80**, p. 3743 (1998).

[42] C. L. Tang and H. Rabin, Selection rules for circularly polarized waves in nonlinear optics, *Phys. Rev. B* **3**, p. 4025 (1971).

[43] X.-S. Kong, H. Liang, X.-Y. Wu, L. Geng, W.-D. Yu and L.-Y. Peng, Manipulation of the high-order harmonic generation in monolayer hexagonal boron nitride by two-color laser field, *J. Chem. Phys.* **156**, p. 074701 (2022).

[44] S. Watanabe, K. Kondo, Y. Nabekawa, A. Sagisaka and Y. Kobayashi, Two-color phase control in tunneling ionization and harmonic generation by a strong laser field and its third harmonic, *Phys. Rev. Lett.* **73**, p. 2692 (1994).

[45] D. A. Telnov, J. Wang and S.-I. Chu, Two-color phase control of high-order harmonic generation in intense laser fields, *Phys. Rev. A* **52**, p. 3988 (1995).

[46] H. Eichmann, A. Egbert, S. Nolte, C. Momma, B. Wellegehausen, W. Becker, S. Long and J. K. McIver, Polarization-dependent high-order two-color mixing, *Phys. Rev. A* **51**, p. R3414 (1995).

[47] I. J. Kim, C. M. Kim, H. T. Kim, G. H. Lee, Y. S. Lee, J. Y. Park, D. J. Cho and C. H. Nam, Highly efficient high-harmonic generation in an orthogonally polarized two-color laser field, *Phys. Rev. Lett.* **94**, p. 243901 (2005).

[48] N. Dudovich, O. Smirnova, J. Levesque, Y. Mairesse, M. Y. Ivanov, D. Villeneuve and P. B. Corkum, Measuring and controlling the birth of attosecond XUV pulses, *Nat. Phys.* **2**, p. 781 (2006).

[49] D. Shafir, H. Soifer, B. D. Bruner, M. Dagan, Y. Mairesse, S. Patchkovskii, M. Y. Ivanov, O. Smirnova and N. Dudovich, Resolving the time when an electron exits a tunnelling barrier, *Nature* **485**, p. 343 (2012).

[50] P. B. Corkum, Plasma perspective on strong field multiphoton ionization, *Phys. Rev. Lett.* **71**, p. 1994 (1993).

[51] X. Zhang, J. Li, Z. Zhou, S. Yue, H. Du, L. Fu and H.-G. Luo, Ellipticity dependence transition induced by dynamical Bloch oscillations, *Phys. Rev. B* **99**, p. 014304 (2019).

[52] G. Vampa, T. J. Hammond, N. Thire, B. E. Schmidt, F. Legare, C. R. McDonald, T. Brabec and P. B. Corkum, Linking high harmonics from gases and solids, *Nature* **522**, p. 462 (2015).

[53] X.-Y. Wu, H. Liang, X.-S. Kong, Q. Gong and L.-Y. Peng, Enhancement of high-order harmonic generation in two-dimensional materials by plasmonic fields, *Phys. Rev. A* **103**, p. 043117 (2021).

[54] S. A. Sato, K. Yabana, Y. Shinohara, T. Otobe, K.-M. Lee and G. F. Bertsch, Time-dependent density functional theory of high-intensity short-pulse laser irradiation on insulators, *Phys. Rev. B* **92**, p. 205413 (2015).

[55] M. Noda, S. Sato, Y. Hirokawa, M. Uemoto, T. Takeuchi, S. Yamada, A. Yamada, Y. Shinohara, M. Yamaguchi, K. Iida, I. Floss, T. Otobe, K. M. Lee, K. Ishimura, T. Boku, G. Bertsch, K. Nobusada and K. Yabana, SALMON: Scalable ab-initio light-matter simulator for optics and nanoscience, *Comput. Phys. Commun.* **235**, p. 356 (2019).

[56] A. D. Rakić, A. B. Djurišić, J. M. Elazar and M. L. Majewski, Optical properties of metallic films for vertical-cavity optoelectronic devices, *Appl. Opt.* **37**, p. 5271 (1998).

[57] G. Le Breton, A. Rubio and N. Tancogne-Dejean, High-harmonic generation from few-layer hexagonal boron nitride: Evolution from monolayer to bulk response, *Phys. Rev. B* **98**, 16 (2018).

[58] N. Tancogne-Dejean and A. Rubio, Atomic-like high-harmonic generation from two-dimensional materials, *Sci. Adv.* **4**, p. eaao5207 (2018).

[59] C. Figueira de Morisson Faria, M. Dörr, W. Becker and W. Sandner, Time-frequency analysis of two-color high-harmonic generation, *Phys. Rev. A* **60**, p. 1377 (1999).

[60] X.-S. Kong, X.-Y. Wu, L. Geng and W.-D. Yu, Strain effects on high-harmonic generation in monolayer hexagonal boron nitride, *Front. Phys.* **10**, p. 1032671 (2022).

[61] R. Qin and Z. Y. Chen, Strain-controlled high harmonic generation with Dirac fermions in silicene, *Nanoscale* **10**, p. 22593 (2018).

[62] G. Pizzi, V. Vitale, R. Arita, S. Blügel, F. Freimuth, G. Géranton, M. Gibertini, D. Gresch, C. Johnson, T. Koretsune *et al.*, Wannier90 as a community code: new features and applications, *J. Phys. Condens. Matter* **32**, p. 165902 (2020).

[63] T. T. Luu and H. J. Wörner, Measurement of the Berry curvature of solids using high-harmonic spectroscopy, *Nat. Commun.* **9**, p. 916 (2018).

https://doi.org/10.1142/9789811279560_0009

# Chapter 9

# High-Order Harmonic Generation in Semiconductors with Excitonic Effects

Matthias Reichelt[1], Ruixin Zuo[1], Xiaohong Song[2], Weifeng Yang[2] and Torsten Meier[1]

[1] *Department of Physics, Paderborn University, Warburger Strasse 100, D-33098 Paderborn, Germany*
[2] *School of Science, Hainan University, Hainan 570288, China*

## 9.1 Introduction

High-order harmonic generation (HHG) may occur when matter is exposed to very intense electromagnetic radiation. It is part of extreme nonlinear optics in the non-perturbative regime and has been observed in gases, liquids, and solids. Initially, HHG was intensively studied in atomic gases for several decades [1–10]. The first observation of HHG in a bulk solid, where harmonics up to the 25th order have been detected from a ZnO crystal, is of more recent date [11]. Since then, numerous studies of HHG in solids have been performed experimentally and theoretically; see, e.g., [12–24]. It has been found that there are certain similarities but also significant differences between HHG generated in atomic gases and in solid-state systems. Generally, it is possible to create very specific light fields, in particular light pulses in the attosecond regime [6, 8, 10, 25, 26] in both types of systems. The intuitive three-step collision picture developed for atomic HHG [4] explains the general mechanisms that lead to the creation of ultrashort pulses. A strong field ionizes the atom, the freed electron is accelerated by the oscillating laser field, and when it returns to the atomic core it can recombine which results in the emission of high-frequency radiation. This process occurs in a similar way in a solid where the ionization and recombination are closely related to the interband transition. The accelaration, however, is different as the electron still remains in the crystal and the hole can also move, leading to a unique intraband contribution to HHG.

In the periodic lattice of crystalline solids the atoms are very close together. The interaction among them is of crucial importance and results, in particular, in the formation of the electronic band structure. As a result, electrons that are accelerated within their bands lead to significant contributions from intraband currents to the emitted radiation [9,27,28]. It is possible that neighboring atoms lead to scattering and thereby strongly modify the electron and hole trajectories [29, 30]. HHG provides detailed information about the ultrafast electronic dynamics [17,31] and also reveals intrinsic material properties like the band structure [32, 33] in situations where they are not easily accessible by other techniques. Analogous to the atomic case the dynamics of the tunneling process in solids has been revealed recently [34]. The direction dependence of the band structure explains the anisotropy of HHG [33,35], i.e., its dependence on the polarization direction of the incident field, and determines whether the excitation and emission have more localized or delocalized contributions, as has been analyzed transparently in mixed Wannier-Bloch approaches [36, 37].

Generally, solids have a much higher density than atomic gases, which can result in an increased photon yield that may allow one to realize compact light sources with very high intensities. Beyond that, a tremendous advantage of using solids is the huge number of design possibilities. By choosing different materials and material combinations, the physical properties of the system can be tailored and optimized for desired effects or functionalities. Spatial nanostructuring and the confinement of carriers in certain dimensions also have a strong impact on the material characteristics, in particular the electronic and optical properties. For instance, thin layers of transition metal dichalcogenides are known to exhibit large exciton binding energies [38] and are thus a suitable platform to explore excitonic effects. Also, without excitonic effects two-dimensional materials can show a significantly different HHG compared to bulk crystals [39]. Semiconductors are often excited with mid-infrared or terahertz (THz) pulses [17, 21, 40]. This far off-resonant excitation with photon energies way below the band gap allows one to apply very high field amplitudes for short times without damaging the material [41]. Besides, THz fields are also well suited for specific spectroscopic experiments since the energetic separation between different exciton states falls within this range [42].

Regarding computational approaches, the most commonly used numerical methods to compute HHG signals are the time-dependent Schrödinger equation [36,43–46], time-dependent density functional theory [47,48], and the semiconductor Bloch equations (SBE) [17,21,27,28,49]. In this chapter we use the SBE which represent a set of coupled differential equations for the electronic coherences between different bands and the occupations within a band, which have been used intensively for several decades to analyze coherent phototransport effects. For instance, they have been used to describe Bloch oscillations in semiconductor superlattices [50, 51] and predicted the signatures of dynamic localization in optical absorption [52, 53]. The generation of charge and/or spin currents via two-color fields has been studied, including different scattering processes such as the carrier-LO-phonon and carrier-carrier scattering on a microscopic basis, as well as excitonic effects [54–57].

The material properties like the band structure and matrix elements, e.g., the transition dipole and the Coulomb matrix elements, that enter the SBE have to be obtained from band structure calculations or models. Besides simple assumptions like, e.g., the tight-binding model which we will use here, different and more involved approaches have been applied in the past. These include, for instance, $\mathbf{k} \cdot \mathbf{p}$ theory [58] or density functional theory [59]. Both approaches involve a numerical diagonalization which can lead to random-valued $\mathbf{k}$-dependent phases of the wave functions. These phases affect the generally complex-valued matrix elements which need to be properly treated [60]. Also, quantities involving $\mathbf{k}$-derivatives cannot be correctly computed when the wave functions carry arbitrary $\mathbf{k}$-dependent phases. Such terms arise when the SBE are formulated in the length gauge since this leads to terms proportional to $\nabla_{\mathbf{k}}$ in the dynamic equations [52,53,56–58]. Typically, however, the length gauge is favorable over the velocity gauge since convergence can be achieved with a smaller number of bands. Furthermore, distinguishing between inter- and intraband contributions is only possible in the length gauge which therefore allows the development of a transparent physical interpretation of the contributions to HHG [58]. It follows that in the general case a construction of smooth matrix elements, for instance based on Berry connections and curvatures or locally smooth structure gauges, is necessary [58,61–63]. Another advantage of the SBE is the possibility to straightforwardly include many-body effects, i.e., interactions among the photoexcited carriers and also with other quasiparticles, e.g., phonons [49,64].

Although the dynamics is intrinsically coupled, in the length gauge the SBE allow one to distinguish between interband and intraband contribution to the total current. In simple terms the interband part resembles somewhat closer the atomic case whereas the intraband part results from electronic motion in the periodic potential in crystals and is more strongly linked to the band structure of the system [33]. The coupling to neighboring atoms of the crystal leads to a revised picture of the atomic re-collision model which was extended to a four-step process [29, 30]. Here, a so-called "pre-acceleration step", i.e., an acceleration within the band that happens before the ionization process [33], plays an important role.

The SBE that allow the inclusion of light fields ranging from THz to optical frequencies are well suited for calculating HHG of semiconductor systems. Depending on the order of the correlations that are taken into account, the number of bands, and the dimensionality of the system, their numerical evaluation can be rather demanding. Since we want to highlight general effects that arise from excitonic resonances in the following, we reduce the approach to a one-dimensional model with two bands and treat the Coulomb interaction on the level of the time-dependent Hartree-Fock (TDHF) approximation. This approximation includes excitonic effects, but higher-order many-body correlations, e.g., biexcitons and Coulomb scattering, are neglected [49, 64]. In the TDHF the many-body hierarchy is truncated by approximating terms with four operators into products of terms containing two operators [49]. The TDHF-SBE are exact in the linear optical regime, but, they provide an approximate description of the nonlinear response. Even though numerous works exist that explore many-body effects with the SBE, so far only a few studies have analyzed HHG including many-body interactions, e.g., [59, 65, 66]. Besides, it was shown that for a strong, resonant excitation of a system with a weak exciton binding energy, the Coulomb interaction can in good approximation be neglected when the light-matter interaction is the dominating energy scale [65]. The situation changes for far off-resonant excitation where excitonic effects may significantly enhance the harmonic emission [59, 66]. In this chapter, we use a rather simple model to compare and discuss results on HHG with and without excitonic effects for certain interesting scenarios.

## 9.2 Semiconductor Bloch Equations and the One-Dimensional Model

Here, we briefly summarize the main steps of the derivation of the SBE in TDHF approximation, show how solutions of the SBE can be used to obtain HHG spectra, and introduce the one-dimensional two-band model system that we use in this chapter. More detailed descriptions of the theory and analyses of several physical effects can be found in, e.g., [13, 27, 28, 33, 49, 65, 66]. The Hamiltonian we consider here contains three terms $\hat{H} = \hat{H}_0 + \hat{H}_{L-M} + \hat{H}_C$. In second quantization the first part $\hat{H}_0$ has the form

$$\hat{H}_0 = \sum_{\lambda,\mathbf{k}} \varepsilon_{\mathbf{k}}^{\lambda} \hat{a}_{\lambda,\mathbf{k}}^{+} \hat{a}_{\lambda,\mathbf{k}} \tag{9.1}$$

where $\hat{a}_{\lambda,\mathbf{k}}^{+}$ is the creation operator of an electron with wave vector $\mathbf{k}$ in band $\lambda$, the annihilation operator is $\hat{a}_{\lambda,\mathbf{k}}$, respectively, and $\varepsilon_{\mathbf{k}}^{\lambda}$ is the band structure.

The light-matter interaction $\hat{H}_{L-M}$ in dipole approximation is described in length gauge by

$$\hat{H}_{L-M} = -\mathbf{E}(t) \cdot \sum_{\lambda,\lambda',\mathbf{k},\mathbf{k}'} e\mathbf{r}_{\lambda,\lambda'}(\mathbf{k},\mathbf{k}') \hat{a}_{\lambda,\mathbf{k}}^{+} \hat{a}_{\lambda',\mathbf{k}} \tag{9.2}$$

with the electric field $\mathbf{E}(t)$. The matrix elements of the position operator $\mathbf{r}_{\lambda,\lambda'}(\mathbf{k},\mathbf{k}') = (i\delta_{\lambda,\lambda'}\nabla_{\mathbf{k}} + \mathbf{d}_{\lambda,\lambda'}(\mathbf{k}))\delta(\mathbf{k}-\mathbf{k}')$ contain the transition dipole matrix element $\mathbf{d}_{\lambda,\lambda'}(\mathbf{k})$.

The many-body part of the Hamiltonian $\hat{H}_C$ accounts for the Coulomb interaction among the photoexcited carriers.

$$\hat{H}_C = \frac{1}{2} \sum_{\substack{\lambda_1,\lambda_2,\lambda_3,\lambda_4 \\ \mathbf{k}_1,\mathbf{k}_2,\mathbf{k}_3,\mathbf{k}_4}} V_{\mathbf{k}_1,\mathbf{k}_2,\mathbf{k}_3,\mathbf{k}_4}^{\lambda_1,\lambda_2,\lambda_3,\lambda_4} \hat{a}_{\lambda_1,\mathbf{k}_1}^{+} \hat{a}_{\lambda_2,\mathbf{k}_2}^{+} \hat{a}_{\lambda_3,\mathbf{k}_3} \hat{a}_{\lambda_4,\mathbf{k}_4} \tag{9.3}$$

Depending on the level of approximations, often not all summations and combinations of band indices are kept in Eq. (9.3). For instance, terms that do not conserve the number of electrons in each band are neglected and interactions within the bands in the ground state of the system are already included effectively in the band structure $\varepsilon_{\mathbf{k}}^{\lambda}$.

At this point it is convenient to transform the operators into the electron-hole picture [49] where a hole describes an electron that is missing in the valence band. In this electron-hole picture no carriers exist in the unexcited state and an optical excitation of the system can create electrons and holes in pairs, i.e., an electron $\hat{c}_{c,\mathbf{k}}^{+} = \hat{a}_{c,\mathbf{k}}^{+}$ and a hole $\hat{d}_{v,-\mathbf{k}}^{+} = \hat{a}_{v,\mathbf{k}}$. The latter definition of the hole operator shows that a hole corresponds to the absence of an electron in the valence band. Since this procedure interchanges creation and annihilation operators in the valence band, the normal ordering in the Hamiltonian has to be restored. Any occurring constant terms are left out as they do not contribute to the system's dynamics. The relevant quantities are the microscopic polarization $p_{\mathbf{k}}(t) = \langle \hat{d}_{v,-\mathbf{k}} \hat{c}_{c,\mathbf{k}} \rangle$, the electron density $n_{\mathbf{k}}^{e}(t) = \langle \hat{c}_{c,\mathbf{k}}^{+} \hat{c}_{c,\mathbf{k}} \rangle$, and the hole density $n^h_{\mathbf{k}}(t) = \langle \hat{d}^+_{v,-\mathbf{k}} \hat{d}_{v,-\mathbf{k}} \rangle$. Their coupled equations of motion are obtained from the Heisenberg equation. This leads to a coupling of the two-point expectation values to four-point expectation values due to the many-body Coulomb interaction. Setting up further equations of motion results in the well-known hierarchy problem. Two commonly used approaches to systematically truncate the relevant correlation functions are the cluster expansion and the dynamics-controlled truncation scheme; see, e.g., [42, 49, 64, 67, 68].

In this article, we apply the dynamical Hartree-Fock approximation and factorize four-operator terms into products of two-operator terms which results in a closed set of differential equations for $p_{\mathbf{k}}$, $n^e_{\mathbf{k}}$, and $n^h_{\mathbf{k}}$. In section 9.3 we will restrict ourselves to a one-dimensional two-band model which includes the relevant phenomena of HHG. Since the tight-binding approach results in a parabolic dispersion at the band edge and can describe the entire Brillouin zone it is suitable to model semiconductors realistically. In real space it corresponds to a one-dimensional chain of coupled atoms. Using only two bands and a linearly polarized electric field, the vector characters of $\mathbf{E}$ and the dipole matrix element $\mathbf{d}$ are unnecessary and $\mathbf{d} \cdot \mathbf{E}$ can be replaced by the projection along the dipole direction $d\,E$. With these simplifications the one-dimensional TDHF-SBE read [13, 27, 28, 49, 65]

$$\begin{aligned}\frac{\partial}{\partial t} p_k = &-\frac{i}{\hbar}\left(\left(\epsilon_{e,k} - \sum_q V_{|k-q|} n^e_q\right) + \left(\epsilon_{h,k} - \sum_q V_{|k-q|} n^h_q\right)\right) p_k \\ &+ \frac{e}{\hbar} E(t) \nabla_k p_k + \frac{i}{\hbar}(1 - n^e_k - n^h_k)\left[d_k E(t) + \sum_q V_{|k-q|} p_q\right] \\ &+ \left.\frac{\partial p_k}{\partial t}\right|_{\mathrm{corr}}, \end{aligned} \tag{9.4}$$

$$\begin{aligned}\frac{\partial}{\partial t} n^{e/h}_k = &-\frac{2}{\hbar}\mathrm{Im}\left(\left[d_k E(t) + \sum_q V_{|k-q|} p_q\right] p^*_k\right) \\ &+ \frac{e}{\hbar} E(t) \nabla_k n^{e/h}_k + \left.\frac{\partial n^{e/h}_k}{\partial t}\right|_{\mathrm{corr}}. \end{aligned} \tag{9.5}$$

The terms marked with "corr" formally describe all terms beyond a mean field approximation. In the following these terms are approximated by a phenomenological dephasing time $T_2$ and a relaxation time $T_1$, i.e., $\left.\frac{\partial p_k}{\partial t}\right|_{\mathrm{corr}} = -\frac{p_k}{T_2}$ and $\left.\frac{\partial n^{e/h}_k}{\partial t}\right|_{\mathrm{corr}} = -\frac{n^{e/h}_k}{T_1}$. The Coulomb matrix element is taken from a one-dimensional quantum wire model with parabolic confinement potential [69] and is given by

$$V_q = a\sqrt{1/|q|}\exp(-2|q|/\pi). \tag{9.6}$$

The prefactor $a$ includes the effective width of the quantum wire and can be adjusted in our model study in order to obtain a certain exciton binding energy. $q$ is a wave vector from the first Brillouin zone. The electric field in Eqs. (9.4)–(9.5) appears in combination with the dipole matrix element $d_k E(t)$ and the gradients $eE(t)\nabla_k$ where $e$ is the electron charge. While the former terms describe interband transitions, the latter represent intraband excitations, i.e., describe the acceleration of carriers within their band. It is noteworthy to mention that the gradient terms can be eliminated from the equations by using the acceleration theorem and transforming into a time-dependent basis in k-space. In this case the wave vector $k$ has to be replaced by a time-dependent $k(t)$ obeying the relation $d/dt\, k(t) = -(e/\hbar)E(t)$.

After solving Eqs. (9.4)–(9.5) the macroscopic interband polarization $P$ and the intraband current $J$ can be computed via [13, 27, 49]

$$P(t) = \sum_k (d_k\, p_k(t) + c.c.) \quad \text{and} \tag{9.7}$$

$$J(t) = \sum_{k,eh} e v_k^{e/h} n_k(t)^{e/h}, \tag{9.8}$$

where $v_k^{e/h}$ is the group velocity which is determined by the derivative of the electron/hole band dispersion. The Fourier transforms of $P$ and $J$ provide the spectrum of the emitted high-harmonic radiation

$$I_{rad}(\omega) \propto |\omega^2 P(\omega) + i\omega J(\omega)|^2. \tag{9.9}$$

For a weak optical excitation one can compute the linear optical polarizations $p_k^{(1)}(t)$ with Eq. (9.4) by setting $n_k^{e/h} = 0$ and neglecting the gradient terms. From such a calculation the linear optical absorption spectrum is obtained by

$$\alpha(\omega) \propto \mathrm{Im}(P^{(1)}(\omega)/E(\omega)). \tag{9.10}$$

In our simulations we use typical parameters for low-dimensional semiconductor systems. Single transition-metal dichalcogenides layers are known to have strong exciton binding energies in the range of several 100 meV up to about 1 eV, the band gaps are several eV, the lattice constants are several Å and the dipole matrix elements have values of some eÅ [38, 70–72]. In our model calculations we choose a k-independent dipole

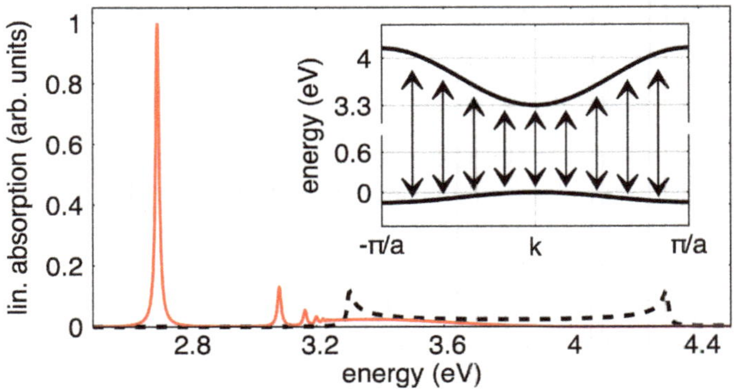


Fig. 9.1 The linear optical absorption spectrum with (solid line) and without (dashed line) excitonic effects. The band gap is 3.3 eV and the 1s exciton peak appears at 2.7 eV which corresponds to a binding energy of 600 meV. The inset displays the tight-binding band structure of the two-band model of a direct semiconductor with a joint band width of 1 eV.

matrix element of 3 eÅ and a lattice constant of $a = 5.65$ Å. The band gap is set to 3.3 eV, the strength of the Coulomb interaction is chosen to result in an exciton binding energy of 0.6 eV, and a two-band dispersion with cosine bands originating from a tight-binding approach is used. The band width of the valence band is 150 meV and of the conduction band 850 meV, i.e., the joint band width amounts to 1 eV, see inset of Fig. 9.1. When the linear absorption Eq. (9.10) is calculated without excitonic effects, i.e., setting $|V_q| = 0$, we obtain the dashed line in Fig. 9.1. In this case, the absorption ranges from 3.3 eV to 4.3 eV and is enhanced at the edges reflecting the corresponding van Hove singularities of the density of states of our one-dimensional model. When the electron-hole attraction is taken into account, i.e., we include the Coulomb matrix elements $V_q$, the spectrum is altered considerably. A very strong 1s exciton peak appears at 2.70 eV, also the next higher excitons (2s and 3s) are visible at 3.08 eV and 3.16 eV, and the absorption in the continuum above the band gap is significantly reduced, see solid line in Fig. 9.1.

## 9.3 Excitonic Enhancement of High Harmonic Generation

In the following we solve the full TDHF-SBE, Eqs. (9.4)–(9.5), with ($V_q$ finite) or vanishing ($V_q = 0$) Coulomb interaction, respectively. In this way it becomes transparent how excitonic effects influence the emitted harmonic radiation. The focus is laid on excitations with multi-THz frequencies, i.e., the emission for incident frequencies well below the band gap is investigated. For instance, the excitation frequency is set to $\omega_L = \omega_{1s}/7 = 2.7\ \text{eV}/7\hbar \approx 0.386\ \text{eV}/\hbar$ which corresponds roughly to 90 THz. The amplitude of the Gaussian-shaped pulse $E(t) = E_0 \exp(-2\ln(2)t^2/\Delta t^2)\cos(\omega_L t)$ is set to 10 MV/cm and the temporal duration to $\Delta t = 300$ fs. The dephasing time is set to $T_2 = 10$ fs [12, 14, 73] and a relaxation time of $T_1 = 500$ fs is used.

In Fig. 9.2(a) a comparison between the radiated high harmonic emission spectra with and without Coulomb interaction is shown. With Coulomb interaction the 7$^{\text{th}}$ harmonic is strongly enhanced by nearly four orders of magnitude. The analogue effect [59, 66] can be seen when exciting with $\omega_L = \omega_{1s}/9 \approx 0.300\ \text{eV}/\hbar$, see Fig. 9.2(b), where the 9$^{\text{th}}$ harmonic is enhanced by more than four orders of magnitude. Generally, if the incident field has a frequency of $\omega_L = \omega_{1s}/n$ with odd $n$, the frequencies of $n$ photons add up to the exciton frequency $\omega_{1s}$ and the 1s exciton state leads to a huge enhancement of the HHG due to its large oscillator strength and its high nonlinearity.

When the excitation frequency is reduced considerably, the effect of excitonic enhancement remains, but the emission pattern changes. In Fig. 9.3 the result for a $\omega_L = \omega_{1s}/27 \approx 0.100\ \text{eV}/\hbar$ pulse is depicted. The 27$^{\text{th}}$ is still enhanced yet neighboring peaks also increase in height.

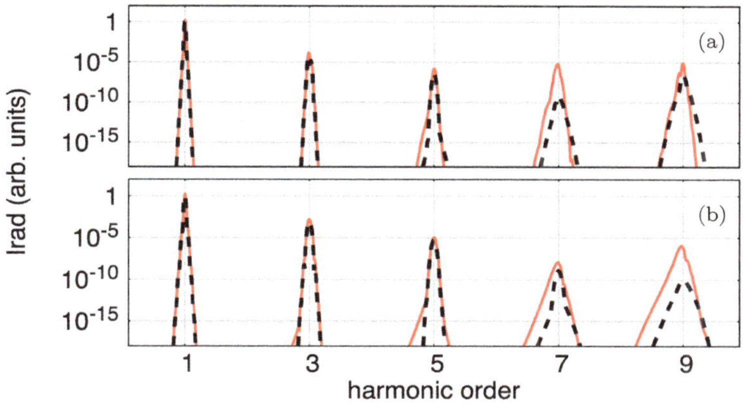


Fig. 9.2 (a) HHG emission with (solid line) and without (dashed line) Coulomb interaction after an excitation with a pulse with a central frequency $\omega_L = \omega_{1s}/7$. The 7$^{\text{th}}$ harmonic is strongly enhanced by excitonic effects. In (b) the result after excitation with a central frequency of $\omega_L = \omega_{1s}/9$ is shown.

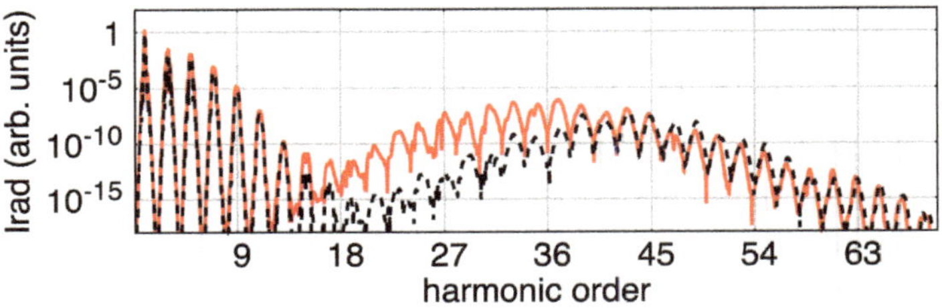


Fig. 9.3 HHG emission with (solid line) and without (dashed line) Coulomb interaction after an excitation with a pulse of frequency $\omega_L = \omega_{1s}/27$. The $15^{\text{th}} \ldots 41^{\text{th}}$ harmonics are enhanced by excitonic effects.

More precisely, the harmonics in the region from $15 \ldots 41$ are enhanced by excitonic effects. Thus for lower THz excitation frequencies corresponding to photon energies significantly smaller than the exciton binding energy the inclusion of the Coulomb interaction substantially increases several harmonic peaks above and below the resonant enhancement. This broad enhancement can be understood to arise from the finite spectral width of each excitation process which broadens the resulting high-order emission.

In HHG the frequencies can mix in a complex way. For instance, the temporal duration of the exciting pulse determines what frequencies are involved in the process. Even though no general behavior of the emission of different harmonics can be predicted *a priori*, some aspects can be ascribed to the pulse spectral. In Fig. 9.2(a) it was found that the $7^{\text{th}}$ harmonic is increased since 7 photons add up to the exciton energy. If instead the frequency of the exciting pulse is slightly reduced from $\omega_{1s}/7$ to $\omega_{1s}/(7+\delta)$, $\delta > 0$, then the sum would be less than $\omega_{1s}$ and should consequently lead to a smaller enhancement than for the exact match. Figure 9.4 displays the $7^{\text{th}}$ harmonic peak for different fractions $f$ of excitation frequencies $\omega_L = \omega_{1s}/f$. The peaks have been obtained from five separate calculations and have been plotted next to each other for comparison. Figure 9.4(a) has been computed for a pulse of temporal width $\Delta t = 1$ ps and shows indeed a slight decrease of the $7^{\text{th}}$ harmonic peak for smaller-excitation frequencies (larger fractions). In Fig. 9.4(b) which has been obtained for $\Delta t = 0.1$ ps the peak height hardly varies with $f$. Shorter pulses contain a broader spectrum and thus the exciton still can be reached quasi-resonantly for all cases shown in Fig. 9.4(b). Even though this reduction effect exists in principle it only becomes recognizable for long-excitation pulses on the order of several 10 cycles and is less essential for short pulses.

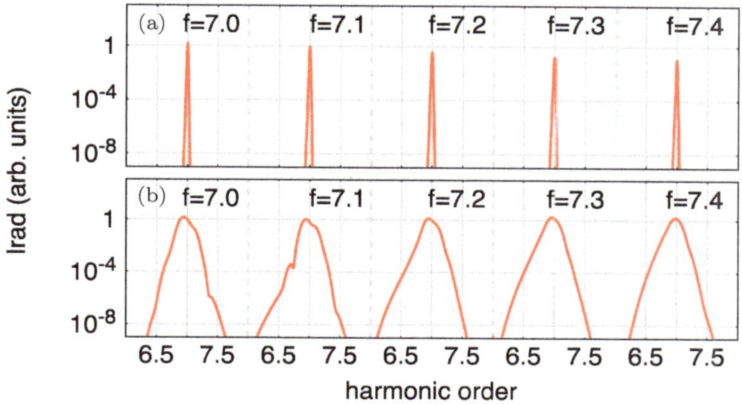


Fig. 9.4 7th harmonic peak when exciting the system with a pulse of frequency $\omega_L = \omega_{1s}/f$. The fraction $f$ has been chosen as $f = 7.0, 7.1, \ldots, 7.4$. For comparison the peaks of harmonic order 7 from individual calculations are displayed for different $f$. The temporal duration of the exciting pulses are (a) $\Delta t = 1$ ps and (b) $\Delta t = 0.1$ ps.

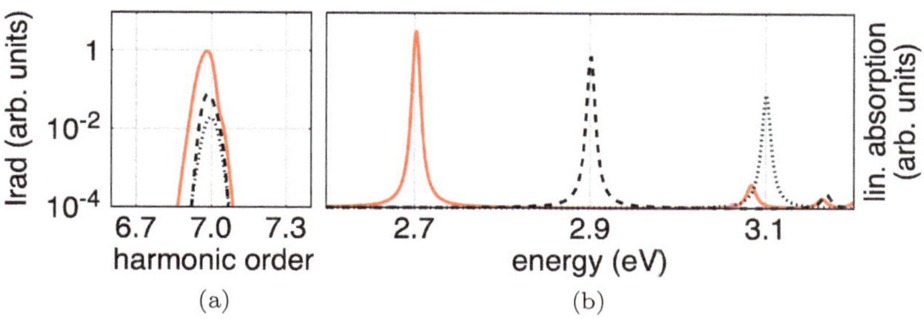


Fig. 9.5 (a) 7th harmonic emission peak for different binding energies of the 1s exciton (solid line 600 meV, dashed line 400 meV, dotted line 200 meV). (b) Corresponding linear absorption spectra. In each calculation the same excitation pulse has been used.

The magnitude of the 1s exciton oscillator strength also contributes to the enhancement factor. The height of the 1s exciton peak is directly linked to the exciton binding energy. When the exciton peak shifts closer to the band gap, its height is also decreased, see Fig. 9.5(b), showing the linear absorption spectra for the binding energies of 600 meV, 400 meV, and 200 meV. Figure 9.5(a) displays the corresponding 7th harmonic peak. Changing the binding energy from 600 to 200 meV reduces the enhancement by a factor of about 50. Moreover, we found that not only the 1s peak alone but also Coulomb nonlinearities are responsible for a part of the enhancement of the signal. In calculations where higher-order Coulomb terms in Eqs. (9.4)–(9.5), i.e., terms proportional to $n^{e/h}V$ and $pp^*V$, have been artifcially switched off, the high harmonic intensity was smaller than for calculations with the full equations. Generally, the HHG is influenced by an intricate interplay between the different Coulomb terms, the binding energy, the band width, and also the excitation conditions. A more precise analysis of the individual contributions will be performed in the future.

All HHG spectra shown so far contain only odd harmonics. This is a consequence of the inversion symmetry of our simple model. Experiments and more realistic models for thin semiconductor layers also exhibit even harmonics [39,74]. Another way to allow the generation of even harmonics is to break the symmetry by additional fields. When an additional static field has a component parallel to the oscillating driving field, even harmonics are actually visible already for relatively weak strengths of the DC field [75]. Here, we consider static electric fields on the order of 1 MV/cm which enter the intraband terms in Eqs. (9.4)–(9.5). In experimental situations quasi-static fields of such magnitude could be realized by using low-frequency THz fields without damaging the material. Figure 9.6 shows high harmonic spectra that are calculated in the presence of a static field with $E_{\text{stat}} = 0.5$ MV/cm. The excitation frequency is $\omega_L = \omega_{1s}/6$ in Fig. 9.6(a) and $\omega_L = \omega_{1s}/8$ in Fig. 9.6(b), respectively. These results demonstrate that an

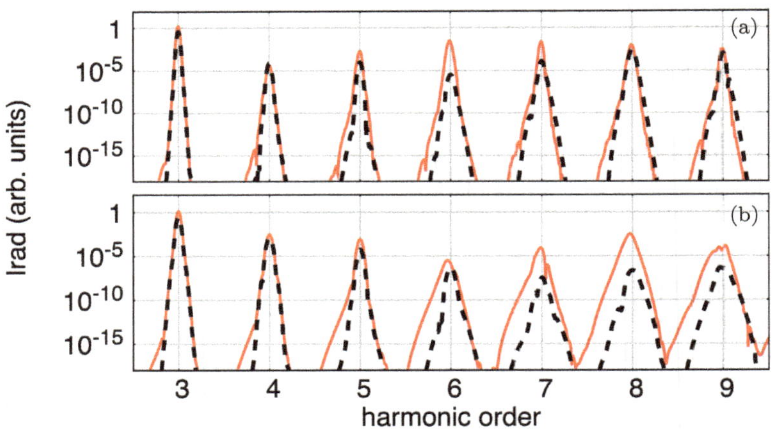


Fig. 9.6 HHG spectra calculated with a static field of 0.5 MV/cm which leads to the generation of even harmonics. The HHG spectra with (solid line) and without (dashed line) Coulomb interaction after an excitation with a pulse of frequency (a) $\omega_L = \omega_{1s}/6$ and (b) $\omega_L = \omega_{1s}/8$, respectively, are shown.

excitonic enhancement also occurs for even harmonics and that it is on the same order of magnitude as for odd harmonics.

## 9.4 Summary and Outlook

We have used the SBE to study the influence of excitonic effects on HHG from semiconductors. We considered a one-dimensional two-band model and solved the equations, including the Coulomb interaction in TDHF approximation. The excitation frequencies were chosen well below the band gap in the THz range and we considered large exciton binding energies as they are realized in transition metal dichalcogenides layers. We find that the Coulomb terms lead to a strong enhancement of the HHG when $n$ photons match the 1s exciton energy. This is true for odd harmonics as well as for even harmonics which appear in the HHG spectra in the presence of an additional static electric field. For small $n$, i.e., when $n$ photons of frequency $\omega_L = \omega_{1s}/n$, or more precisely the frequencies involved in the incident pulse, add up to the exciton energy, an individual peak is predominantly strongly enhanced by several orders of magnitude. For larger $n$, i.e., smaller excitation frequencies, a broader range of neighboring peaks is enhanced.

In the future, it will be interesting to investigate the influence of Coulomb correlation effects, i.e., many-body effects beyond the Hartree-Fock approximation, on the HHG. They are necessary in order to properly describe, e.g., intra-excitonic transitions which can be induced by THz excitation [42]. The SBE can be further extended to also include the carrier-phonon coupling on a microscopic level. All these interaction mechanisms noticeably affect the dephasing and relaxation processes and could result in a more precise prediction of high harmonic radiation than is possible by using phenomenological $T_2$ and $T_1$ times, respectively.

## Acknowledgements

This work was supported by the Deutsche Forschungsgemeinschaft (DFG, German Research Foundation) (project number 231447078 TRR 142/3) (project A10), the National Natural Science Foundation of China (Grant Nos. 12074240, 12264013, 12204136, 12204135 and 91950101), Hainan Provincial Natural Science Foundation of China (Grant Nos. 122CXTD504, 123MS002, 123QN180, 123QN179 and 122QN217), and the Sino-German Mobility Program (Grant No. M-0031). We thank the PC$^2$ (Paderborn Center for Parallel Computing) for a computing time grant.

## References


[1] A. McPherson, G. Gibson, H. Jara, U. Johann, T. S. Luk, I. A. McIntyre, K. Boyer and C. K. Rhodes, Studies of multiphoton production of vacuum-ultraviolet radiation in the rare gases, *J. Opt. Soc. Am. B* **4**, p. 595 (1987).

[2] M. Ferray, A. L'Huillier, X. F. Li, L. A. Lompre, G. Mainfray and C. Manus, Multiple-harmonic conversion of 1064 nm radiation in rare gases, *J. Phys. B* **21**, p. L31 (1988).

[3] J. L. Krause, K. J. Schafer and K. C. Kulander, High-order harmonic generation from atoms and ions in the high intensity regime, *Phys. Rev. Lett.* **68**, p. 3535 (1992).

[4] P. B. Corkum, Plasma perspective on strong field multiphoton ionization, *Phys. Rev. Lett.* **71**, p. 1994 (1993).

[5] M. Lewenstein, P. Balcou, M. Y. Ivanov, A. L'Huillier and P. B. Corkum, Theory of high-harmonic generation by low-frequency laser fields, *Phys. Rev. A* **49**, p. 2117 (1994).

[6] T. Brabec and F. Krausz, Intense few-cycle laser fields: Frontiers of nonlinear optics, *Rev. Mod. Phys.* **72**, p. 545 (2000).

[7] P. M. Paul, E. S. Toma, P. Breger, G. Mullot, F. Augé, P. Balcou, H. G. Muller and P. Agostini, Observation of a train of attosecond pulses from high harmonic generation, *Science* **292**, p. 1689 (2001).

[8] A. Baltuška, T. Udem, M. Uiberacker, M. Hentschel, E. Goulielmakis, C. Gohle, R. Holzwarth, V. S. Yakovlev, A. Scrinzi, T. W. Hänsch and F. Krausz, Attosecond control of electronic processes by intense light fields, *Nature* **421**, p. 611 (2003).

[9] M. Wegener, *Extreme Nonlinear Optics*. Springer (2005).

[10] F. Krausz and M. Ivanov, Attosecond physics, *Rev. Mod. Phys.* **81**, p. 163 (2009).

[11] S. Ghimire, A. D. DiChiara, E. Sistrunk, P. Agostini, L. F. DiMauro and D. A. Reis, Observation of high-order harmonic generation in a bulk crystal, *Nat. Phys.* **7**, p. 138 (2011).

[12] G. Vampa, C. R. McDonald, G. Orlando, D. D. Klug, P. B. Corkum and T. Brabec, Theoretical analysis of high-harmonic generation in solids, *Phys. Rev. Lett.* **113**, p. 073901 (2014).

[13] O. Schubert, M. Hohenleutner, F. Langer, B. Urbanek, C. Lange, U. Huttner, D. Golde, T. Meier, M. Kira, S. W. Koch and R. Huber, Sub-cycle control of terahertz high-harmonic generation by dynamical bloch oscillations, *Nat. Photonics* **8**, p. 119 (2014).

[14] G. Vampa, C. R. McDonald, G. Orlando, P. B. Corkum and T. Brabec, Semiclassical analysis of high harmonic generation in bulk crystals, *Phys. Rev. B* **91**, p. 064302 (2015).

[15] T. T. Luu, M. Garg, S. Y. Kruchinin, A. Moulet, M. T. Hassan and E. Goulielmakis, Extreme ultraviolet high-harmonic spectroscopy of solids, *Nature* **521**, p. 498 (2015).

[16] G. Vampa, T. J. Hammond, N. Thiré, B. E. Schmidt, F. Légaré, C. R. McDonald, T. Brabec and P. B. Corkum, Linking high harmonics from gases and solids, *Nature* **522**, p. 462 (2015).
[17] F. Langer, M. Hohenleutner, C. P. Schmid, C. Poellmann, P. Nagler, T. Korn, C. Schüller, M. S. Sherwin, U. Huttner, J. T. Steiner, S. W. Koch, M. Kira and R. Huber, Lightwave-driven quasiparticle collisions on a subcycle timescale, *Nature* **533**, p. 225 (2016).
[18] G. Ndabashimiye, S. Ghimire, M. Wu, D. A. Browne, K. J. Schafer, M. B. Gaarde and D. A. Reis, Solid-state harmonics beyond the atomic limit, *Nature* **534**, p. 520 (2016).
[19] M. F. Ciappina, J. A. Pérez-Hernández, A. S. Landsman, W. A. Okell, S. Zherebtsov, B. Förg, J. Schötz, L. Seiffert, T. Fennel, T. Shaaran, T. Zimmermann, A. Chacón, R. Guichard, A. Zaïr, J. W. G. Tisch, J. P. Marangos, T. Witting, A. Braun, S. A. Maier, L. Roso, M. Krüger, P. Hommelhoff, M. F. Kling, F. Krausz and M. Lewenstein Attosecond physics at the nanoscale, *Rep. Prog. Phys.* **80**, p. 054401 (2017).
[20] G. Vampa, Y. S. You, H. Liu, S. Ghimire and D. A. Reis, Observation of backward high-harmonic emission from solids, *Opt. Express* **26**, p. 12210 (2018).
[21] F. Langer, C. P. Schmid, S. Schlauderer, M. Gmitra, J. Fabian, P. Nagler, C. Schüller, T. Korn, P. G. Hawkins, J. T. Steiner, U. Huttner, S. W. Koch, M. Kira and R. Huber, Lightwave valleytronics in a monolayer of tungsten diselenide, *Nature* **557**, p. 76 (2018).
[22] S. Ghimire and D. A. Reis, High-harmonic generation from solids, *Nat. Phys.* **15**, p. 10 (2019).
[23] J. Li, J. Lu, A. Chew, S. Han, J. Li, Y. Wu, H. Wang, S. Ghimire and Z. Chang, Attosecond science based on high harmonic generation from gases and solids, *Nat. Commun.* **11**, p. 2748 (2020).
[24] E. Goulielmakis and T. Brabec, High harmonic generation in condensed matter, *Nat. Photonics* **16**, p. 411 (2022).
[25] P. B. Corkum and F. Krausz, Attosecond science, *Nat. Phys.* **3**, p. 381 (2007).
[26] X. Song, S. Yang, G. Wang, J. Lin, L. Wang, T. Meier and W. Yang, Control of the electron dynamics in solid-state high harmonic generation on ultrafast time scales by a polarization-skewed laser pulse, *Opt. Express* **31**, p. 18862 (2023).
[27] D. Golde, T. Meier and S. W. Koch, High harmonics generated in semiconductor nanostructures by the coupled dynamics of optical inter- and intraband excitations, *Phys. Rev. B* **77**, p. 075330 (2008).
[28] D. Golde, M. Kira, T. Meier and S. W. Koch, Microscopic theory of the extremely nonlinear terahertz response of semiconductors, *Phys. Status Solidi B* **248**, p. 863 (2011).
[29] L. Li, P. Lan, X. Zhu, T. Huang, Q. Zhang, M. Lein and P. Lu, Reciprocal-space-trajectory perspective on high-harmonic generation in solids, *Phys. Rev. Lett.* **122**, p. 193901 (2019).

[30] X. Song, S. Yang, R. Zuo, T. Meier and W. Yang, Enhanced high-order harmonic generation in semiconductors by excitation with multicolor pulses, *Phys. Rev. A* **101**, p. 033410 (2020).

[31] R. E. F. Silva, I. V. Blinov, A. N. Rubtsov, O. Smirnova and M. Ivanov, High-harmonic spectroscopy of ultrafast many-body dynamics in strongly correlated systems, *Nat. Photonics* **12**, p. 266 (2018).

[32] G. Vampa, T. J. Hammond, N. Thiré, B. E. Schmidt, F. Légaré, C. R. McDonald, T. Brabec, D. D. Klug and P. B. Corkum, All-optical reconstruction of crystal band structure, *Phys. Rev. Lett.* **115**, p. 193603 (2015).

[33] R. Zuo, A. Trautmann, G. Wang, W.-R. Hannes, S. Yang, X. Song, T. Meier, M. Ciappina, H. T. Duc and W. Yang, Neighboring atom collisions in solid-state high harmonic generation, *Ultrafast Science* **2021**, p. 1 (2021).

[34] R. Zuo, X. Song, S. Ben, T. Meier and W. Yang, Revealing the nonadiabatic tunneling dynamics in solid-state high harmonic generation, *Phys. Rev Res.* **5**, p. L022040 (2023).

[35] Y. S. You, D. Reis and S. Ghimire, Anisotropic high-harmonic generation in bulk crystals, *Nat. Phys.* **13**, p. 345 (2017).

[36] E. N. Osika, A. Chacón, L. Ortmann, N. Suárez, J. A. Pérez-Hernández, B. Szafran, M. F. Ciappina, F. Sols, A. S. Landsman and M. Lewenstein, Wannier-Bloch approach to localization in high-harmonics generation in solids, *Phys. Rev. X* **7**, p. 021017 (2017).

[37] A. M. Parks, G. Ernotte, A. Thorpe, C. R. McDonald, P. B. Corkum, M. Taucer and T. Brabec, Wannier quasi-classical approach to high harmonic generation in semiconductors, *Optica* **7**, p. 1764 (2020).

[38] F. Xia, H. Wang, D. Xiao, M. Dubey and A. Ramasubramaniam, Two-dimensional material nanophotonics, *Nat. Photonics* **8**, p. 899 (2014).

[39] H. Liu, Y. Li, Y. S. You, S. Ghimire, T. F. Heinz and D. A. Reis, High-harmonic generation from an atomically thin semiconductor, *Nat. Phys.* **13**, p. 262 (2017).

[40] N. Yoshikawa, K. Nagai, K. Uchida, Y. Takaguchi, S. Sasaki, Y. Miyata and K. Tanaka, Interband resonant high-harmonic generation by valley polarized electron–hole pairs, *Nat. Commun.* **10**, p. 3709 (2019).

[41] C. Schmidt, J. Bühler, A.-C. Heinrich, J. Allerbeck, R. Podzimski, D. Berghoff, T. Meier, W. G. Schmidt, C. Reichl, W. Wegscheider, D. Brida and A. Leitenstorfer, Signatures of transient Wannier-Stark localization in bulk gallium arsenide, *Nat. Commun.* **9**, p. 2890 (2018).

[42] M. Kira and S. W. Koch, Many-body correlations and excitonic effects in semiconductor spectroscopy, *Prog. Quantum Electron.* **30**, p. 155 (2006).

[43] M. F. Ciappina, J. Biegert, R. Quidant and M. Lewenstein, High-order-harmonic generation from inhomogeneous fields, *Phys. Rev. A* **85**, p. 033828 (2012).

[44] M. Wu, S. Ghimire, D. A. Reis, K. J. Schafer and M. B. Gaarde, High-harmonic generation from Bloch electrons in solids, *Phys. Rev. A* **91**, p. 043839 (2015).

[45] P. G. Hawkins, M. Y. Ivanov and V. S. Yakovlev, Effect of multiple conduction bands on high-harmonic emission from dielectrics, *Phys. Rev. A* **91**, p. 013405 (2015).

[46] T. Ikemachi, Y. Shinohara, T. Sato, J. Yumoto, M. Kuwata-Gonokami and K. L. Ishikawa, Trajectory analysis of high-order-harmonic generation from periodic crystals, *Phys. Rev. A* **95**, p. 043416 (2017).

[47] N. Tancogne-Dejean, O. D. Mücke, F. X. Kärtner and A. Rubio, Impact of the electronic band structure in high-harmonic generation spectra of solids, *Phys. Rev. Lett.* **118**, p. 087403 (2017).

[48] I. Floss, C. Lemell, G. Wachter, V. Smejkal, S. A. Sato, X.-M. Tong, K. Yabana and J. Burgdörfer, Ab initio multiscale simulation of high-order harmonic generation in solids, *Phys. Rev. A* **97**, p. 011401 (2018).

[49] H. Haug and S. W. Koch, *Quantum Theory of the Optical and Electronic Properties of Semiconductors*, 5th edition. World Scientific (2009).

[50] J. Feldmann, K. Leo, J. Shah, D. A. B. Miller, J. E. Cunningham, T. Meier, G. von Plessen, A. Schulze, P. Thomas and S. Schmitt-Rink, Optical investigation of Bloch oscillations in a semiconductor superlattice, *Phys. Rev. B* **46**, p. 7252 (1992).

[51] F. Rossi, T. Meier, P. Thomas, S. W. Koch, P. E. Selbmann and E. Molinari, Ultrafast carrier relaxation and vertical-transport phenomena in semiconductor superlattices: A Monte Carlo analysis, *Phys. Rev. B* **51**, p. 16943 (1995).

[52] T. Meier, F. Rossi, P. Thomas and S. W. Koch, Dynamic localization in anisotropic Coulomb systems: Field induced crossover of the exciton dimension, *Phys. Rev. Lett.* **75**, p. 2558 (1995).

[53] T. Meier, G. von Plessen, P. Thomas and S. W. Koch, Coherent effects induced by static and time-dependent electric fields in semiconductors, *Phys. Rev. B* **51**, p. 14490 (1995).

[54] R. Atanasov, A. Haché, J. L. P. Hughes, H. M. van Driel and J. E. Sipe, Coherent control of photocurrent generation in bulk semiconductors, *Phys. Rev. Lett.* **76**, p. 1703 (1996).

[55] R. D. R. Bhat and J. E. Sipe, Optically injected spin currents in semiconductors, *Phys. Rev. Lett.* **85**, p. 5432 (2000).

[56] H. T. Duc, T. Meier and S. W. Koch, Microscopic analysis of the coherent optical generation and the decay of charge and spin currents in semiconductor heterostructures, *Phys. Rev. Lett.* **95**, p. 086606 (2005).

[57] H. T. Duc, Q. T. Vu, T. Meier, H. Haug and S. W. Koch, Temporal decay of coherently optically injected charge and spin currents due to carrier–lo-phonon and carrier-carrier scattering, *Phys. Rev. B* **74**, p. 165328 (2006).

[58] L. H. Thong, C. Ngo, H. T. Duc, X. Song and T. Meier, Microscopic analysis of high harmonic generation in semiconductors with degenerate bands, *Phys. Rev. B* **103**, p. 085201 (2021).

[59] J. Hader, J. Neuhaus, J. V. Moloney and S. W. Koch, Coulomb enhancement of high harmonic generation in monolayer transition metal dichalcogenides, *Opt. Lett.* **48**, p. 2094 (2023).

[60] S. Jiang, J. Chen, H. Wei, C. Yu, R. Lu and C. D. Lin, Role of the transition dipole amplitude and phase on the generation of odd and even high-order harmonics in crystals, *Phys. Rev. Lett.* **120**, p. 253201 (2018).

[61] K. S. Virk and J. E. Sipe, Semiconductor optics in length gauge: A general numerical approach, *Phys. Rev. B* **76**, p. 035213 (2007).

[62] D. Vanderbilt, *Berry Phases in Electronic Structure Theory: Electric Polarization, Orbital Magnetization and Topological Insulators*. Cambridge University Press (2018).

[63] L. Yue and M. B. Gaarde, Structure gauges and laser gauges for the semiconductor Bloch equations in high-order harmonic generation in solids, *Phys. Rev. A* **101**, p. 053411 (2020).

[64] M. Kira and S. W. Koch, *Semiconductor Quantum Optics*. Cambridge University Press (2011).

[65] D. Golde, T. Meier and S. W. Koch, Microscopic analysis of extreme nonlinear optics in semiconductor nanostructures, *J. Opt. Soc. Am. B* **23**, p. 2559 (2006).

[66] A. Trautmann, R. Zuo, G. Wang, W.-R. Hannes, S. Yang, L. H. Thong, C. Ngo, J. T. Steiner, M. Ciappina, M. Reichelt, H. T. Duc, X. Song, W. Yang and T. Meier, Microscopic simulations of high harmonic generation from semiconductors, *Proc. SPIE* **11999**, p. 1199909 (2022).

[67] V. M. Axt and A. Stahl, A dynamics-controlled truncation scheme for the hierarchy of density matrices in semiconductor optics, *Z. Phys. B* **93**, p. 195 (1994).

[68] M. Lindberg, Y. Z. Hu, R. Binder and S. W. Koch, $\chi^{(3)}$ formalism in optically excited semiconductors and its applications in four-wave-mixing spectroscopy, *Phys. Rev. B* **50**, p. 18060 (1994).

[69] W. Hoyer, *Quantentheorie zu Exzitonbildung und Photolumineszenz in Halbleitern*, Ph.D. thesis, Philipps-Universität Marburg (2002).

[70] Q. H. Wang, K. Kalantar-Zadeh, A. Kis, J. N. Coleman and M. S. Strano, Electronics and optoelectronics of two-dimensional transition metal dichalcogenides, *Nat. Nanotechnol.* **7**, p. 699 (2012).

[71] D. Y. Qiu, F. H. da Jornada and S. G. Louie, Optical spectrum of $MoS_2$: Many-body effects and diversity of exciton states, *Phys. Rev. Lett.* **111**, p. 216805 (2013).

[72] L. Meckbach, *Microscopic Theory of the Linear and Nonlinear Optical Properties of TMDCs*, Ph.D. thesis, Philipps-Universität Marburg (2020).

[73] T. T. Luu and H. J. Wörner, High-order harmonic generation in solids: A unifying approach, *Phys. Rev. B* **94**, p. 115164 (2016).

[74] J. Cao, F. Li, Y. Bai, P. Liu and R. Li, Inter-half-cycle spectral interference in high-order harmonic generation from monolayer $MoS_2$, *Opt. Express* **29**, p. 4830 (2021).

[75] G. Vampa, T. J. Hammond, M. Taucer, X. Ding, X. Ropagnol, T. Ozaki, S. Delprat, M. Chaker, N. Thiré, B. E. Schmidt, F. Légaré, D. D. Klug, A. Y. Naumov, D. M. Villeneuve, A. Staudte and P. B. Corkum, Strong-field optoelectronics in solids, *Nat. Photonics* **12**, p. 465 (2018).

https://doi.org/10.1142/9789811279560_0010

## Chapter 10

# Solid-State Harmonic Generation Governed by Crystal Symmetry

Chen Qian and Ruifeng Lu

*Institute of Ultrafast Optical Physics, Nanjing University of Science and Technology, China*

## 10.1 Introduction

A large number of statistically significant photons exhibit strong wave property, and many phenomena in strong-field physics can be attributed to the interference of light waves. The nonlinear photocurrent in a material driven by strong laser fields can coherently emit discrete high-order harmonic generation (HHG) [1,2]. Crystal symmetry, including point-group symmetry and time-reversal symmetry, regulates the band structure, chirality [3, 4], ferroelectricity, ferromagnetism and topological geometry [5, 6] of electrons. Furthermore, the nonlinear optical effect is very sensitive to the crystal symmetry [7, 8]. Space-time symmetry of the applied field combined with the crystal symmetry provide strict coherence conditions for photocurrent, which can be recorded by HHG that conforms to the selection rules. The contribution of photocurrent can be divided into interband currents generated by the polarization of electron-hole pairs and intraband currents generated by excited electron wave packets moving in the lattice. In the absence of external fields, the symmetry of band structures and wave functions is prescribed by the crystal symmetry. With the addition of ultrafast oscillating laser fields, a handful of electrons are excited into the conduction band and form paired electric dipoles with holes in the valence band. The dipoles are forced to oscillate in the electric potential formed by the Coulomb and laser fields, coherently producing harmonic radiation. Much intrinsic information about crystal bands can be traced by these moving dipoles.

In this chapter, we derive the selection rule of HHG via the band structure, shift vector [9, 10] and transition dipole moment, and reveal their roles in the photocurrent interference. In addition, we will discuss the orientation dependence, polarization properties, time-frequency analysis and ellipticity dependence of harmonics determined by crystal symmetry [11]. We will clarify universal rules of HHG governed by the crystal symmetry with a fundamental picture.

## 10.2 Theory

Based on the single-electron and dipole approximations, we can directly write the time-dependent Hamiltonian of matter interacting with an external laser field (atomic units are used throughout unless otherwise stated),

$$\hat{H}(t) = \hat{H}_0(t) + \hat{H}_l(t), \tag{10.1}$$

$$\hat{H}_0 = -\frac{1}{2}\nabla_{\mathbf{r}}^2 + \hat{V}(\mathbf{r}), \tag{10.2}$$

$$\hat{H}_l(t) = \hat{\mathbf{r}} \cdot \mathbf{E}(t), \tag{10.3}$$

where $\nabla_{\mathbf{r}}^2$ is the Laplace operator with respect to the electronic coordinate operator $\hat{\mathbf{r}}$, $V(\mathbf{r})$ is the Coulomb potential, and $\mathbf{E}(t)$ is the applied electric field. It can be seen that the laser field only acts on the coordinate operators of a single electron without changing the Hamiltonian $\hat{H}_0$ of the initial system. In crystals, $V(\mathbf{r})$ has the translational symmetry, and electrons in the periodic lattice potential can be described by a wave packet composed of Bloch waves,

$$\psi(\mathbf{r}, t) = \frac{1}{N_c^{1/2}} \sum_m \int_{\mathrm{BZ}} d\mathbf{k} a_{m,\mathbf{k}}(t)\phi_{m,\mathbf{k}}(\mathbf{r}), \tag{10.4}$$

with $\phi_{m,\mathbf{k}}(\mathbf{r}) = e^{i\mathbf{k}\mathbf{r}} u_{m,\mathbf{k}}(\mathbf{r})$ (where $u_{m,\mathbf{k}}(\mathbf{r})$ is the periodic part of the Bloch wave function), $N_c$ is the total number of unit cells, and $a_{m,\mathbf{k}}(t)$ is the time-dependent probability amplitude of the Bloch wave. By using the time-dependent Schrödinger equation $i\frac{\partial \psi(\mathbf{r},t)}{\partial t} = H(t)\psi(\mathbf{r}, t)$, we can obtain

$$i\frac{\partial a_{n,\mathbf{k}'}(t)}{\partial t} = \varepsilon_{n,\mathbf{k}'} a_{n,\mathbf{k}'}(t) + \mathbf{E}(t) \cdot \sum_m \int_{\mathrm{BZ}} d\mathbf{k} a_{m,\mathbf{k}}(t) \langle \phi_{n,\mathbf{k}'} | \hat{\mathbf{r}} | \phi_{m,\mathbf{k}} \rangle, \tag{10.5}$$

where the position operator under Bloch basis can be rewritten as $\langle \phi_{n,\mathbf{k}'} | \hat{\mathbf{r}} | \phi_{m,\mathbf{k}} \rangle = \delta(\mathbf{k}' - \mathbf{k}) \left( -i\delta_{n,m} \nabla_{\mathbf{k}'} + \mathbf{d}_{nm}(\mathbf{k}) \right)$; here the transition dipole matrix element $\mathbf{d}_{nm}(\mathbf{k}) = i \langle u_{n,\mathbf{k}} | \nabla_{\mathbf{k}} | u_{m,\mathbf{k}} \rangle$ is introduced to describe the polarization of electron-hole pairs. Equation (10.5) can be converted to the Houston basis after gauge transformations,

$$i\frac{\partial a_{n,\mathbf{k}(t)}(t)}{\partial t} = \mathbf{E}(t)\cdot\sum_{m}\mathbf{d}_{nm}(\mathbf{k}(t))a_{m,\mathbf{k}(t)}(t)e^{i\int_{-\infty}^{t}\varepsilon_{nm}(\mathbf{k}(\tau))d\tau}, \tag{10.6}$$

where $\varepsilon_{nm}(\mathbf{k}) = \varepsilon_n(\mathbf{k}) - \varepsilon_m(\mathbf{k})$ is the energy difference between bands. The quasi momentum $\mathbf{k}(t)$ of electrons changes adiabatically with the laser field, and the evolution relationship is $\mathbf{k}(t) = \mathbf{K} + \mathbf{A}(t)$. $\mathbf{K}$ is the canonical momentum of the crystal in the absence of field. This is a multi-band coupling equation, in which the probability amplitude of the electron in the $n$th eigenstate is related to other states through the transition dipole moments. Equation (10.6) can be regarded as a linear superposition of these transition processes, which form a statistical ensemble.

We introduce density matrix $\rho_{nm}(\mathbf{K},t) = a^{\dagger}_{n,\mathbf{k}(t)}(t)a_{m,\mathbf{k}(t)}(t)$ to describe the time-dependent evolution of the electronic population. For simplicity, here we consider the population transition that occur mainly between two bands, which can always be described by a two-band equation. The densities of interband and intraband currents obey (we ignore the dephasing time related to coupling between particles):

$$\begin{aligned}\dot{\rho}_{nm}(\mathbf{K},t) = -i\mathbf{E}(t)\cdot\Big[&(\mathbf{d}_{mm}(\mathbf{k}(t)) - \mathbf{d}_{nn}(\mathbf{k}(t)))\,\rho_{nm}(\mathbf{K},t)\\ &+ \mathbf{d}_{mn}(\mathbf{k}(t))f_{nm}(\mathbf{K},t)e^{i\int_{-\infty}^{t}\varepsilon_{mn}(\mathbf{k}(\tau))d\tau}\Big],\end{aligned} \tag{10.7}$$

$$\dot{\rho}_{nn}(\mathbf{K},t) = -i\mathbf{E}(t)\cdot\mathbf{d}_{nm}(\mathbf{k}(t))\rho_{nm}(\mathbf{K},t)e^{i\int_{-\infty}^{t}\varepsilon_{nm}(\mathbf{k}(\tau))d\tau} - \text{c.c.}, \tag{10.8}$$

where $f_{nm}(\mathbf{K},t) = \rho_{nn}(\mathbf{K},t) - \rho_{mm}(\mathbf{K},t)$ is the difference of Fermi-Dirac distribution, and the band index $n \neq m$. Equations (10.7) and (10.8) are the result of the coupling between adiabatic evolution and non-adiabatic tunneling process of the electron density. The time-dependent photocurrent can be divided into interband and intraband components.

$$\mathbf{J}_{nm}(t) = -\frac{1}{N_c}\sum_{\mathbf{K}\in\mathrm{BZ}}\rho_{nm}(\mathbf{K},t)\mathbf{p}_{nm}(\mathbf{k}(t)), \tag{10.9}$$

$$\mathbf{J}_{nn}(t) = -\frac{1}{N_c}\sum_{\mathbf{K}\in\mathrm{BZ}}\rho_{nn}(\mathbf{K},t)\mathbf{p}_{nn}(\mathbf{k}(t)), \tag{10.10}$$

where the momentum operator can be given by $\hat{\mathbf{p}}(\mathbf{k},t) = \partial_{\mathbf{k}}\hat{H}(\mathbf{k},t)$. We have assumed that the evolution of the electron population has no effect on the Coulomb potential. Therefore, the original Hilbert space does not change with the addition of laser fields. The momentum matrix element takes the form $\mathbf{p}_{nm}(\mathbf{k}) = \langle u_{n,\mathbf{k}}|\hat{\mathbf{p}}|u_{m,\mathbf{k}}\rangle$, which can be calculated by

$$\mathbf{p}_{nm}(\mathbf{k}) = i\varepsilon_{nm}(\mathbf{k})\mathbf{d}_{nm}(\mathbf{k}),\ n \neq m, \tag{10.11}$$

$$\mathbf{p}_{nn}(\mathbf{k}) = \nabla_{\mathbf{k}}\varepsilon_n(\mathbf{k}). \tag{10.12}$$

The anomalous velocity induced by the Berry curvature has been included in Eq. (10.9), which comes from first-order correction in the electron transition [5]. The interband and intraband currents can be given as

$$\begin{aligned}\mathbf{J}^a_{nm}(t) = &-\frac{1}{N_c}\sum_{\mathbf{K}\in\mathrm{BZ}}\int_{-\infty}^{t} dt'\varepsilon_{nm}(\mathbf{k}(t))|\mathbf{d}^a_{nm}(\mathbf{k}(t))|\\ &\times\left[\mathbf{E}^b(t')\cdot|\mathbf{d}^b_{mn}(\mathbf{k}(t'))|\right] f_{nm}(\mathbf{K},t)\\ &\times e^{-i\left[S_{\mathrm{dyn}}(\mathbf{K},t,t')+S_{\mathrm{shift}}(\mathbf{K},t,t')+S_{\Delta\mathrm{TDP}}(\mathbf{k}(t))\right]},\end{aligned} \tag{10.13}$$

$$\begin{aligned}\mathbf{J}^a_{nn}(t) = &\frac{1}{N_c}\sum_{\mathbf{K}\in\mathrm{BZ}}\int_{-\infty}^{t} dt'\int_{-\infty}^{t'} dt''\nabla_{\mathbf{k}_a}\varepsilon_n(\mathbf{k}(t))\left[\mathbf{E}^b(t')\cdot|\mathbf{d}^b_{nm}(\mathbf{k}(t'))|\right]\\ &\times\left[\mathbf{E}^b(t'')\cdot|\mathbf{d}^b_{mn}(\mathbf{k}(t''))|\right] f_{nm}(\mathbf{K},t')\\ &\times e^{-i\left[S_{\mathrm{dyn}}(\mathbf{K},t',t'')+S_{\mathrm{shift}}(\mathbf{K},t',t'')+S_{\Delta\mathrm{TDP}}(\mathbf{k}(t'))\right]} + \mathrm{c.c.},\end{aligned} \tag{10.14}$$

where a, b are Cartesian indices, labeling the directions of the currents and the electric field, respectively. $S_{\mathrm{dyn}}(\mathbf{K},t,t') = \int_{t'}^{t}\varepsilon_{mn}(\mathbf{k}(\tau))d\tau$ is the dynamical phase. A shift phase is introduced here as

$$S_{\mathrm{shift}}(\mathbf{K},t,t') = \int_{t'}^{t}\mathbf{E}^b(\tau)\cdot\mathbf{R}^{b,b}_{mn}(\mathbf{k}(\tau))d\tau, \tag{10.15}$$

where the shift vector $\mathbf{R}^{b,b}_{mn}(\mathbf{k}) = \mathbf{d}^b_{mm}(\mathbf{k}) - \mathbf{d}^b_{nn}(\mathbf{k}) - \nabla_{\mathbf{k}_b}\phi^b_{mn}(\mathbf{k})$, formed by the Berry connections and transition dipole phase (TDP), represents the offset of charge centers of different bands. $\phi^a_{mn}(\mathbf{k})$ is the TDP as $\mathbf{d}^a_{mn}(\mathbf{k}) = |\mathbf{d}^a_{mn}(\mathbf{k})|\, e^{i\phi^a_{mn}(\mathbf{k})}$. $|\mathbf{d}^a_{mn}(\mathbf{k})|$ is the transition dipole amplitude (TDA) and denotes the polarization intensity. $S_{\Delta\mathrm{TDP}}(\mathbf{k}) = \phi^a_{mn}(\mathbf{k}) - \phi^b_{mn}(\mathbf{k})$ denotes the difference of transition dipoles phases ($\Delta$TDP) along a and b directions, which comes from the deflection of non-collinear currents relative to the driving field.

Each of the three phase factors is gauge independent. The total current is $\mathbf{J}^a = \sum_{n,m}(\mathbf{J}^a_{nm} + \mathbf{J}^a_{nn})$. From Eqs. (10.13) and (10.14), we notice that the phases of interband and intraband currents have the same form;

thus, the dependence of their coherence on the crystal symmetry is always consistent.

## 10.3 Selection Rules of HHG

One of the most widely studied and robust laws in HHG is its selection rules, which mainly depends on the crystal symmetry and can be divided into two categories. The first type appears in the typical Floquet systems and results from periodic oscillations of laser fields. Photoexcited carriers can display dynamical symmetry and coherently generate harmonic radiation [8]. The other comes directly from the crystal symmetry that is not broken by applied fields. The photocurrents cancel each other out leading to no harmonic generation in some particular directions.

Under the framework of the perturbation theory, the polarization can be expanded by an applied electric field:

$$\mathbf{P}(t) = \epsilon_0 \sum_i \chi_\omega^{(i)} [\mathbf{E}(t)]^i \tag{10.16}$$

with $\epsilon_0$ the vacuum permittvity. $\chi_\omega^{(i)}$ is the susceptibility of order $i$ and only depends on the frequency of the fundamental fields. This primary formula shows that the electric field $\mathbf{E}(t)$ is completely extracted and that the susceptibility is time independent. Thus, Eq. (10.16) cannot describe the time-dependent dynamical evolution of electrons in the lattice.

When an external electric field can be compared with the Coulomb field in crystals, it cannot be regarded as a perturbation. In this case, we need to consider the influence of the time-dependent population of charge density on the nonlinear process. Therefore, compared with the perturbation approximation, a broader theory is expected to treat crystals under external strong laser fields.

The crystal symmetry we considered here include the spatial-inversion, mirror, rotational and time-reversal symmetries. Due to the periodic translational symmetry, crystals are limited to 32 point groups and 122 magnetic point groups. In addition, the laser fields could also contain abundant time-space symmetry. We mainly consider the common monochrome light with different ellipticities. By combining the symmetries of crystals and light fields, we can obtain a wide variety of selection rules for HHG.

Consider a point-group symmetry operation $\hat{G}$ on the Bloch state of electrons,

$$\hat{G}|u_{n,\mathbf{k}}\rangle = |u_{n,G^{-1}\mathbf{k}}\rangle. \tag{10.17}$$

For the transition dipole matrix element, we have

$$\begin{aligned}\hat{G}\mathbf{d}_{nm}^{a}(\mathbf{k})\hat{G}^{\dagger} &= i\hat{G}\langle u_{n,\mathbf{k}}|\partial_{\mathbf{k}_a}|u_{m,\mathbf{k}}\rangle\hat{G}^{\dagger}\\ &= i\langle u_{n,G^{-1}\mathbf{k}}|\hat{G}\partial_{\mathbf{k}_a}\hat{G}^{\dagger}|u_{m,G^{-1}\mathbf{k}}\rangle\\ &= i\langle u_{n,G^{-1}\mathbf{k}}|\partial_{(G^{-1}\mathbf{k})_{a'}}|u_{m,G^{-1}\mathbf{k}}\rangle\\ &= \mathbf{d}_{nm}^{a'}(G^{-1}\mathbf{k}),\end{aligned} \tag{10.18}$$

where $a$, $a' = G^{-1}a$ denote the directions of the transition dipole moments. If the crystal has $G$ symmetry, its Hamiltonian satisfies $\hat{G}\hat{H} = \hat{H}\hat{G}$, and the Bloch wave is thus the eigenstate of $\hat{G}$ as well. Since $\hat{G}$ is unitary, its eigenvalues are complex numbers of modulo 1:

$$\hat{G}|u_{n,\mathbf{k}}\rangle = e^{i\phi_G}|u_{n,\mathbf{k}}\rangle. \tag{10.19}$$

Thus, we obtain $|u_{n,G^{-1}\mathbf{k}}\rangle = e^{i\phi_G}|u_{n,\mathbf{k}}\rangle$. Then,

$$\begin{aligned}\mathbf{d}_{nm}^{a'}(G^{-1}\mathbf{k}) &= i\left\langle u_{n,\mathbf{k}}\left|\partial_{(G^{-1}\mathbf{k})_{a'}}\right|u_{m,\mathbf{k}}\right\rangle\\ &= G_{a'a}\mathbf{d}_{nm}^{a}(\mathbf{k}),\end{aligned} \tag{10.20}$$

in which $G_{a'a}$ is just the ΔTDP between $a$ and $a'$ direction under $\hat{G}$. Moreover, scalar quantities such as the dynamical phase and shift phase are invariant under $\hat{G}$.

With the addition of the light field, we firstly discuss the physical origin of selection rules caused by the dynamical symmetries. Let's combine $\hat{G}$ and the time translation operator:

$$\hat{X} = \hat{G}\cdot\hat{\tau}_N, \tag{10.21}$$

with $\hat{\tau}_N t \equiv t \pm \frac{T_0}{N}$, $T_0 = \frac{2\pi}{\omega_0}$ is the period of the laser field of frequency $\omega_0$. We apply the dynamical symmetry operation to the photocurrent,

$$\hat{X}\mathbf{J}^{a}(\mathbf{K},t)\hat{X}^{\dagger} = \mathbf{J}^{G^{-1}a}(G^{-1}\mathbf{K},\hat{\tau}_N t). \tag{10.22}$$

Every wave vector $\mathbf{K}$ in the lattice is a candidate for harmonic peaks unless symmetry forbids it. For the time-dependent quasi momentum $\mathbf{k}(t) \equiv \mathbf{K} + \mathbf{A}(t)$,

$$\hat{X}\mathbf{k}(t) = G^{-1}\mathbf{K} + \mathbf{A}(\hat{\tau}_N t). \tag{10.23}$$

The action of the laser field may disrupt initial symmetries of crystals, but new dynamical symmetry can be induced. Next, we give some typical cases of the crystal symmetry or dynamical symmetry of systems under monochromatic lights. We will reveal in a unified way the microscopic origin of selection rules in HHG.

### 10.3.1 *Mirror Symmetry*

The mirror reflection associates two symmetric dipoles about the mirror plane. A two-fold dynamical symmetry can be obtained by exciting the dipole pairs with oscillating electric field which is symmetric about the mirror plane [see Fig. 10.1(a)]. Adjoining order-2 temporal operator $\hat{\tau}_2$ with the mirror reflection $\hat{M}_c$, we define the dynamical symmetry operation:

$$\hat{F} = \hat{M}_c \cdot \hat{\tau}_2, \tag{10.24}$$

where $c$ is the normal direction of the mirror plane. Monochromatic lights reverse along $c$ under $\hat{\tau}_2$ (i.e., $\hat{\tau}_2\mathbf{E}^c(t) = -\mathbf{E}^c(t)$), so we have $\hat{F}\mathbf{k}(t) = M_c^{-1}\mathbf{K} + \mathbf{A}(\hat{\tau}_2 t) = M_c^{-1}\mathbf{k}(t)$. Based on the transformation rule given by Eqs. (10.20) and (10.22), we can derive that the interband current can transform as

$$\begin{aligned}\hat{F}\mathbf{J}^a_{nm}(\mathbf{K},t)\hat{F}^\dagger &= \mathbf{J}^{M_c^{-1}a}_{nm}(M_c^{-1}\mathbf{K},\hat{\tau}_2 t)\\ &= -i\varepsilon_{nm}(M_c^{-1}\mathbf{k}(t))\mathbf{d}^{M_c^{-1}a}_{nm}(M_c^{-1}\mathbf{k}(t))\rho_{nm}(M_c^{-1}\mathbf{K},\hat{\tau}_2 t)\\ &= -i\varepsilon_{nm}(\mathbf{k}(t))\mathbf{d}^a_{nm}(\mathbf{k}(t))e^{i\left[\phi^{M_c^{-1}a}_{nm}(\mathbf{k}(t))-\phi^a_{nm}(\mathbf{k}(t))\right]}\rho_{nm}(\mathbf{K},t).\end{aligned} \tag{10.25}$$

When $a \parallel c$, the $\Delta$TDP is $\phi^{M_c^{-1}a}_{nm}(\mathbf{k}) - \phi^a_{nm}(\mathbf{k}) = \pi$. The electron density is invariable under the point-group symmetry operation, so we can obtain

$$\hat{F}\mathbf{J}^a_{nm}(\mathbf{K},t)\hat{F}^\dagger = e^{i\pi}\mathbf{J}^a_{nm}(\mathbf{K},t). \tag{10.26}$$

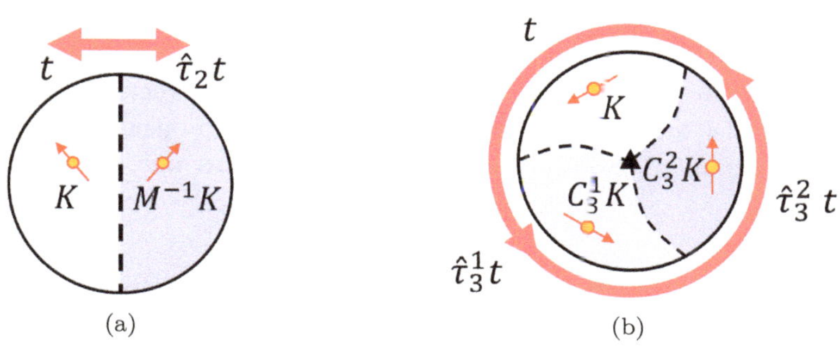


Fig. 10.1 (a) Mirror symmetry combined with order-2 temporal symmetry of light. (b) Rotational symmetry combined with order-3 temporal symmetry of light. Reproduced with permission from [11].

After integration over the entire Brillouin Zone (BZ), the photocurrent satisfies

$$\hat{F}\mathbf{J}^a_{nm}(t)\hat{F}^\dagger = -\mathbf{J}^a_{nm}(t). \tag{10.27}$$

Such reversal of the current only comes from the reflected dipole moment $\mathbf{d}^a_{nm}$. This Floquet system can be reduced to two pairs of electric dipoles with opposing polarization directions and separated by half an optical cycle in the time domain. The photons with frequency $\omega$ become interference enhanced if the current satisfies $\hat{F}\mathbf{J}^a_{nm}(\omega)\hat{F}^\dagger = \mathbf{J}^a_{nm}(\omega)$. Using the Fourier transform that $\mathbf{J}^a_{nm}(\omega) = \int_{-\infty}^{+\infty} dt\mathbf{J}^a_{nm}(t)e^{i\omega t}$, we have

$$e^{i\omega\frac{T_0}{2}} = e^{i\pi} \Rightarrow \omega = (2l+1)\omega_0, \quad l \in \mathbf{N}, \tag{10.28}$$

where $\mathbf{N}$ denotes natural numbers. Considering similar transformation for Eq. (10.14) corresponding to intraband current, we can easily reach the same conclusion. Therefore, when the laser field is perpendicular to the mirror plane, only odd-order harmonics emit in the direction parallel to the driving field. When $a\perp c$, the $\Delta$TDP is $\phi^{M_c^{-1}a}_{nm}(\mathbf{k}) - \phi^a_{nm}(\mathbf{k}) = 0$. The photocurrent satisfies

$$\mathbf{J}^{M_c^{-1}a}(\hat{\tau}_2 t) = \mathbf{J}^a(t). \tag{10.29}$$

This derives only even-order harmonic generation in the direction perpendicular to the driving field ($\omega = 2l\omega_0, l \in \mathbf{N}$).

Appling linearly polarized laser parallel to the mirror plane does not break the mirror symmetry of the system, i.e., $\hat{M}_c\hat{H}(t)\hat{M}_c^\dagger = \hat{H}(t)$. The current perpendicular to the mirror plane meets $\hat{M}_c\mathbf{J}^a(\mathbf{K},t)\hat{M}_c^\dagger = \mathbf{J}^{M_c^{-1}a}(M_c^{-1}\mathbf{K},t) = -\mathbf{J}^a(\mathbf{K},t)$. Here, the reverse current also comes from the reflection of the transition dipole moment. Since the currents on both sides of the mirror are opposite, it is concluded that when the driving light is parallel to the mirror plane, no harmonics are generated in the normal direction of the mirror plane (relevant rules are presented in Table 10.1).

Table 10.1 Selection rule on mirror symmetry.

| | $a \parallel c$ | $a\perp c$ |
|---|---|---|
| $b \parallel c$ | $\Delta$TDP $= \pi$, $\omega = (2l+1)\omega_0$ | $\Delta$TDP $= 0$, $\omega = 2l\omega_0$ |
| $b\perp c$ | no harmonics | integer order harmonics |

### 10.3.2 Rotational Symmetry

Rotation usually takes place in a two-dimensional plane; there are only 5 types of rotation axes in crystals (1,2,3,4,6-fold) due to periodic translational symmetry of lattices. Combining order-$N$ temporal operator $\hat{\tau}_N$ and $N$-fold rotational symmetry operator $\hat{C}_N$, we define a dynamical symmetry operation $\hat{R}_N = \hat{C}_N \cdot \hat{\tau}_N$. When a laser field with $R_N$ symmetry acts on a crystal with $C_N$ symmetry (considering the plane of electric field is always perpendicular to the $C_N$ axis), we have $\hat{R}_N\mathbf{k}(t) = C_N^{-1}\mathbf{K} + \mathbf{A}(\hat{\tau}_N t) = C_N^{-1}\mathbf{k}(t)$, this system can form a dynamical symmetry. The interband current transforms as

$$\begin{aligned}\hat{R}_N\mathbf{J}^a_{nm}(\mathbf{K},t)\hat{R}_N^{\dagger} &= \mathbf{J}^{C_N^{-1}a}_{nm}(C_N^{-1}\mathbf{K},\hat{\tau}_N t)\\ &= -i\varepsilon_{nm}(C_N^{-1}\mathbf{k}(t))\mathbf{d}^{C_N^{-1}a}_{nm}(C_N^{-1}\mathbf{k}(t))\rho_{nm}(C_N^{-1}\mathbf{K},\hat{\tau}_N t)\\ &= -i\varepsilon_{nm}(\mathbf{k}(t))\mathbf{d}^a_{nm}(\mathbf{k}(t))e^{i\left[\phi^{C_N^{-1}a}_{nm}(\mathbf{k}(t))-\phi^a_{nm}(\mathbf{k}(t))\right]}\rho_{nm}(\mathbf{K},t).\end{aligned} \tag{10.30}$$

Let's first consider the in-plane polarization. Since the rotational symmetry operator has two eigenvalues $e^{\pm\, i\frac{2\pi}{N}}$, the ΔTDP should be $\phi^{C_N^{-1}a}_{nm}(\mathbf{k}) - \phi^a_{nm}(\mathbf{k}) = \pm\frac{2\pi}{N}$, denoting the rotation angle of dipoles under $\hat{C}_N$ [see Fig. 10.1(b) for the case of $N = 3$]. Thus, integrating over the entire BZ, we have $\hat{R}_N\mathbf{J}^a_{nm}(t)\hat{R}_N^{\dagger} = e^{\pm\, i\frac{2\pi}{N}}\mathbf{J}^a_{nm}(t)$. Using the interference form of the Fourier transform that $\hat{R}_N\mathbf{J}^a_{nm}(\omega)\hat{R}_N^{\dagger} = \mathbf{J}^a_{nm}(\omega)$, we know that the frequency of photons emitted perpendicular to the rotation axis can only be

$$\omega = (Nl \pm 1)\omega_0, l \in \mathbf{N}, \tag{10.31}$$

which corresponds to co-rotating and counter-rotating photons relative to driving lasers, respectively. A same result can be obtained for the intraband current. For the out-of-plane polarization, we have $\hat{R}_N\mathbf{J}^a(\mathbf{K},t)\hat{R}_N^{\dagger} = \mathbf{J}^a(\mathbf{K},t)$. Thus, the frequency of photons emitted parallel to the rotation axis can only be

$$\omega = Nl\omega_0, l \in \mathbf{N}. \tag{10.32}$$

When the electric field is parallel to the rotation axis, the rotational symmetry of the system is always maintained, i.e., $\hat{C}_N\hat{H}(t)\hat{C}_N^{\dagger} = \hat{H}(t)$. The photocurrent perpendicular to the axis satisfies $\hat{C}_N\mathbf{J}^a(\mathbf{K},t)\hat{C}_N^{\dagger} = \mathbf{J}^{C_N^{-1}a}(C_N^{-1}\mathbf{K},t) = e^{\pm\, i\frac{2\pi}{N}}\mathbf{J}^{C_N^{-1}a}(C_N^{-1}\mathbf{K},t)$. Then we get $e^{\pm\, i\frac{2\pi}{N}} = 1 \Rightarrow N = 1$. In other words, when the electric field is always parallel to the

Table 10.2 Selection rule on rotational symmetry.

| | $a \perp c$ | $a \parallel c$ |
|---|---|---|
| $b \perp c$ | $\Delta\text{TDP} = \pm\frac{2\pi}{N}$, $\omega = (Nl \pm 1)\omega_0$ | $\Delta\text{TDP} = 0$, $\omega = Nl\omega_0$ |
| $b \parallel c$ | no harmonics | integer order harmonics |

$N$-fold ($N \geq 2$) rotation axis, no current generates perpendicular to the axis (relevant rules are presented in Table 10.2).

### 10.3.3 *Inversion Symmetry*

Inversion symmetry is not limited by axes or planes. The second-order nonlinear effect is one of the most common methods to determine the inversion symmetry of crystals. We know that when a monochromatic light field is applied to a system with inversion symmetry, the even-order harmonics are canceled by destructive interference, while the odd-order harmonics show constructive interference.

Let $\hat{P} = \hat{I} \cdot \hat{\tau}_2$, where $\hat{I}$ is the spatial-inversion operation. The interband current under $\hat{P}$ transforms as

$$\begin{aligned}\hat{P}\mathbf{J}^a_{nm}(\mathbf{K},t)\hat{P}^\dagger &= \mathbf{J}^{I^{-1}a}_{nm}(I^{-1}\mathbf{K},\hat{\tau}_2 t)\\ &= -i\varepsilon_{nm}(I^{-1}\mathbf{k}(t))\mathbf{d}^{I^{-1}a}_{nm}(I^{-1}\mathbf{k}(t))\rho_{nm}(I^{-1}\mathbf{K},\hat{\tau}_2 t)\\ &= -i\varepsilon_{nm}(\mathbf{k}(t))\mathbf{d}^a_{nm}(\mathbf{k}(t))e^{i\left[\phi^{I^{-1}a}_{nm}(\mathbf{k}(t))-\phi^a_{nm}(\mathbf{k}(t))\right]}\rho_{nm}(\mathbf{K},t).\end{aligned} \tag{10.33}$$

It follows that

$$\hat{P}\mathbf{J}^a_{nm}(\mathbf{K},t)\hat{P}^\dagger = e^{i\pi}\mathbf{J}^a_{nm}(\mathbf{K},t). \tag{10.34}$$

Integrating over the entire BZ, we then have $\mathbf{J}^{I^{-1}a}_{nm}(\hat{\tau}_2 t) = -\mathbf{J}^a_{nm}(t)$. The reversal of the current comes from the inversed dipole moment. After applying Fourier transform, we know that the radiated photon frequency is $\omega = (2l+1)\omega_0, l \in \mathbf{N}$. The same conclusion can be reached by analyzing the intraband current.

### 10.3.4 *Time-Reversal Symmetry*

We now derive the optical response of crystals with time-reversal symmetry. Considering time-reversal operation $\hat{T}$ on the electron Bloch state,

$$\hat{T}\left|u_{n,\mathbf{k}}\right\rangle = \left|u_{n,-\mathbf{k}}\right\rangle^*. \tag{10.35}$$

Thus, for transition dipole matrix elements,

$$\begin{aligned}\hat{T}\mathbf{d}_{nm}^{a}(\mathbf{k})\hat{T}^{\dagger} &= \hat{T}\left\langle u_{n,\mathbf{k}}\right|i\partial_{\mathbf{k}_a}\left|u_{m,\mathbf{k}}\right\rangle\hat{T}^{\dagger} \\ &= -i\left\langle\partial_{-\mathbf{k}_a}u_{m,-\mathbf{k}}\middle|u_{n,-\mathbf{k}}\right\rangle \\ &= i\left\langle u_{m,-\mathbf{k}}\right|\partial_{-\mathbf{k}_a}\left|u_{n,-\mathbf{k}}\right\rangle \\ &= \mathbf{d}_{mn}^{a}(-\mathbf{k}).\end{aligned} \tag{10.36}$$

If the system has time-reversal symmetry (i.e., $\hat{T}\hat{H} = \hat{H}\hat{T}$), the Bloch wave is the eigenstate of $\hat{T}$ as well. Due to the antiunitarity of $\hat{T}$, its eigenvalues are complex numbers of modulo 1:

$$\hat{T}\left|u_{n,\mathbf{k}}\right\rangle = e^{i\phi_{\mathbf{k}}}\left|u_{n,\mathbf{k}}\right\rangle. \tag{10.37}$$

Therefore, we have $\left|u_{n,-\mathbf{k}}\right\rangle^{*} = e^{i\phi_{\mathbf{k}}}\left|u_{n,\mathbf{k}}\right\rangle$. Then,

$$\begin{aligned}\mathbf{d}_{mn}^{a}(-\mathbf{k}) &= -i\left\langle\partial_{-\mathbf{k}_a}u_{m,-\mathbf{k}}\middle|u_{n,-\mathbf{k}}\right\rangle \\ &= i\left\langle u_{n,\mathbf{k}}\right|\partial\mathbf{k}_a\left|u_{m,\mathbf{k}}\right\rangle \\ &= \mathbf{d}_{nm}^{a}(\mathbf{k}).\end{aligned} \tag{10.38}$$

Similarly, we can obtain the constraints of time-reversal symmetry on the band dispersion and the shift vector, respectively:

$$\hat{T}\varepsilon_{n}(\mathbf{k})\hat{T}^{\dagger} = \varepsilon_{n}(-\mathbf{k}) = \varepsilon_{n}(\mathbf{k}), \tag{10.39}$$

$$\hat{T}\mathbf{R}_{nm}^{b,b}(\mathbf{k})\hat{T}^{\dagger} = \mathbf{R}_{nm}^{b,b}(-\mathbf{k}) = \mathbf{R}_{nm}^{b,b}(\mathbf{k}). \tag{10.40}$$

Similar to exploring the case of inversion symmetry, we define an operator $\hat{U}$ that performs $\hat{T}$ on the crystals, and performs an order-2 temporal operator $\hat{\tau}_2$ on the time. The time-dependent quasi momentum has $\hat{U}\mathbf{k}(t) = -\mathbf{k}(t)$. The dynamical phase, shift phase and ΔTDP transform as

$$\hat{U}S_{\text{dyn}}(\mathbf{K},t,t')\hat{U}^{\dagger} = S_{\text{dyn}}(-\mathbf{K},\hat{\tau}_2 t,\hat{\tau}_2 t') = S_{\text{dyn}}(\mathbf{K},t,t'), \tag{10.41}$$

$$\hat{U}S_{\text{shift}}(\mathbf{K},t,t')\hat{U}^{\dagger} = S_{\text{shift}}(-\mathbf{K},\hat{\tau}_2 t,\hat{\tau}_2 t') = -S_{\text{shift}}(\mathbf{K},t,t\ ), \tag{10.42}$$

$$\hat{U}S_{\Delta\text{TDP}}(\mathbf{k}(t))\hat{U}^{\dagger} = S_{\Delta\text{TDP}}(-\mathbf{k}(t)) = -S_{\Delta\text{TDP}}(\mathbf{k}(t)). \tag{10.43}$$

The interband current under $\hat{U}$ takes

$$\begin{aligned}&\hat{U}\mathbf{J}_{nm}^{a}(\mathbf{K},t)\hat{U}^{\dagger} = \mathbf{J}_{nm}^{a}(-\mathbf{K},\hat{\tau}_2 t) \\ &= -\int_{-\infty}^{\hat{\tau}_2 t} d\hat{\tau}_2 t'\varepsilon_{nm}(-\mathbf{k}(t))|\mathbf{d}_{nm}^{a}(-\mathbf{k}(t))|\left[\mathbf{E}^{b}(\hat{\tau}_2 t')\cdot\left|\mathbf{d}_{mn}^{b}(-\mathbf{k}(t'))\right|\right] \\ &\quad\times f_{nm}(-\mathbf{K},\hat{\tau}_2 t)e^{-i\left[S_{\text{dyn}}\left(-\mathbf{K},\hat{\tau}_2 t,\hat{\tau}_2 t'\right)+S_{\text{shift}}(-\mathbf{K},\hat{\tau}_2 t,\hat{\tau}_2 t')+S_{\Delta\text{TDP}}(-\mathbf{k}(t))\right]}\end{aligned}$$

$$
\begin{aligned}
&= \int_{-\infty}^{t} dt' \varepsilon_{nm}(\mathbf{k}(t)) \left|\mathbf{d}_{mn}^{a}(\mathbf{k}(t))\right| \left[\mathbf{E}^{b}(t') \cdot \left|\mathbf{d}_{nm}^{b}(\mathbf{k}(t'))\right|\right] \\
&\quad \times f_{nm} e^{-i\left[S_{\mathrm{dyn}}(\mathbf{K},t,t') - S_{\mathrm{shift}}(\mathbf{K},t,t') - S_{\Delta\mathrm{TDP}}(\mathbf{k}(t))\right]} \\
&= -\mathbf{J}_{nm}^{a}(\mathbf{K},t) e^{2i\left[S_{\mathrm{shift}}(\mathbf{K},t,t') + S_{\Delta\mathrm{TDP}}(\mathbf{k}(t))\right]}. \qquad (10.44)
\end{aligned}
$$

By comparing the phases on the exponents of Eq. (10.44) and Eq. (10.34), it can be found that the impact of $\hat{P}$ and $\hat{U}$ on the photocurrent differs by the shift phase and ΔTDP. However, by time-reversal symmetry alone, a phase mismatch between the dynamical phase and shift phase as well as between dynamical phase and ΔTDP will be caused, and completely destructive interference cannot be formed. Since the inversion symmetry can support pure coherent odd-order harmonic generation, the shift phase and ΔTDP should be the source of even-order harmonics in crystals with time-reversal symmetry.

A channel of the three-step model in strong-field physics includes excitation, acceleration and re-collision processes of an electron-hole pair. Let's further consider two interference channels driven by a monochromatic laser field as Fig. 10.2 shows. When inversion symmetry exists, the two channels are identical with an interval of half an optical cycle (indicated by

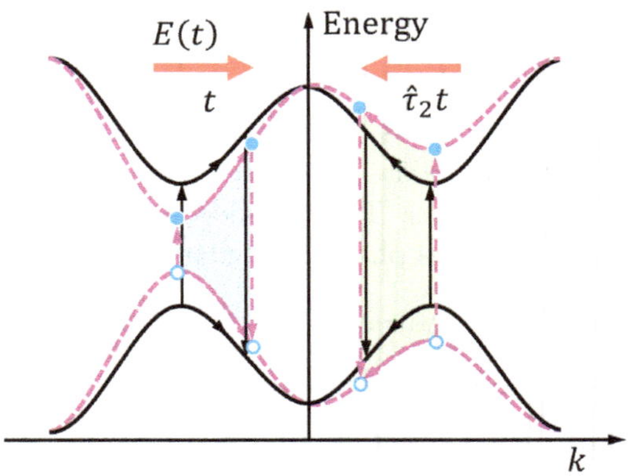


Fig. 10.2 Interference pattern of two channels. The solid black lines are symmetric bands and channels protected by the inversion symmetry. The magenta dashed line is the effective bands deformed by the laser field, its asymmetry coming from the inversion symmetry breaking of crystals. The arrows between or along the bands indicate excitation, acceleration, and recollision of electrons and holes. Different color areas surrounded by the arrows highlight the asymmetry of the channels. Reproduced with permission from [11].

black arrows), their interference leading to pure odd-order harmonics. If the system only possesses time-reversal symmetry, then these two channels are going to be different (indicated by magenta arrows). The band dispersion and TDA remain symmetric in **k**-space due to the protection of time-reversal symmetry. However, due to the existence of the shift vector in non-centrosymmetric crystals, electrons need to do extra work in the photoelectric field when they take interband transitions [10]. Thus, the external light field equivalently modulate the energy curve of electrons, forming the laser-dressed effective bands with symmetry breaking in **k**-space shown by magenta solid curves in Fig. 10.2. The interference condition of pure odd-order harmonics is broken and even-order harmonics can be produced. The movement of electron-hole pairs on the effective bands could accumulate in the dynamical phase and shift phase. In previous works of strong-field physics, only the dynamical phase induced by Coulomb barriers has been considered, but the shift phase induced by external electric barriers has not been discussed. The shift phase and the dynamical phase have fully equivalent effects on photocurrent, and together constitute the photocurrent phase:

$$S_J(\mathbf{K}, t, t') = S_{\text{dyn}}(\mathbf{K}, t, t') + S_{\text{shift}}(\mathbf{K}, t, t') + S_{\Delta\text{TDP}}(\mathbf{k}(t)), \qquad (10.45)$$

which takes the influences of the band structure, shift vector and TDP into account.

This fundamental image of interference involving two channels can be easily generalized to multiple channels. In conclusion, the photocurrent phase and selection rules are determined by the effective band gap between the excited electron-hole pair.

### 10.3.5 *Using Circular Dichroism to Discriminate Time-Reversal Symmetry Breaking*

The inversion symmetry breaking of crystals can be judged by even-order harmonic generation, which arises from interference of non-equivalent currents between two adjacent half-cycles. Similarly, we can use the helicity reversal of laser fields to find evidence of time-reversal symmetry breaking. The helicity of elliptically polarized light can be flipped by the time-reversal operation (without considering the Poynting vector of lights). It is found that the transport processes of charge carriers are different in magnetic materials driven by lasers with different helicity. In experiments, the magnetic circular dichroism of nonlinear optical response has been used to record the magnetic switching of materials [12,13]. The circular dichroism caused by laser fields also have access to selective excitation of spin, valley and chirality of electron states [4, 14, 15].

In the following, we will show that the circular dichroism of HHG can be directly used to identify magnetic materials. Under the time-reversal transformation, the initial right-hand helically polarized laser ($\sigma_+$) is changed to left-hand helically polarized laser ($\sigma_-$), which have $\mathbf{E}_{\sigma_+}(t) = \mathbf{E}_{\sigma_-}(-t)$, and $\mathbf{A}_{\sigma_+}(t) = -\mathbf{A}_{\sigma_-}(-t)$. Each component of the photocurrent phase has the following transformation relation,

$$\begin{aligned}\hat{T}\mathbf{d}_{nm}(\mathbf{K}+\mathbf{A}_{\sigma_+}(t))\hat{T}^\dagger &= \mathbf{d}_{nm}(-\mathbf{K}+\mathbf{A}_{\sigma_-}(-t))\\ &= \mathbf{d}^*_{nm}(\mathbf{K}+\mathbf{A}_{\sigma_+}(t)),\end{aligned} \tag{10.46}$$

$$\begin{aligned}\hat{T}S_{\text{dyn}}(\mathbf{K},\mathbf{A}_{\sigma_+},t,t')\hat{T}^\dagger &= S_{\text{dyn}}(-\mathbf{K},\mathbf{A}_{\sigma_-},-t,-t')\\ &= -S_{\text{dyn}}(\mathbf{K},\mathbf{A}_{\sigma_+},t,t'),\end{aligned} \tag{10.47}$$

$$\begin{aligned}\hat{T}S_{\text{shift}}(\mathbf{K},\mathbf{A}_{\sigma_+},t,t')\hat{T}^\dagger &= S_{\text{shift}}(-\mathbf{K},\mathbf{A}_{\sigma_-},-t,-t')\\ &= -S_{\text{shift}}(\mathbf{K},\mathbf{A}_{\sigma_+},t,t').\end{aligned} \tag{10.48}$$

The dynamical phase, shift phase and TDP all reverse signs under $\hat{T}$, but this cannot be achieved by any pure point-group symmetry. Therefore, the interband current generated by helically polarized lasers transforms as

$$\begin{aligned}&\hat{T}\mathbf{J}^a_{nm,\sigma_+}(\mathbf{K},t)\hat{T}^\dagger = \mathbf{J}^a_{nm,\sigma_-}(-\mathbf{K},-t)\\ &= -\int_{+\infty}^{-t} dt'\varepsilon_{nm}(-\mathbf{K}+\mathbf{A}_{\sigma_-}(-t))\left|\mathbf{d}^a_{nm}(-\mathbf{K}+\mathbf{A}_{\sigma_-}(-t))\right|\\ &\quad\times\left[\mathbf{E}^b_{\sigma_-}(t')\cdot\left|\mathbf{d}^b_{mn}(-\mathbf{K}+\mathbf{A}_{\sigma_-}(t'))\right|\right] f_{nm}(-\mathbf{K},-t)\\ &\quad\times e^{-i\left[S_{\text{dyn}}(-\mathbf{K},\mathbf{A}_{\sigma_-},-t,t')+S_{\text{shift}}(-\mathbf{K},\mathbf{A}_{\sigma_-},-t,t')+S_{\Delta\text{TDP}}(-\mathbf{K}+\mathbf{A}_{\sigma_-}(-t))\right]}\\ &= \int_{-\infty}^{t} dt'\varepsilon_{nm}(\mathbf{K}+\mathbf{A}_{\sigma_+}(t))\left|\mathbf{d}^a_{mn}(\mathbf{K}+\mathbf{A}_{\sigma_+}(t))\right|\\ &\quad\times\left[\mathbf{E}^b_{\sigma_+}(t')\cdot\left|\mathbf{d}^b_{nm}(\mathbf{K}+\mathbf{A}_{\sigma_+}(t'))\right|\right] f_{nm}(-\mathbf{K},-t)\\ &\quad\times e^{-i\left[-S_{\text{dyn}}(\mathbf{K},\mathbf{A}_{\sigma_+},t,t')-S_{\text{shift}}(\mathbf{K},\mathbf{A}_{\sigma_+},t,t')-S_{\Delta\text{TDP}}(\mathbf{K}+\mathbf{A}_{\sigma_+}(t))\right]}\\ &= -\mathbf{J}^{a,*}_{nm,\sigma_+}(\mathbf{K},t)\frac{f_{nm}(-\mathbf{K},-t)}{f_{nm}(\mathbf{K},t)}.\end{aligned} \tag{10.49}$$

If we assume that the difference of Fermi-Dirac distribution is time-reversal invariant (i.e., $f_{nm}(-\mathbf{K},-t) = f_{nm}(\mathbf{K},t)$), the total interband current has $\hat{T}\mathbf{J}^a_{nm,\sigma_+}(t)\hat{T}^\dagger = -\mathbf{J}^{a,*}_{nm,\sigma_+}(t)$. This assumption can work in the perturbation region with low-order changing rate of electron distribution. For the intraband current, the same conclusion can be derived. Then we have $\hat{T}\mathbf{J}^a_{\sigma_+}(t)\hat{T}^\dagger = \mathbf{J}^a_{\sigma_-}(-t) = -\mathbf{J}^a_{\sigma_+}(t)$, so lasers with opposite helicity can produce HHG with the same intensity. We deduce that, under the protection of the time-reversal symmetry of crystals, the intensity of low-order nonlinear optical response does not show the circular dichroism. In contrast, circular dichroism emerges when magnetic materials are considered. It allows us to use helically polarized laser fields to detect the magnetism of crystals. See Sec. 10.4 for more numerical verifications.

## 10.4 Model Calculations

We have theoretically revealed the role of the ΔTDP and the shift phase in selection rules of HHG. Now we perform numerical calculations based on tight-binding models including graphene, hexagonal boron nitride (h-BN) and Haldane model [16]. In this section, we propose to analyze the rules of orientation dependence, polarization characteristics, time-frequency spectra and ellipticity dependence of HHG prescribed by the crystal symmetry.

### 10.4.1 *Orientation Dependence and Polarization Characteristics*

Our discussion focuses on tight-binding models with honeycomb lattice due to its universality. Considering hoppings to nearest-neighbor sites, the Hamiltonian is

$$H = t_1 \sum_{\langle i,j \rangle} c_i^\dagger c_j, \tag{10.50}$$

where $i$, $j$ denote different sublattices. This is a simplest two-band Hamiltonian used to describe graphene, which is subject to $D_{6h}$ point-group and time-reversal symmetries. For calculations, we set the lattice constant to 2.5 Å, and nearest-neighbor hopping to 2.33 eV. The peak intensity of the driving laser we selected is $1.2 \times 10^{12}$ W/cm$^2$, wavelength is 1.9 $\mu$m, and full width at half maximum is 55 fs in a Gaussian envelope.

Figure 10.3 shows the polarization characteristics of harmonics parallel and perpendicular to linearly polarized laser field as a function of crystal orientation. The orientation angle is set to 0° when the laser is along $\Gamma - K$ direction. Pure odd-order harmonics generate due to the inversion symmetry of graphene [see Figs. 10.3(a) and 10.3(b)]. The $C_6$ axis out of plane leads to the orientation periodicity of 60° for all harmonics.

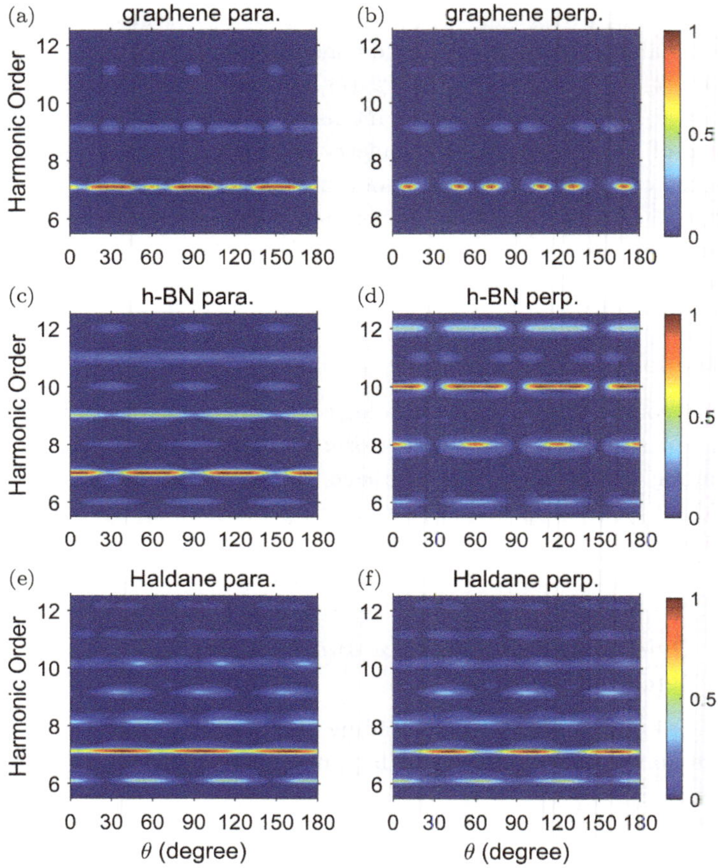


Fig. 10.3 Orientation dependence and polarization characteristics. Parallel and perpendicular components, respectively, of the HHG for (a, b) graphene, (c, d) h-BN, and (e, f) Haldane model. Reproduced with permission from [11].

Let's break the inversion symmetry by introducing different on-site energy for adjacent atoms. The Hamiltonian is

$$H = t_1 \sum_{\langle i,j \rangle} c_i^\dagger c_j + M_0 \sum_i \epsilon_i c_i^\dagger c_i, \tag{10.51}$$

in which $\epsilon_i = \pm 1$ for different atoms. It can be used to describe h-BN, its point-group symmetry reduced to $D_{3h}$. Different from graphene, even harmonics are generated due to the inversion symmetry breaking [see Figs. 10.3(c) and 10.3(d)]. The inversion symmetry breaking term $M_0$

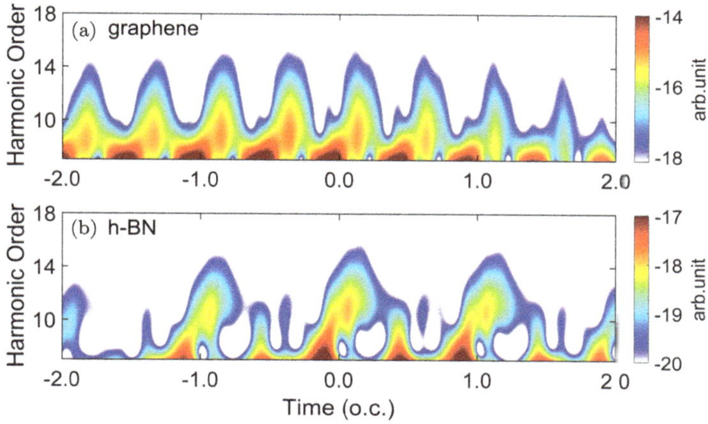


Fig. 10.4 Time-frequency analysis diagram for (a) graphene and (b) h-BN. The linear driving laser is along $\Gamma - M$ direction, and the polarization of harmonics is parallel to the driving laser. Reproduced with permission from [11].

is set to 1.96 eV. The interference processes in time domain can be reflected by time-frequency analysis spectra. Combining Figs. 10.4(a) and 10.4(b), we now know that the even-order harmonic generation results from different harmonic radiations between adjacent half-cycles. As we analyzed above, the inverted laser field every half optical cycle cannot change the dynamical phase and TDA. However, the shift phases formed by two unequal interference channels are different, which is responsible for the different radiations in non-centrosymmetric crystals.

In addition, the mirror symmetry can also induce destructive interference of harmonics as we derived before. When the driving laser is oriented parallel to the mirror plane ($\theta = 30° \pm 60° l$, $l \in \mathbf{N}$), there are no harmonics of perpendicular polarization because the currents cancel each other out. When the laser field is perpendicular to the mirror plane ($\theta = 60° \pm 60° l$, $l \in \mathbf{N}$), it strictly follows that only odd-order harmonics are generated for the parallel polarization and even-order harmonics for the perpendicular polarization. In addition, notice that although h-BN only has an in-plane $C_3$ symmetry, the harmonics can form 6-fold orientation periodicity. This can be completely attributed to the multi-cycle driving field we used; the laser field equivalently shifting half a cycle can hardly change the intensity of harmonics. If a few-cycle driving field is used, such extra 2-fold orientation periodicity of harmonics cannot be observed. Therefore, we need to use a few-cycle driving probe to determine the rotation axis that the crystal has.

To further discuss the effect of the mirror symmetry and time-reversal symmetry breaking, we consider the Haldane model with magnetic phases on the next-nearest neighbor sites. The Hamiltonian is expanded as

$$H = t_1 \sum_{\langle i,j \rangle} c_i^\dagger c_j + M_0 \sum_i \epsilon_i c_i^\dagger c_i + t_2 \sum_{\langle\langle i,j \rangle\rangle} e^{-iv_{ij}\varphi} c_i^\dagger c_j, \tag{10.52}$$

where $v_{ij} = \pm 1$, depending on the kind of atoms that the hopping takes place between. Here, we do not care about the optical behavior of its topological properties, the complex hopping strength considered only to break the mirror symmetry and time-reversal symmetry. The basic point-group symmetry of this Hamiltonian is reduced to $C_{3h}$. The complex hopping strength $t_2 = 0.63$ eV, its phase $\varphi = \frac{\pi}{2}$. Since the mirrors are broken, the stable destructive interference at some orientation angles that occurs in the h-BN case vanishes in the Haldane model [see $\theta = 30^\circ \pm 30^\circ l$, $l \in \mathbf{N}$ in Figs. 10.3(e) and 10.3(f)]. The spectra keep the orientation periodicity of 60°, which arises from the $C_3$ axis and the multi-cycle driving field.

### 10.4.2 *Ellipticity Dependence and Circular Dichroism*

We can apply circularly polarized laser by modulating the ellipticity of driving lasers. The helicity dependence of HHG can be used to probe molecular chirality, which is attributed to the circular dichroism of chiral molecules [17]. This similarly happens in crystals, the mirror symmetry protecting the harmonics from the circular dichroism. Let's apply a helically polarized laser to a crystal with a mirror symmetry. Under the mirror reflection, the applied right-hand helically polarized ($\sigma_+$) laser is changed to left-hand helically polarized ($\sigma_-$) laser. The photocurrent has

$$\hat{M}_c \mathbf{J}^a_{\sigma_+}(t) \hat{M}_c^\dagger = \mathbf{J}^{M_c^{-1}a}_{\sigma_-}(t) = M_{ca} \mathbf{J}^a_{\sigma_+}(t), \tag{10.53}$$

where c is the normal direction of the mirror plane, and $M_{ca} = e^{i\left(\phi_{nm}^{M_c^{-1}a} - \phi_{nm}^a\right)}$. In other words, the mirror reflection of the laser-crystal system directly causes the reflection of photocurrent. Therefore, there is no circular dichroism of HHG in crystals with mirror symmetry. As Fig. 10.5(a) shows, the harmonic spectra of h-BN driven by circularly polarized lights with inverse helicities have the same intensity. Due to the $C_3$ symmetry of h-BN and the circularly polarized driving laser we used here, only $3l \pm 1 (l \in \mathbf{N})$ harmonic orders are allowed.

However, an asymmetric profile of elliptically dependent HHG from cubic crystals has been demonstrated in Refs. [18–20]. Here, we attribute

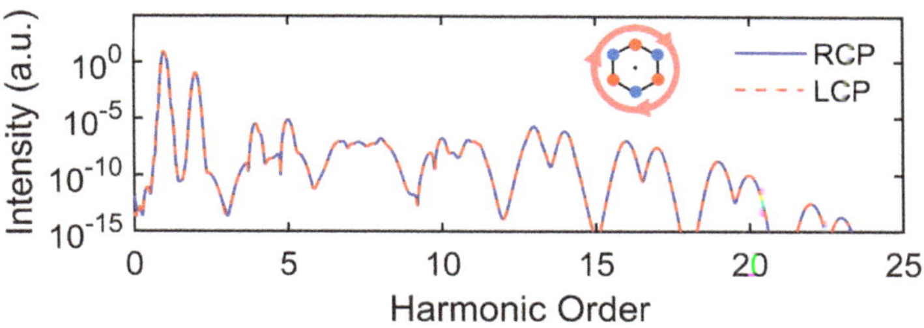


(a)

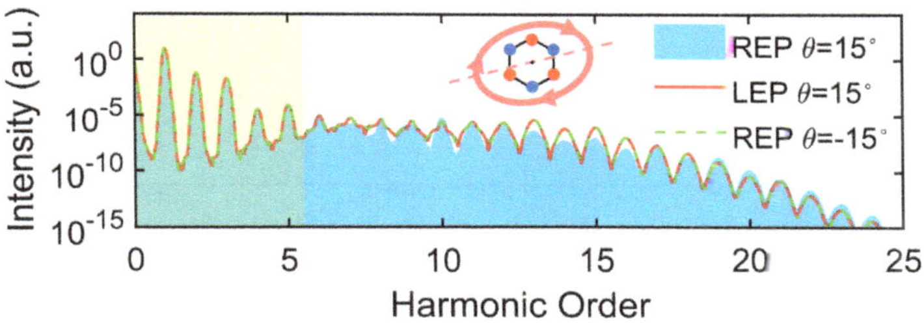


(b)

Fig. 10.5 (a) Solid-state harmonic generation of h-BN driven by right-handed circularly polarized (RCP) and left-handed circularly polarized (LCP) light. (b) Solid-state harmonic generation of h-BN driven by elliptically polarized light (the ellipticity of REP is 0.5, and LEP is $-0.5$), where $\theta$ is the orientation angle between the main axis of ellipse and $\Gamma - K$ direction. The harmonic intensity driven by left-handed elliptically polarized (LEP) light with $\theta = 15°$ (red solid line) is the same as that driven by right-handed elliptically polarized (REP) light with $\theta = -15°$ (green dashed line), but the REP case with $\theta = 15°$ (blue area) is offset from them in the high-order region. Reproduced with permission from [11].

this asymmetry to the mirror reflection mismatch between light fields and crystals. The mirror reflection of the laser-crystal system not only causes the helicity inversion of the driving light, but also leads to the reflection of its orientation angle. Therefore, the circular dichroism of HHG may be induced if only the helicity of the elliptically polarized light is reversed regardless of orientation dependence [see the difference in high-order region between the blue area and red solid lines in Fig. 10.5(b)]. The circular dichroism vanishes when we reflect the laser-crystal system simultaneously [see the red solid line and green dashed line in Fig. 10.5(b)].

The time-reversal symmetry can also reverse the helicity of laser fields. We can use the circular dichroism of harmonics to identify the time-reversal symmetry of crystals. We note the area shaded yellow in Fig. 10.5(b) that the low-order harmonics, nearly below band gap, never exhibit circular dichroism in h-BN. However, a completely different phenomenon emerges in the Haldane model with breaking time-reversal symmetry.

We deduce from Eq. (10.49) that the time-reversal symmetry of crystals prevents the low-order harmonics from circular dichroism. In order to exclude the effect of mirror symmetry, we use an elliptically polarized laser and set the orientation angle to 15°. Nevertheless, the low-order harmonics of h-BN keep perfect ellipticity-dependent symmetry [see Fig. 10.6(a)]. However, due to the absence of the time-reversal symmetry, the circular dichroism of low-order harmonics can be clearly seen in the Haldane model [see Fig. 10.6(b)]. Thus, the circular dichroism of harmonics can be used to identify magnetic materials.

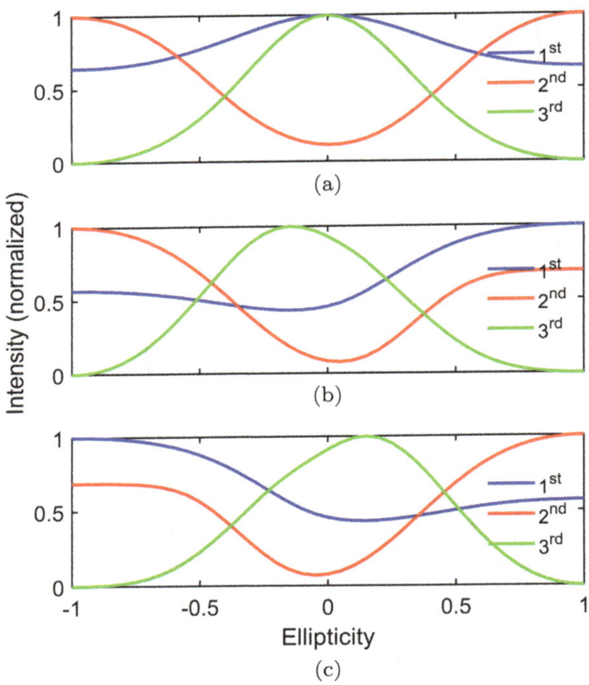


Fig. 10.6 Ellipticity dependence of low-order harmonic intensity for h-BN (a) and Haldane model with $\varphi = \frac{\pi}{2}$ (b), $\varphi = -\frac{\pi}{2}$ (c). The orientation angle $\theta$ is set to be 15°, other parameters of the models and laser being the same as Fig. 10.3. The legends refer to harmonic orders. The intensity is normalized for each harmonic. Reproduced with permission from [11].

Then we consider the time-reversal enantiomer by flipping the magnetic flux of the Haldane model. As Figs. 10.6(b) and 10.6(c) show, when the phase of the flux is inverted, the ellipticity dependence is reversed exactly as well. This is equivalent to performing $\hat{T}$ on the laser-crystal system, but the intensity of low-order harmonics is unaltered. Combined with ultrafast time-resolved spectra, we can expectantly use the low-order harmonics to observe the magnetization degree of materials or ultrafast spin dynamics.

## 10.5 Conclusion

In summary, the expression of photocurrent under the dipole approximation and a systematic picture of the HHG in solids have been obtained. Firstly, we point out here that the selection rule of HHG is determined by the phase difference of transition dipole moments under point-group symmetry operation. Secondly, similar to the dynamical phase caused by the Coulomb field, the shift phase is induced by the instantaneous potential of the oscillating laser, and the motion of electrons on the laser-dressed effective band is a complementary strong-field physical image to study the HHG. For example, when we reconstruct the band structure of non-centrosymmetric crystals and consider the propagation effect of HHG or its phase-matching condition, the effects of the shift phase and $\Delta$TDP are non-negligible. This framework can help us to understand the microscopic mechanism of the selection rules, orientation dependence, polarization characteristics, time-frequency analysis and ellipticity dependence of HHG. The two-channel image can also be used to explain the cause of the valley polarization and valley Hall effect. Thirdly, we point out that the low-order harmonics do not have circular dichroism in non-magnetic systems, which is important for judging the magnetization degree of materials based on strong-field nonlinear optics. Strong-field ultrafast nonlinear effect is a vital tool for exploring crystal structure and electron spin dynamics. This theory and conclusions can be extended to discuss optical phenomena about the magnetic point groups. In addition, the discussion is limited to the photocurrent under the electric dipole approximation, which can be analogized to the electric quadrupole effect and the magnetic dipole effect. Last but not least, the single-electron framework is no longer suitable for systems with strongly correlated interactions, thus establishing an efficient model involving quasiparticle transitions becomes particularly urgent. All of this chapter is supported by the original article on https://arxiv.org/abs/2304.10109.

## References


[1] H. Liu, Y. Li, Y. S. You, S. Ghimire, T. F. Heinz and D. A. Reis, High-harmonic generation from an atomically thin semiconductor, *Nature Physics* **13**, 3, pp. 262–265 (2017).

[2] A. H. Chin, O. G. Calderón and J. Kono, Extreme midinfrared nonlinear optics in semiconductors, *Physical Review Letters* **86**, pp. 3292–3295 (2001).

[3] H. Weng, C. Fang, Z. Fang, B. A. Bernevig and X. Dai, Weyl semimetal phase in noncentrosymmetric transition-metal monophosphides, *Physical Review* **5**, p. 011029 (2015).

[4] R. Yu, H. Weng, Z. Fang, H. Ding and X. Dai, Determining the chirality of Weyl fermions from circular dichroism spectra in time-dependent angle-resolved photoemission, *Physical Review B* **93**, p. 205133 (2016).

[5] D. Xiao, M.-C. Chang and Q. Niu, Berry phase effects on electronic properties, *Reviews of Modern Physics* **82**, pp. 1959–2007 (2010).

[6] T. Morimoto and N. Nagaosa, Topological nature of nonlinear optical effects in solids, *Science Advances* **2**, 5 (2016).

[7] O. E. Alon, V. Averbukh and N. Moiseyev, Selection rules for the high harmonic generation spectra, *Physical Review Letters* **80**, pp. 3743–3746 (1998).

[8] O. Neufeld, D. Podolsky and O. Cohen, Floquet group theory and its application to selection rules in harmonic generation, *Nature Communications* **10**, 1, p. 405 (2019).

[9] J. E. Sipe and A. I. Shkrebtii, Second-order optical response in semiconductors, *Physical Review B* **61**, pp. 5337–5352 (2000).

[10] C. Qian, C. Yu, S. C. Jiang, T. Zhang, J. C. Gao, S. Shi, H. Q. Pi, H. M. Weng and R. F. Lu, Role of shift vector in high-harmonic generation from noncentrosymmetric topological insulators under strong laser fields, *Physical Review X* **12**, 2, p. 021030 (2022).

[11] C. Qian, S. Jiang, T. Wu, H. Weng, C. Yu and R. Lu, On the theory of solid-state harmonic generation governed by crystal symmetry, arXiv:2304.10109 [physics.optics] (2023).

[12] F. Willems, C. T. L. Smeenk, N. Zhavoronkov, O. Kornilov, I. Radu, M. Schmidbauer, M. Hanke, C. von Korff Schmising, M. J. J. Vrakking and S. Eisebitt, Probing ultrafast spin dynamics with high-harmonic magnetic circular dichroism spectroscopy, *Physical Review B* **92**, p. 220405 (2015).

[13] T. Fan, P. Grychtol, R. Knut, C. Hernández-García, D. D. Hickstein, D. Zusin, C. Gentry, F. J. Dollar, C. A. Mancuso, C. W. Hogle, O. Kfir, D. Legut, K. Carva, J. L. Ellis, K. M. Dorney, C. Chen, O. G. Shpyrko, E. E. Fullerton, O. Cohen, P. M. Oppeneer, D. B. Milošević, A. Becker, A. A. Jaroń-Becker, T. Popmintchev, M. M. Murnane and H. C. Kapteyn, Bright circularly polarized soft X-ray high harmonics for X-ray magnetic circular dichroism, *Proceedings of the National Academy of Sciences* **112**, 46, pp. 14206–14211 (2015).

[14] D. Xiao, G.-B. Liu, W. Feng, X. Xu and W. Yao, Coupled spin and valley physics in monolayers of $MoS_2$ and other group-VI dichalcogenides, *Physical Review Letters* **108**, p. 196802 (2012).
[15] Q. Ma, S.-Y. Xu, C.-K. Chan, C.-L. Zhang, G. Chang, Y. Lin, W. Xie, T. Palacios, H. Lin, S. Jia, P. A. Lee, P. Jarillo-Herrero and N. Gedik, Direct optical detection of Weyl fermion chirality in a topological semimetal, *Nature Physics* **13**, 9, pp. 842–847 (2017).
[16] F. D. M. Haldane, O(3) nonlinear $\sigma$ model and the topological distinction between integer- and half-integer-spin antiferromagnets in two dimensions, *Physical Review Letters* **61**, pp. 1029–1032 (1988).
[17] O. Neufeld, D. Ayuso, P. Decleva, M. Y. Ivanov, O. Smirnova and O. Cohen, Ultrasensitive chiral spectroscopy by dynamical symmetry breaking in high harmonic generation, *Physical Review X* **9**, p. 031002 (2019).
[18] N. Klemke, N. Tancogne-Dejean, G. M. Rossi, Y. Yang, F. Scheiba, R. E. Mainz, G. Di Sciacca, A. Rubio, F. X. Kartner and O. D. Mucke, Polarization-state-resolved high-harmonic spectroscopy of solids, *Nature Communications* **10** (2019).
[19] N. Tancogne-Dejean, O. D. Mucke, F. X. Kartner and A. Rubio, Ellipticity dependence of high-harmonic generation in solids originating from coupled intraband and interband dynamics, *Nature Communications* **8**, p. 745 (2017).
[20] Y. S. You, D. A. Reis and S. Ghimire, Anisotropic high-harmonic generation in bulk crystals, *Nature Physics* **13**, 4, pp. 345–349 (2017).

https://doi.org/10.1142/9789811279560_0011

# Chapter 11

# High-Order Harmonic Generation in Solids: Dependence on Target and Laser Properties

Francisco Navarrete[1,5], Marcelo F. Ciappina[2,3,4] and Uwe Thumm[5]

[1] *Institute of Physics, University of Rostock, 18051 Rostock, Germany*
[2] *Department of Physics, Guangdong Technion — Israel Institute of Technology, 241 Daxue Road, Shantou, Guangdong, China, 515063*
[3] *Technion — Israel Institute of Technology, Haifa, 32000, Israel*
[4] *Guangdong Provincial Key Laboratory of Materials and Technologies for Energy Conversion, Guangdong Technion — Israel Institute of Technology, 241 Daxue Road, Shantou, Guangdong, China, 515063*
[5] *Department of Physics, Kansas State University, Manhattan, Kansas 66506, USA*

## 11.1 Introduction

Illuminated by intense laser fields, gases and solids emit radiation that is strongly enhanced at or near multiples of the driving laser frequency. In the last three decades, extensive research has been conducted on the high-order harmonic generation (HHG) process in atomic gases, leading to a comprehensive understanding of the underlying mechanism, i.e., the emission of radiation by rescattered electrons that are driven by the laser electric field [1, 2]. For more than a decade, theoretical studies considered the substantial electronic density in solids as a potential means for designing high-intensity harmonic radiation sources [3, 4]. After its experimental demonstration more than a decade ago [5] the HHG process in solids was investigated in diverse theoretical [6–15], as well as experimental [16–26], studies. Nevertheless, questions remain about the physical mechanisms responsible for many distinctive features in measured HHG spectra [27].

The electron dynamical processes involved in the harmonic up-conversion in solids have been described by different numerical models, such as the semiconductor Bloch equations (SBE) [6], the time-dependent Schrödinger equation (TDSE) [7] using either a static or adiabatic (in the driving laser field) basis, and time-dependent density functional theory [10]. We here develop a model that includes initial electronic states in the entire Brillouin zone (BZ). This model applies to materials with a fully occupied valence band (e.g., dielectrics). We compare our analytical and numerical results for HHG in solids, including initial electronic states in the entire first BZ, with predictions obtained by only considering initial states at the minimum valence-conduction band gap (at the Γ point of the BZ). We distinguish between approximations needed for carrying out the calculations and those made for interpretation purposes only. We start this chapter with general considerations for the prototypical case of a pristine crystal exposed to a monochromatic laser source, before discussing applications to HHG by two-colour driving laser pulses [13] and in doped solids [21].

## 11.2 Crystal Momentum-Resolved Contributions

In this section, we discuss a crystal momentum-resolved numerical model for HHG in solids, using as example the interaction of intense few-cycle mid-infrared driving laser pulses with MgO crystals. We employ the single-active-electron (SAE) approximation and expand the active electron's wave function in a basis of adiabatic Houston states [7]. We model the bulk electronic band structure in terms of a Kronig-Penney model potential, which matches adequately the characteristics of the conduction and valence bands of MgO. Importantly, we distinguish contributions to the HHG yield from initial SAE states at the Γ point in HHG spectra we obtained by including initial SAE valence states in the entire BZ of the target's band structure (Sec. 11.2.5.1). To facilitate the interpretation of HHG spectra, we compare our numerical spectra with the prediction of an analytical saddle point approximation (Sec. 11.2.5.2).

### 11.2.1 *Theoretical Formulation*

To analyze HHG in solids, we describe the dynamics of an electron in a one-dimensional infinite periodic potential $V(x) = V(x + a)$. The TDSE of this system can be written as

$$\left[\frac{1}{2}\left(\hat{p} + A(t)\right)^2 + V(x)\right] |\psi_k(t)\rangle = i\frac{\partial}{\partial t}|\psi_k(t)\rangle \tag{11.1}$$

where $k$ is the Bloch (crystal) momentum and $\hat{p} = -i\frac{\partial}{\partial x}$ the electron-momentum operator. We use atomic units ($e = \hbar = m_e = 1$) throughout this chapter, unless specified otherwise. The vector potential $A(t)$ is related to the electric field $E(t)$ by

$$A(t) = -\int_0^t E(t')\, dt'.$$

Here, we have employed the dipole approximation $A(x, t) \approx A(t)$, because the driving laser wavelengths we are interested in are much larger than the lattice constant. We note that, even though we are presenting here a 1D model, crystal orientation orientation effects can be simulated, by changing the band structure [i.e., by adjusting $V(x)$].

### 11.2.2 *Single-Active-Electron Solution of the Time-Dependent Schrödinger Equation in the Houston Basis*

In this section we describe the method we employ to solve Eq. (11.1). For both numerical calculations and physical interpretation, we expand the wave function as

$$|\psi_k(t)\rangle = e^{-iA(t)x} \sum_n B_{nk}(t) \exp\left[-i \int_0^t \varepsilon_{n\kappa(t')} dt'\right] |\phi_{n\kappa(t)}\rangle, \qquad (11.2)$$

where the eigenfunctions $|\phi_{n\kappa(t)}\rangle$ are Bloch states that are adiabatic in $A(t)$, i.e., with regard to the comparatively slow time dependence of the driving-laser electric field. They are known as "Houston states" [7] and solutions of the time-independent Schrödinger equation

$$\left[\frac{\hat{p}^2}{2} + V(x)\right] |\phi_{n\kappa(t)}\rangle = \varepsilon_{n\kappa(t)} |\phi_{n\kappa(t)}\rangle, \qquad (11.3)$$

which depends adiabatically on time through the Houston-state crystal momentum

$$\kappa(t) = k + A(t).$$

The electronic evolution driven by the infrared pulse is thus parameterized by $\kappa(t)$, starting from an initial state with Bloch-state crystal momentum $k$ (Fig. 11.1). Similar to ordinary Bloch functions [28], the Hamiltonian in Eq. (11.3) does not couple Houston states with different band indices $n$ or initial (field-free) crystal momenta $k$. Adiabatic Bloch states thus explore each electronic band separately for every given initial momentum $k$, i.e., for each "$k$-channel" [11].

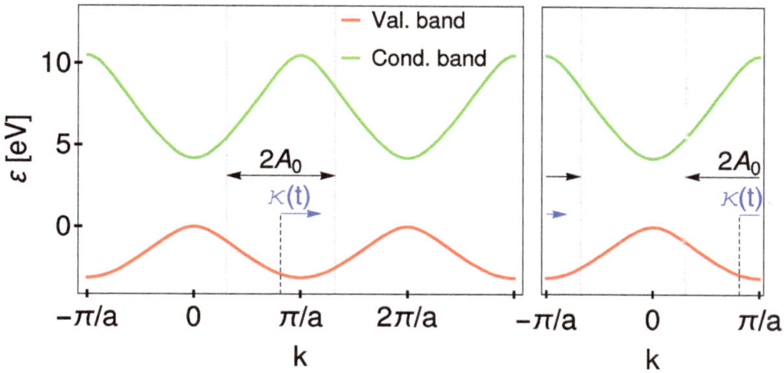


Fig. 11.1 The two lowest dispersion curves of a two-band system in the repeated-zone scheme (left) and in the first Brillouin zone (right). The vertical black dashed line indicates a particular $k$-channel. The blue arrow represents the adiabatic momentum $\kappa(t)$ at time $t$, while the double-headed black arrow denotes the maximum range ($2A_0$) covered by $\kappa(t)$ within the first Brillouin zone. Reproduced with permission from [11].

#### 11.2.2.1 *Equations of Motion*

Substituting Eq. (11.2) in (11.1) leads to the set of coupled differential equations

$$i\dot{B}_{nk}(t) = -E(t) \sum_{n' \neq n} B_{n'k}(t) D^{nn'}_{\kappa(t)} \exp\left[ i \int_0^t \Delta\varepsilon^{nn'}_{\kappa(t)} dt' \right], \tag{11.4}$$

with the band gap energy $\Delta\varepsilon^{nn'}_{\kappa(t)} = \varepsilon_{n\kappa(t)} - \varepsilon_{n'\kappa(t)}$ between Houston states [7, 29] and transition dipole moments (TDMs)

$$D^{nn'}_{\kappa(t)} = \frac{i\, P^{nn'}_{\kappa(t)}}{\Delta\varepsilon^{nn'}_{\kappa(t)}} \quad (n \neq n'), \quad D^{nn}_{\kappa(t)} = 0. \tag{11.5}$$

The TDMs are given in terms of the momentum-operator matrix elements [30]

$$P^{nn'}_{\kappa(t)} = \frac{1}{a} \int_0^a \phi^*_{n\kappa(t)}(x) \frac{1}{i} \frac{\partial}{\partial x} \phi_{n'\kappa(t)}(x)\, dx,$$

with diagonal elements

$$P^{nn}_{\kappa(t)} = \frac{\partial \varepsilon_{n\kappa(t)}}{\partial \kappa(t)}$$

that can be interpreted as the group velocity of the active electron's wave packet in band $n$.

#### 11.2.2.2 *Density Matrix Formulation*

We describe the electronic evolution in Eq. (11.4) in terms of the density operator,

$$\hat{\rho}_k(t) = |\psi_k(t)\rangle\langle\psi_k(t)|,$$

where $|\psi_k(t)\rangle$ is a pure state given by Eq. (11.2). The matrix elements in the basis of Houston states $\{|\phi_{n,\kappa(t)}\rangle\}$,

$$\begin{aligned}\rho_k^{nn'}(t) &= B_{nk}(t)B^*_{n'k}(t)\\ &= \langle\phi_{n\kappa(t)}|\hat{\rho}_k|\phi_{n'\kappa(t)}\rangle e^{i\int_0^t \Delta\varepsilon^{nn'}_{\kappa(t')}dt'},\end{aligned}$$

represent band populations for $n = n'$. We obtain their time evolution by solving the Liouville equation

$$\begin{aligned}i\dot{\rho}_k^{nn'}(t) = E(t)\sum_{n''}[&D^{n''n}_{\kappa(t)}e^{i\int_0^t \Delta\varepsilon^{nn''}_{\kappa(t')}dt'}\rho_k^{n''n'}(t)\\ &- D^{n'n''}_{\kappa(t)}e^{i\int_0^t \Delta\varepsilon^{n''n'}_{\kappa(t')}dt'}\rho_k^{nn''}(t)].\end{aligned} \tag{11.6}$$

The density matrix formulation, apart from being equivalent to solving the TDSE, allows for an ad-hoc introduction of relaxation effects and decoherence in a natural way. This procedure leads to the SBE [31], that we will use in Sec. 11.4. Within the density matrix formulation, it is straightforward to relate two broadly employed calculation methods of HHG in solids, the solution of the TDSE and the SBE.

#### 11.2.2.3 *Laser-Driven Electron Currents*

The $k$-resolved current density $J_k(t) = J_k^{ra}(t) + J_k^{er}(t)$ has intra- and interband contributions,

$$J_k^{ra}(t) = -\sum_n \rho_k^{nn}(t)\,P^{nn}_{\kappa(t)} \tag{11.7}$$

$$J_k^{er}(t) = -\sum_{n'>n}\sum_n e^{i\int_0^t \Delta\varepsilon^{nn'}_{\kappa(t')}dt'}\rho_k^{n'n}(t)P^{nn'}_{\kappa(t)} + \text{c.c.}, \tag{11.8}$$

respectively, from which the spectral HHG yield is obtained as

$$Y_k(\omega) = \left|\int_{-\infty}^{\infty} dt\, e^{-i\omega t}J_k(t)\right|^2 \equiv |\hat{J}_k(\omega)|^2.$$

In the numerical application in Sec. 11.2.5 below, we will consider dielectric solids for which the Fermi energy lies in the middle of the band gap between the valence and first conduction band. Thus the total yield is obtained by integrating over all the initially populated states (the full BZ of the valence band),

$$Y(\omega) = \left| \int_{-\infty}^{\infty} dt\, e^{-i\omega t} \int_{BZ} dk\, J_k(t) \right|^2 .$$

### 11.2.3 *Two-Band HHG Mechanism*

Restricting our analysis to the valence ($v$) and first conduction band ($c$), the intra- and interband currents in Eqs. (11.7) and (11.8) simplify to

$$J_k^{ra}(t) = -\rho_k^{vv}(t)\, P_{\kappa(t)}^{vv} - \rho_k^{cc}(t)\, P_{\kappa(t)}^{cc} \tag{11.9}$$

$$J_k^{er}(t) = ie^{-iS(k,\,t)} \rho_k^{cv}(t) D_{\kappa(t)}^{vc} \Delta\epsilon_{\kappa(t)}^{vc} + \text{c.c.}, \tag{11.10}$$

with the action

$$S(k,\, t) = \int_0^t \Delta\varepsilon_{\kappa(t')}^{cv} dt'.$$

The intraband HHG yield is given by

$$Y_k^{ra}(\omega) = \left| \int_{-\infty}^{\infty} dt\, e^{-i\omega t} \left[ P_{\kappa(t)}^{vv} - \rho_k^{cc}(t)\, \Delta P_{\kappa(t)}^{vc} \right] \right|^2 ,$$

with

$$\Delta P_{\kappa(t)}^{vc} = P_{\kappa(t)}^{vv} - P_{\kappa(t)}^{cc}.$$

For low laser intensities, the population of the conduction band is very small ($\rho_k^{cc}(t) \ll 1$), and the dominant contribution in each $k$-channel comes from $P_{\kappa(t)}^{vv}$. After integrating over the BZ, because $P_k^{vv}$ is an odd function of $k$, the first term in Eq. (11.9) vanishes, such that the total intraband yield reduces to

$$Y^{ra}(\omega) = \left| \int_{-\infty}^{\infty} dt\, e^{-i\omega t} \int_{BZ} dk\, \rho_k^{cc}(t) \Delta P_{\kappa(t)}^{vc} \right|^2 .$$

### 11.2.4 *Semiclassical Analysis of the Interband Yield*

By means of a semiclassical analysis, we can derive analytical expressions for the interband yield. These allow us to better understand the underlying electron dynamics. For simplicity, we restrict our analysis to vector-potential amplitudes $A_0 < \pi/a$, since for higher amplitudes there will be more than one pathway to get to the $\Gamma$ point from any initial crystal momentum (see Fig. 11.1). This does not mean that we will restrict our analysis to the Γ point contribution.

As we show in this chapter, a vector potential $A_0 \approx \pi/(2a)$ exhibits a number of harmonics that represent a large portion of the total extension of the first plateau. For instance, for ZnO [32] crystals, a laser wavelength of 3250 nm and electric field strengths up to 0.6 V/Å, it was shown experimentally that the whole extent of the first plateau is explored. Thus, up to that field strength, we assume a two-band model to be sufficient for modeling HHG. At higher intensities, the semiclassical model can be extended to include electronic transitions to higher conduction bands. This results in new plateaus in the HHG spectra and coherent electron dynamics in more than one conduction band.

### 11.2.4.1 *Saddle Point Approximation*

By applying the frozen valence-band approximation $|\Delta\rho_k^{vv}(t)| \approx 1$ [33] and setting the TDM in Eq. (11.5) equal to its value at the Γ point [6, 34], $D^{vc}_{\kappa(t)} \approx D^{vc}_0$,[1] Eq. (11.10) simplifies to

$$\hat{J}_k^{er}(\omega) \approx -\,\omega\,|D_0^{vc}|^2 \int_{-\infty}^{\infty} dt \int_0^t dt'\, e^{i\sigma_\omega(k,\,t,\,t')} E(t'), \tag{11.11}$$

where

$$\sigma_\omega(k,\,t,\,t') = -\omega t - \int_{t'}^{t} \Delta\varepsilon^{cv}_{\kappa(t'')} dt''. \tag{11.12}$$

Since in Eq. (11.11) the factor $e^{i\sigma_\omega(k,\,t,\,t')}$ oscillates rapidly as a function of $t$ and $t'$, the integrals are dominated by contributions at times $t = t_e$ and $t' = t_s$, when the exponent is stationary. At these times we get

$$\Delta\varepsilon^{cv}_{\kappa(t_e)} = \omega \tag{11.13}$$

$$\Delta\varepsilon^{cv}_{\kappa(t_s)} = 0. \tag{11.14}$$

From Eq. (11.13), we obtain the cutoff frequency for each $k$-channel,

$$\omega^c(k,\,A_0) = \begin{cases} \Delta\varepsilon^{cv}_{|k|+A_0}, & |k| + A_0 < \pi/a \\ \Delta\varepsilon^{cv}_{\pi/a}, & |k| + A_0 \geq \pi/a. \end{cases} \tag{11.15}$$

[1]While in our calculations we do not assume a constant $D^{vc}_{\kappa(t)}$, here we use it because it allows us to obtain analytical results.

Because the band gap is non-zero across the entire BZ, Eq. (11.14) can only be fulfilled for complex-valued times, energies, and crystal momenta.

For a continuous-wave driving field, $A(t) = A_0 \sin(\omega_0 t)$, the interband HHG yield within this semiclassical analysis is

$$
\begin{aligned}
Y_k^{er}(\omega) &\approx \exp\left[-\frac{\sqrt{2}\,\Delta\varepsilon_0^{cv}}{E_0\,|D_0^{vc}|\,\sqrt{1-(k/A_0)^2}}\right] \\
&\quad \times 2\,(\pi\omega\,|D_0^{vc}|)^2 \left(\frac{E_0\,|D_0^{vc}|\,\sqrt{1-(k/A_0)^2}}{\sqrt{2}\Delta\varepsilon_0^{cv}}\right) \\
&\quad \times \left|\sum_{t_e} \frac{e^{-i\omega t_e}\left[e^{i\left\{\Re[S(k,\,t_s)]+S(k,\,t_e)+\frac{\pi}{2}\right\}}+\text{c.c.}\right]}{\left|E(t_e)\Delta P^{vc}_{\kappa(t_e)}\right|^{1/2}}\right|^2 .
\end{aligned}
\tag{11.16}
$$

The first exponential factor in this equation modulates the amplitude. It shows that the interband HHG yield increases with decreasing band gap $\Delta\varepsilon_0^{cv}$ and increasing TDM $|D_0^{vc}|$, as expected. The real-valued reduced effective mass,

$$
m_k^* = \frac{m_{vk}^* m_{ck}^*}{m_{vk}^* - m_{ck}^*}, \tag{11.17}
$$

is evaluated at the band center ($\Gamma$ point),

$$
m_0^* = m_k^*|_{k=0}.
$$

It is given in terms of the valence- and conduction-band effective masses by

$$
\frac{1}{m_{nk}^*} = \frac{\partial^2 \varepsilon_{nk}}{\partial k^2}, \qquad n = v,\, c.
$$

Analyzing the process in the repeated-zone scheme, which explicitly shows the periodic nature of the BZ, with a period $2\pi/a$, we see that for $A_0 < k < \pi/a$, we have $0 < k \pm A_0 < 2\pi/a$, and thus $0 < \kappa(t) < 2\pi/a$ (Fig. 11.1). Here we considered for simplicity that $k > 0$, but an equivalent analysis applies to a crystal momentum of value $-k$.

Thus, for $|k| > A_0$, $|\kappa(t)|$ never reaches the limits 0 and $2\pi/a$ (in a repeated zone scheme this means the electron will not reach the $\Gamma$ point, neither to the left nor to the right), and Eq. (11.14) is only fulfilled in the interval $k \in (-A_0, A_0)$. This can be seen in our numerical calculations in Fig. 11.2(b) that we will discuss in more detail further below in this chapter.

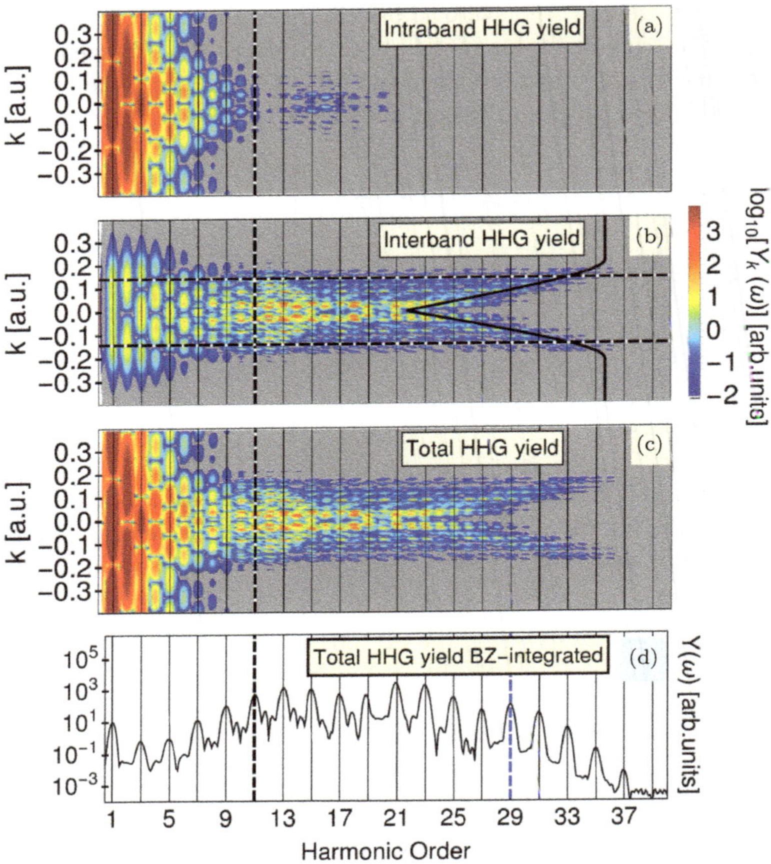


Fig. 11.2 HHG in MgO driven by a 10-cycle 3250 nm pulse with a peak electric field strength of 0.13 V/Å (corresponding to a peak intensity of $2.24 \times 10^{11}$ W/cm$^2$). Thin vertical lines mark odd harmonics. The vertical black dashed line indicates the harmonic order ($\Delta\varepsilon_0^{cv}/\omega_0 = 11$) corresponding to the minimum band gap energy $\Delta\varepsilon_0^{cv}$ at $k = 0$. (a) Intraband HHG spectrum. (b) Interband HHG spectrum. The horizontal dashed black lines indicate crystal momenta for which the $k$-resolved yield decreases five orders of magnitude relative to its maximum value. The V-shaped black line shows the maximal band gap energy $\omega^c(k, A_0)$ as given by Eq. (11.15). (c) Total HHG spectrum. The yields in (a–c) are given as functions of the lattice momentum $k$ over the entire first Brillouin zone (BZ) and on the same logarithmic scale. (d) Total HHG yield, integrated over the first BZ. Above harmonic order 29, the yield drops exponentially, defining the cutoff energy indicated by the vertical blue dashed line. Reproduced with permission from [11].

Taking into account that an upper limit for the integrated interband yield is given by

$$Y^{er}(\omega) = \left| \int_{BZ} dk\, \hat{J}_k^{er}(\omega) \right|^2 \leq \left( \int_{BZ} dk \left| \hat{J}_k^{er}(\omega) \right| \right)^2, \tag{11.18}$$

and since $\left[1 - (k/A_0)^2\right]^{1/2} \leq 1$ in Eq. (11.16), the contribution of each $k$-channel to the interband yield is limited by

$$Y_k^{er}(\omega) \propto \exp\left[ -\frac{\sqrt{2}\Delta\varepsilon_0^{cv}}{E_0 \left|D_0^{vc}\right| \sqrt{1 - (k/A_0)^2}} \right] = \exp\left[ -\frac{\sqrt{8m_0^*}\,(\Delta\varepsilon_0^{cv})^{3/2}}{E_0 \sqrt{1 - (k/A_0)^2}} \right].$$

This shows that even though the maximal contribution to the interband yield occurs at the $\Gamma$ point, there is a broader range of size $2A_0$ which contributes substantially to it. Nevertheless, the $\Gamma$ point is an upper bound to all contributions in the BZ. Thus, from Eq. (11.18) we find

$$Y^{er}(\omega) \propto \exp\left[ -\frac{\sqrt{8m_0^*}\,(\Delta\varepsilon_0^{cv})^{3/2}}{E_0} \right]. \tag{11.19}$$

#### 11.2.4.2 *Cutoff Frequency of the Integrated Interband Yield*

From Eqs. (11.6) and (11.8) the interband current, integrated over the BZ, becomes

$$\begin{aligned} \hat{J}^{er}(t) = \int_{BZ} dk \Bigg[ & \omega \int_{-\infty}^{\infty} dt\, e^{-i\omega t} \rho_k^{cv}(t) D_{\kappa(t)}^{vc} e^{-iS(k,t)} \\ & - \int_{-\infty}^{\infty} dt\, e^{-i\omega t} \Delta J_k^{er}(t) \Bigg]. \end{aligned} \tag{11.20}$$

Assuming constant TDMs and a frozen valence-band population in Eq. (11.20), we arrive at the saddle point conditions

$$\frac{\partial \sigma_\omega(k_r, t_s, t_e)}{\partial k} = \frac{\partial \sigma_\omega(k_r, t_s, t_e)}{\partial t} = \frac{\partial \sigma_\omega(k_r, t_s, t_e)}{\partial t'} = 0,$$

with $\sigma_\omega(k, t', t)$ being defined in Eq. (11.12). This implies

$$\Delta\varepsilon_{\kappa_r(t_s)}^{cv} = 0 \tag{11.21}$$

$$\int_{t_s}^{t_e} \Delta P_{\kappa_r(t)}^{vc} dt = 0 \tag{11.22}$$

$$\Delta\varepsilon_{\kappa_r(t_e)}^{cv} = \omega, \tag{11.23}$$

where $\kappa_r(t) = k_r + A(t)$, allowing us to determine the roots $k_r$, $t_s$, and $t_e$.

We note that the pair of Eqs. (11.21) and (11.23) is equivalent to Eqs. (11.14) and (11.13). On the other hand, Eq. (11.22) arises due to the integration over the BZ. This helps us elucidate the photoemission dynamics: the excited photoelectron wave packet moves with group velocity $P^{cc}_{k_r+A(t)}$, while the residual hole wave packet propagates with group velocity $P^{vv}_{k_r+A(t)}$. Both recombine at time $t_e$ emitting a photon with energy $\Delta\varepsilon^{cv}_{\kappa_r(t_e)}$ at time $t_0 = \Re[t_s]$. Since the model lattice potential $V(x)$ is inversion symmetric, the matrix elements $P^{nn}_k$ are odd functions of $k$, and $\Im[t_s] \propto \omega_0\sqrt{2\Delta\varepsilon^{cv}_0 m^*_0}$, so Eq. (11.22) can be approximated as

$$\int_{t_0}^{t_e} \Delta P^{vc}_{\kappa_r(t)} dt \approx 0. \tag{11.24}$$

Thus, we integrate along the real axis in Eq. (11.22), with real-valued integrand and integration limits. The cutoff frequency is now given by the $k$-channel at $k = k^c_r$, which yields the maximum HHG frequency

$$\omega^c_{BZ}(A_0) = \varepsilon_{k^c_r+A(t_e)}.$$

### 11.2.5 *HHG Spectra in the Two-Band Approximation*

Next, we validate the semiclassical analysis of the previous sections against numerical calculations. For this purpose we represent the solid electronic structure by two bands, just one conduction and the valence band. Furthermore, we assume "moderate" vector-potential amplitudes of the driving laser pulse that are smaller than the extent of half of the BZ ($A_0 < \pi/a$). While being limited, such moderate driving-laser intensities are sufficiently intense to prohibit their perturbative inclusion in our model. Here, we discuss results of the numerical solution of Eq. (11.4) for a model MgO solid with an initially fully occupied valence band.

#### 11.2.5.1 *Crystal Momentum-Resolved Contributions to HHG*

Our calculations were performed for flat-top-envelope pulses with a field strength of 0.13 V/Å, laser wavelength of 3250 nm and pulse duration of 10 optical cycles. We use the Kronig-Penney model potential discussed in Sec. 11.2 above, with positive delta-shaped spikes and an interlayer spacing of $a = 8$ a.u. [35] to model the electronic structure of MgO. Each spike represents a crystal plane in this 1D model. We adjust the potential strength to $V_0 = 22.345$ eV to match the 4.19 eV minimum band gap at the $\Gamma$ point and BZ boundary. Figure 11.2 shows spectra we obtained by numerically solving the 1D TSDE. Figures 11.2(a)–11.2(c) display the $k$-resolved two-band HHG spectra, while Fig. 11.2(d) shows the total integrated spectrum. The comparison of Figs. 11.2(a) and 11.2(b) reveals that intraband harmonics dominate the spectrum for harmonic photon energies below the band gap (which corresponds to $\Delta\varepsilon_0^{cv} \approx 11\omega_0$), while above the band gap energy the largest contribution to the total yield is due to interband recombination. Because of the low depletion of the valence band at the field strength used for this calculation, the overall shape of the intraband spectrum is dominated by the Fourier transform of $P^{vv}_{\kappa(t)}$ [11], i.e., the electronic motion in the valence band. We can see that along the BZ the spectrum shows even and odd harmonics, while at the $\Gamma$ point and BZ boundary the spectrum exhibits predominantly odd harmonics.

The interband spectra in Figs. 11.2(a) and 11.2(b) exhibit a well-defined structure of lobes in the below-the-band-gap region, with a significantly lower interband yield. The two spectra exhibit a different distribution of yield nodes along the BZ. The interband spectrum above the interband gap threshold dominates the shape of the total yield. It consists of a rich structure of even and odd harmonics and is contained in a region limited in $k$-space by the peak vector potential $A_0$ (horizontal dashed lines), while the highest energies are limited by $\omega^c(k, A_0)$ (solid black line), as predicted by the semiclassical analysis [see Eq. (11.15)]. The interband spectrum is V-shaped close to the center of the BZ. The total $k$-resolved HHG spectrum along the BZ in Fig. 11.2(c) exhibits a complex structure of even and odd harmonics. Contrary to assumptions made in early studies [7, 32, 36], contributions to the integrated yield in Fig. 11.2(d) are not limited to the vicinity of the $\Gamma$ point at $k = 0$. For an inversion-symmetric crystal illuminated by a strong continuous-wave driving field, the total BZ-integrated yield would be composed of strictly odd harmonics [37]. In contrast, for the few-cycle driving laser pulses we consider, the spectral pulse profile contains a range of frequencies close to the central frequency, which introduces additional (yet lower in intensity) harmonics between odd-frequency contributions.

#### 11.2.5.2 *Field-Strength Dependence of HHG Spectra*

Figure 11.3 shows calculated HHG yields, integrated over the BZ, for field strengths ranging from 0.05 to 0.30 V/Å. Figures 11.3(a) and 11.3(b) correspond to the intra- and interband contributions, respectively, while Fig. 11.3(c) gives the total yield. Figure 11.3 reveals that, for the entire driving-laser peak intensity range, the part of the harmonic spectra below

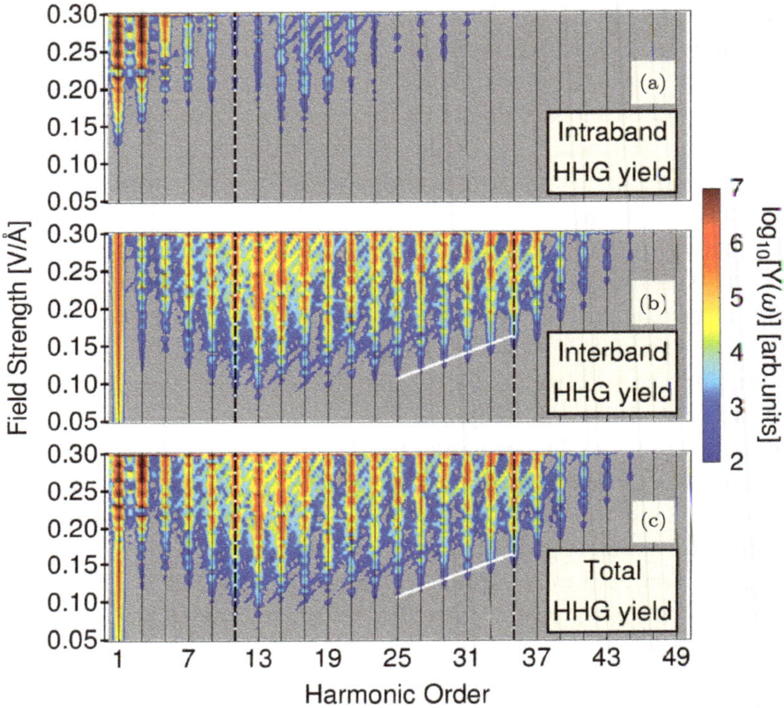


Fig. 11.3 Field-strength dependence of the BZ-integrated HHG yield in MgO driven by a 3250 nm 10-cycle pulse. (a) Intraband, (b) interband, and (c) total spectrum. Odd harmonics are indicated with thin vertical lines. Vertical black dashed lines show the maximal and minimal band gap energies. Cutoff harmonics are indicated as a white line in (b, c). Reproduced with permission from [11].

the interband gap threshold is mainly due to intraband emission, while above this threshold harmonic emission is primarily related to interband recombination. Relative to a weak background of non-odd integer contributions, the spectra are characterized by high-contrast odd harmonics.

In Fig. 11.3 we show the field-strength dependence of the cutoff harmonic order (white lines) as the last harmonic before the yield starts to decrease exponentially. Its dependence on the field strength is almost linear, which can be confirmed within the semiclassical two-band analysis in Secs. 11.2.4.1 and 11.2.4.2, by solving Eq. (11.24) for $t_e$ and subsequently Eq. (11.23) for each $k_r$. This yields the $\Delta\varepsilon^{cv}_{\kappa_r(t_e)}$ as a function of the driving-laser peak field strength, from which the maximal frequency $\omega^c_{BZ}$ is obtained.

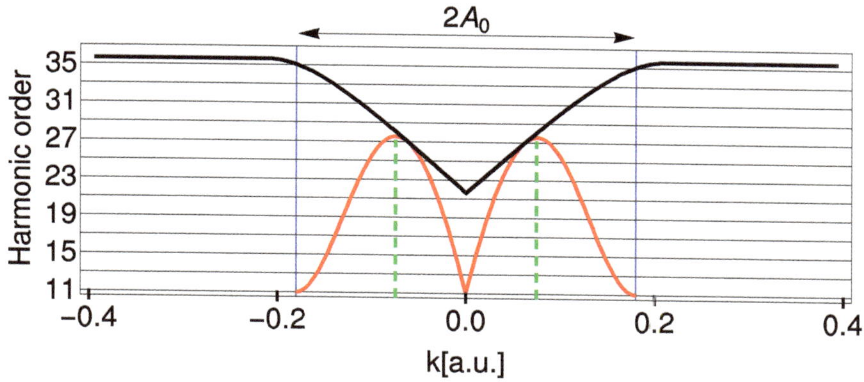


Fig. 11.4 Determination of the cutoff HHG order for a driving-laser peak intensity of 0.13 V/Å and a model MgO crystal with interlayer spacing $a = 8$ a.u. The black solid line represents the maximum local vertical band gap energy at the field-dressed crystal momentum $\kappa(t) = k + A(t)$ during one optical cycle of the driving laser as a function of the field-free crystal momentum $k \in [-\pi/a = -0.39, \pi/a = 0.39]$ a.u. The vertical thin blue lines indicate crystal momenta $\pm k_r$, determined in the saddle point approximation, that maximize the interband HHG yield $Y^{er}(\omega)$. The red line shows the interband HHG cutoff in each $k$-channel. The vertical green dotted lines indicate the highest cutoff energies reached at crystal momenta $k_r^c$, which do not necessarily coincide with the maximum vertical band gap at the first BZ edge ($k = \pi/a$). Reproduced with permission from [11].

The solid red line in Fig. 11.4 represents the harmonic energy $\Delta\varepsilon^{cv}_{\kappa_r(t_e)}$ as a function of crystal momenta $k_r$ that contribute to the BZ-integrated interband yield for a peak field strength of 0.13V/Å. As anticipated in Sec. 11.2.4.2, we can see that $|k_r| < A_0$, as indicated by vertical blue lines. We refer to the crystal momentum for which the maximum harmonic energy is obtained as $k_r^c$, which is indicated by green dotted lines. The black solid line is V-shaped around the $\Gamma$ point and shows the maximum energy $\omega^c(k, A_0)$ in each $k$-channel according to Eq. (11.15) [c.f., Fig. 11.2(b)]. This energy is equal to the maximum local band gap in each channel and exceeds the cutoff energy of the BZ-integrated yield $\Delta\varepsilon^{cv}_{\kappa_r(t_e)}$ (red line). Furthermore, since $\omega^c(k_r, A_0) \approx \Delta\varepsilon^{cv}_{\kappa_r(t_e)}$, the $k = \pm k_r^c$ channels have the largest contribution to the HHG yield.

## 11.3 Two Color-Driven Enhanced High-Order Harmonic Generation in Solids

In this section we explore HHG by bichromatic few-cycle driving pulses, which consist of a mid-infrared fundamental pulse and its second harmonic, combined phase-coherently. We will work again with a model MgO crystal and assess how we can control and enhance different areas of the HHG spectrum by adjusting the temporal profile of the bichromatic driving-laser electric field. We modify the shape of the driving pulse by varying the amplitude ratio and delay between its fundamental and second harmonic pulse components, while keeping the energy of the driving laser pulse constant. A complete analysis of the physics along the BZ allows us to understand the different roles the electric field and vector potential play in determining the HHG spectrum, in particular, its plateau region and cut-off energy.

According to semiclassical models for HHG in solids, the efficiency of the interband excitation in direct band gap materials can be enhanced by driving a large fraction of electrons to the vicinity of the minimum band gap in the BZ (steered by the vector potential), at a time when the electric field is as strong as possible [34,38]. This fine-tuning of the spectral amplitudes and phases maximizes the tunneling excitation rate and is a promising way to optimize interband HHG yields.

Even though synthesizing pulses using a large number of harmonics might look attractive for both, tailoring the HHG spectrum and distinguishing the effects of the electric field and the driving-pulse vector potential, upconversion to the third or higher odd harmonics is significantly less efficient than second harmonic generation [39]. We therefore consider bichromatic driving fields synthesized by frequency-doubling a fraction $R$ of a primary driver with a central frequency $\omega_0$ and pulse energy $U_{\omega_0}^{\text{prim}}$ (Fig. 11.5). Thus, in our HHG-control scheme the driving field consists of a linearly polarized bichromatic pulse with a fundamental pulse of energy $U_{\omega_0}$ and a co-propagating phase-coherent second harmonic pulse of energy $U_{2\omega_0}$. The centers of the two monochromatic constituents are assumed to have a controllable delay $\tau$. Such pulses can be conveniently produced in the laboratory and have been employed in molecular dissociation [40] and solid HHG [18] experiments.

By means of numerical simulations of the HHG spectrum as a function of the energy conversion parameter

$$R = U_{2\omega_0}/U_{\omega_0}^{\text{prim}}$$

and chromatic delay $\tau$ for fixed primary pulse energies $U_{\omega_0}^{\text{prim}}$, we analyze different ways to tailor the yield and spectral shape of the harmonic emission [13]. Even though our numerical calculations include excitation to and recombinations from many conduction bands (as many as required

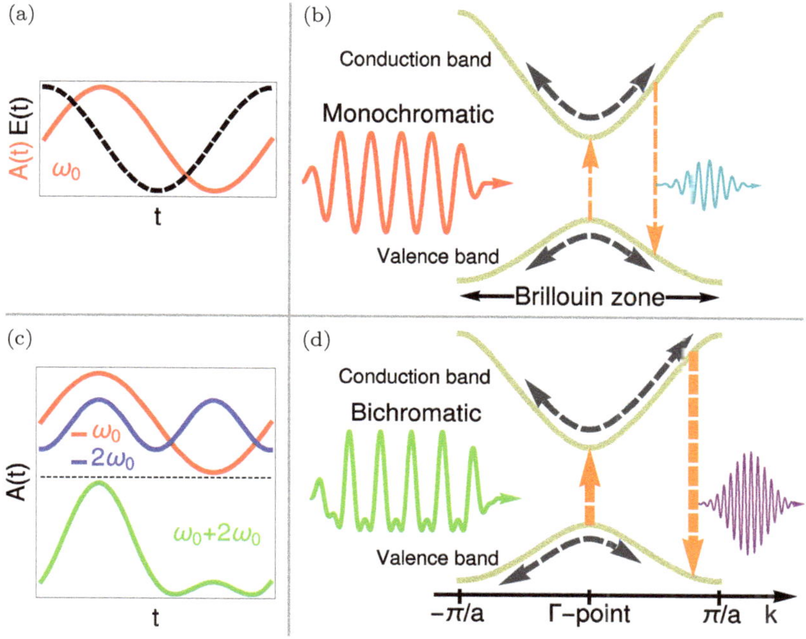


Fig. 11.5 Mechanisms for HHG in solids using (a, b) monochromatic and (c, d) bichromatic pulses. (a) Vector potential $A(t)$ (solid red line) and electric field $E(t)$ (dashed black line) of a monochromatic driving laser pulse with a carrier frequency $\omega_0$, and period $T_{\omega_0} = 2\pi/\omega_0$. (b) Schematic of HHG by a monochromatic driving pulse (red line) interacting with a two-band model solid represented by the dispersion curves of its valence and conduction bands as a function of the crystal momentum $k$ in the first BZ. The induced intraband motion in both bands is indicated as dashed black arrows. Laser-driven interband excitation and recombination are represented as orange arrows. Electronic intra- and interband currents contribute to the emission of HH radiation (blue line). (c) Vector potentials for one period of the fundamental pulse (red line) and phase-coherent harmonic pulse (blue line) with frequencies $\omega_0$ and $2\omega_0$, and their superposition into a bichromatic pulse (green line) with a harmonic conversion ratio $R = 0.5$ and chromatic delay $\tau = T_{\omega_0}/8$. (d) Schematic of HHG in a bichromatic driving pulse (green line). The asymmetry and non-sinusoidal profile of the driving pulse modify intra- and interband currents as compared to a monochromatic driver of identical pulse energy. This alteration affects the HHG spectrum (purple line) and can significantly increase the HHG yield and cutoff frequency. Reproduced with permission from [13].

for numerical convergence), we focus on the HHG yield in the first plateau, because yields in higher plateaus are orders of magnitude smaller [11,16,41] and thus not suitable for practical applications [42, 43]. The crystal momentum-resolved analysis of the spectrum is useful for analyzing the parity of the harmonics as well as obtaining semiclassical approximations of the yield in the plateau region.

### 11.3.1 *Bichromatic Driving Electric Field*

The assumed bichromatic electric field is the superposition of two co-propagating pulses with frequencies $\omega_0$ of the "fundamental" and $2\omega_0$ of the "harmonic" pulse, i.e.,

$$E(t) = E_{\omega_0}(t) + E_{2\omega_0}(t).$$

The profiles of the fundamental and harmonic components,

$$E_i(t) = E_{0,i} f(t - t_i) \cos\left[n_i \omega_0 (t - t_i)\right],$$

are modeled with the same "flat-top" electric-field envelope $f(t)$. The parameters $R = U_{2\omega_0}/U_{\omega_0}^{\mathrm{prim}}$ and $\tau = t_{2\omega_0} - t_{\omega_0}$ shape the bichromatic field. We quantify the departure of the combined field from that of each spectral component by defining the asymmetry parameter,

$$\mathcal{A} = \frac{\int_{t_0}^{t_0+T_{\omega_0}} dt\,\left[A_+^2(t) - A_-^2(t)\right]}{\int_{t_0}^{t_0+T_{\omega_0}} dt\,\left[A_+^2(t) + A_-^2(t)\right]},$$

where $A_\pm(t) = \Theta[\pm A(t)][\pm A(t)]$ are the positive and negative parts of the vector potential, $\Theta$ being the Heaviside step function. This definition is more appropriate than considering just the maxima of $A_\pm(t)$, because even when the maximum and minimum of $A(t)$ are equal in magnitude, the pulse can still be asymmetric.

### 11.3.2 *HHG Mechanism for Bichromatic Pulses*

Figure 11.5 compares the HHG mechanism for mono- and bichromatic driving fields for a two-band solid. It illustrates the potential to enhance the harmonic yield by using a two-color scheme. This enhancement can be understood by means of a semiclassical estimation [13] of the yield,

$$Y_k^{er}(\omega) \propto \exp\left[-\frac{\left(\sqrt{8m_0^*}\Delta\varepsilon_0^{cv}\right)^{3/2}}{E_0 \tilde{E}[\tilde{A}^{-1}(-k/A_0)]}\right],$$

where $\tilde{E}$ and $\tilde{A}$ are the bichromatic electric field and corresponding vector potential normalized by $E_0$ and $A_0$, respectively. The crystal momentum channels that contribute to the HHG spectrum are restricted to $k \in (-A_0\tilde{A}^{\text{peak}}, A_0\tilde{A}^{\text{peak}})$. As shown in [13], the semiclassical estimate of the BZ-integrated interband yield is given by

$$Y^{er}(\omega) \propto \exp\left[-\frac{\sqrt{8m_0^*}\,(\Delta\varepsilon_0^{cv})^{3/2}}{E_0\tilde{E}^{\text{peak}}(R,\,\tau)}\right], \tag{11.25}$$

where $\tilde{E}^{\text{peak}}$ is the absolute maximum of $E/E_0$ within the pulse duration. This equation shows that tailoring the peak electric field can enhance the harmonic yield.

### 11.3.3 *HHG Spectra for Two-Color Pulses with Adjustable Temporal Profile*

Since $\tilde{E}^{\text{peak}}$ and $\tilde{A}^{\text{peak}}$ reach their maximum for $\tau = 0$ and $\tau = T_{\omega_0}/8$, respectively, these values are particularly attractive for analyzing the HHG spectrum of the bichromatic field. Figure 11.6 shows numerical HHG spectra for monochromatic driving pulses with the fundamental frequency $\omega_0$ and electric-field amplitude $E_0 = 0.15\,\text{V/Å}$ (for which R = 0) compared to bichromatic pulses with delays $\tau = 0$ [Figs. 11.6(a, c)] and $\tau = T_{\omega_0}/8$ [Figs. 11.6(b, d)]. The spectrum shown in Fig. 11.6(a) is calculated for the maximal electric field strength $E^{\max} = 0.21$ V/Å, which is achieved for the parameters $\tau = 0$ and $R = 0.5$. The spectrum in Fig. 11.6(d) is computed for bichromatic pulses with $A^{\max} = 0.23$ a.u., which is the maximal vector potential amplitude, obtained for $\tau = T_{\omega_0}/8$ and $R = 0.2$.

From Fig. 11.6 it is evident that bichromatic outperform the monochromatic driving pulses in terms of producing higher total integrated harmonic yields. These particular bichromatic pulses result in high harmonics occurring at even multiples of the fundamental frequency $\omega_0$. The emergence of even harmonics is not restricted to odd multiples of the harmonic frequency $2\omega_0$ ($n\omega_0$, $n = 2, 6, 10, \ldots$). Even harmonics of the fundamental frequency $\omega_0$ appear since the distinctive asymmetrical shape of the coherent bichromatic pulse breaks the symmetry of the HHG process. This occurs, for example, when $\tau$ is set to $T_{\omega_0}/8$. Our analytical investigation [13] reveals that for $0 < R < 1$ the most significant contributions to the HHG yields originate from $k$-channels that possess non-zero initial crystal momenta $k_0$. Furthermore, when using driving pulses with asymmetric temporal profiles, the adiabatic crystal momentum $\kappa(t) = k_0 + A(t)$ does not symmetrically "explore" the first BZ, due to the asymmetry of the pulse. We anticipate a similar pattern for all chromatic delays $\tau$ that yield an asymmetric driving pulse ($\mathcal{A} \neq 0$).

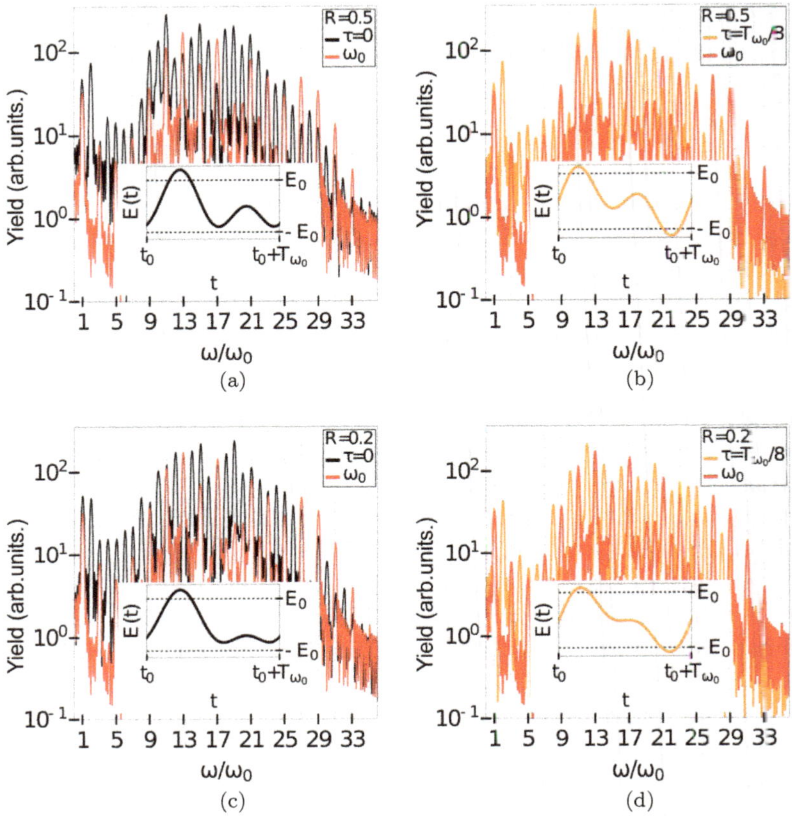


Fig. 11.6 HHG spectra generated by $\omega_0 + 2\omega_0$ bichromatic pulses (black and orange lines) in comparison with spectra generated by the monochromatic primary pulse (red line) with fundamental frequency $\omega_0$ (corresponding to a wavelength of 3200 nm) of equal pulse energy. The electric-field amplitude of the monochromatic pulse is 0.15 V/Å (corresponding to a peak intensity of $3 \times 10^{11}$ W/cm$^2$). The chromatic delays and energy conversion ratio of the bichromatic pulses are: (a) $\tau = 0$, $R = 0.5$ (maximizing the peak bichromatic electric field strength), (b) $\tau = T_{\omega_0}/8$, $R = 0.5$, (c) $\tau = 0$, $R = 0.2$, and (d) $\tau = T_{\omega_0}/8$, $R = 0.2$ (maximizing the peak bichromatic vector-potential amplitude). Corresponding electric fields over one optical period $T_{\omega_0}$ in the flat part of the pulse profile are shown as insets. Reproduced with permission from [13].

Figure 11.6(a) shows a higher total integrated yield for the two-color HHG spectrum than Fig. 11.6(b). This observation aligns with the fact that the bichromatic driver exhibits a stronger maximal electric field and a distinctive characteristic of HHG within the plateau spectral range, as can be inferred from the semiclassical analysis of Eq. (11.25). Determining the cutoff frequency for the spectra generated by bichromatic pulses in Fig. 11.6 is more intricate than for monochromatic driving pulses. We tentatively assign nearly identical cutoff orders (relative to the fundamental frequency $\omega_0$) to the bichromatic pulses in Figs. 11.6(a) and 11.6(b), 27 and 26, respectively. An apparent advantage of maximizing the peak electric field of bichromatic pulses is a substantial enhancement of the plateau harmonic yields.

Similarly, Fig. 11.6(c) displays a greater total integrated yield for the two-color HHG spectrum than Fig. 11.6(d). This is again in line with the stronger maximal electric field of the bichromatic driver. However, even though the peak vector potential is maximized for $R = 0.2$ and $\tau = T_{\omega_0}/8$, in this case we cannot detect an extension of the cutoff energy. This might be attributed to the fact that the increase in the peak vector potential of the bichromatic pulse is only 10% relative to the primary pulse. The most noticeable change in the cutoff frequency occurs for $0.5 < R < 1.0$. Over this $R$ range, the peak vector potential decreases to 50% of its maximum value for all values of $R$, and the cutoff energy closely follows its variation with $R$.

The integrated yield in Fig. 11.6(c) is computed for the driver parameters that maximize the field strength. It exceeds the yield in Fig. 11.6(d), obtained by maximizing the vector potential. This implies that augmenting the field strength through pulse shaping at a constant pulse energy is the most effective approach for enhancing the harmonic yield. Despite the peak vector potential of a monochromatic pulse affecting the cutoff frequency, solely maximizing the peak magnitude of the bichromatic vector potential does not guarantee the largest achievable cutoff frequency (for the given constraint of constant pulse energy).

## 11.4 HHG in Doped Solids

In recent years, there has been a growing theoretical search for the effect of crystal impurities on HHG mechanism [44–47]. It was recently found experimentally by Nefedova *et al.* [21] by comparison of a pristine sample of MgO and a doped MgO:Cr crystal (at a 0.5% concentration of the Cr dopant) that the secondary HHG plateau of the doped target is enhanced. In these experiments a 1.55 $\mu$m central wavelength driving-laser pulse with peak intensities ranging from 9 to 20 TW/cm$^2$ was used. It was also observed that the angular distributions of the spectra for different laser polarizations are similar for pristine and doped crystals. In this section we will analyze how a semiclassical analytical approach of the electron dynamics along the BZ can reveal the main features of these HHG spectra.

### 11.4.1 *Semiconductor Bloch Equations*

We numerically solve the SBE, obtained by adding an extra *ad hoc* term to Eq. (11.6). This leads to

$$i\dot{\rho}_k^{nn'}(t) = -i\frac{\rho_k^{nn'}(t)}{T_2} + E(t)\sum_{n''}[D_{\kappa(t)}^{n''n} e^{i\int_0^t \Delta\varepsilon_{\kappa(t')}^{nn''}dt'} \rho_k^{n''n'}(t)$$

$$- D_{\kappa(t)}^{n'n''} e^{i\int_0^t \Delta\varepsilon_{\kappa(t')}^{n''n'}dt'} \rho_k^{nn''}(t)].$$

These equations are identical with Eq. (11.6), except for the phenomenolocial dephasing term (first term on the r.h.s.). The phenomenological dephasing time $T_2$ is introduced to account for polarization damping [48] and effects beyond the SAE picture. The dephasing term in the equations of motion has the effect of generating greater contrast between odd and non-integer harmonics [6]. When numerically modelling experimental results, the conditions under which experiments are conducted can be difficult to reproduce without introducing a dephasing time, i.e., implying the involvement of relaxation effects because the contrast between odd and non-integer harmonics decreases as we use shorter and more intense pulses in the calculations. Additionally, it was shown that by adequately tuning $T_2$ it is possible to take into account macroscopic effects, such as the suppression of the long electron trajectories [49]. As an example, we mention the well-established interpretation of the first experimental HHG spectra in solids [32] by Vampa *et al.* [6], who observed that without introducing a dephasing time, the interband spectrum obtained from SBE calculations exhibits an unstructured mix of non-integer harmonics. This contrasts the clearly defined odd harmonics observed in the experimental spectrum. On the contrary, for a dephasing time of at least a quarter of the central pulse period, the calculations exhibit well-defined odd harmonics, in agreement with the experimental spectrum.

### 11.4.2 *Modeling MgO:Cr*

With the aim of theoretically supporting the experimental results of Nefedova *et al.* [21], we use in our model sample and laser parameters similar to those employed in the experiment. We perform our calculations in reduced dimensionality, representing the crystal as a 2D square lattice. We model the effective electronic interaction in terms of a separable potential

$$V(x, y) = V_x(x) + V_y(y),$$

where $x$ and $y$ are spatial coordinates. We describe the band structure of MgO in each spatial dimension using a cosine potential [6, 7, 50],

$$V_j(r_j) = V_0 \cos(a r_j), \quad j = x, y,$$

with $r_x = x$ and $r_y = y$. Here, $V_0$ is set to 0.37 a.u., and the lattice parameter is $a = 8$ a.u. This potential reproduces the band gap between the valence and first conduction band of MgO in the Γ-$X$ direction in comparison with results from density functional theory [35]. The band gap region is approximately cosine shaped, and its value at the center and zone boundaries agree with a difference of less than 10%. However, it results in an unrealistic minimum band gap of 4.2 eV. To achieve agreement between our model and the experimental value of 7.8 eV [21, 51], we energetically shift the valence band downward as suggested in [52] (see Fig. 11.7).

We model the band structure of MgO:Cr with 0.5% Cr doping including both the substitution of Mg ions with Cr ions and the presence of Mg vacancies. We achieve this by modifying the potential of the pristine MgO sample, following Refs. [44–47, 53]. To account for the impact of vacancies, we increase the amplitude of oscillation at the position $n_{\text{vac}}a$. Additionally, to address the substitution of $Mg^{2+}$ by $Cr^{3+}$ ions of the doped solid, we reduce the oscillation amplitude at a neighboring lattice position $n_{\text{imp}}a$.

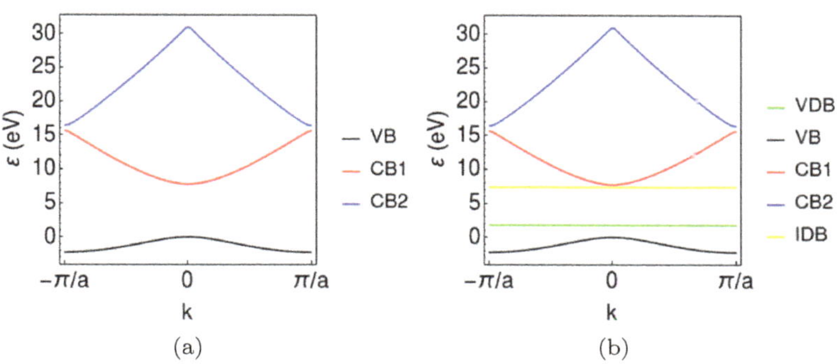


Fig. 11.7 Valence (VB) conduction bands CB1 and CB2 for (a) pristine MgO and (b) MgO:Cr. Panel (b) features additional bands: vacancy defect (VDB) and impurity defect (IDB) of MgO:Cr. Reproduced with permission from [21].

This results in the adjusted electronic potential

$$V_j(r_j) = V_0\cos(ar_j) + \begin{cases} V_{\text{vac}}\cos(ar_j), & n_{\text{vac}}a \le x \le (n_{\text{vac}}+1)a \\ V_{\text{imp}}\cos(ar_j), & n_{\text{imp}}a \le x \le (n_{\text{imp}}+1)a \end{cases}.$$

$V_0$, $V_{\text{vac}}$, and $V_{\text{imp}}$ represent the potential strengths of the pristine solid, vacancies, and impurities, respectively. To match the observed 1.8 eV reduction in the band gap of the MgO:Cr [21], we set $V_{\text{imp}}$ to 0.11 a.u. and $V_{\text{vac}}$ to $-0.11$ a.u.

The low dopant concentration allows us to approximate defect states as non-dispersive levels within the band gap, characterized by infinite effective masses.[2] These states interact across the entire BZ, exciting conduction band levels [53]. This interaction is enabled by the presence of impurities, which break the translational symmetry of the pristine lattice.[3] However, due to the low defect concentration, the electronic band structure of MgO is only marginally altered. Thus, we simplify the coupling between bands as strictly vertical. We thus model the band structure of MgO:Cr used in our calculations primarily in terms of dispersive valence and conduction bands of the pristine crystal, with two key modifications: (i) an adjusted band gap energy and (ii) the inclusion of two flat bands representing non-dispersive vacancy and impurity states, as illustrated in Fig. 11.7.

### 11.4.3 *Semiclassical Analysis*

Figure 11.8 provides an overview of the key processes involved in HHG in pure and doped MgO. For the doped sample, the Cr donor-dopants play a dual role. They replace $Mg^{2+}$ ions in the MgO lattice with $Cr^{3+}$ ions, introducing additional electrons [see Figs. 11.8(a) and 11.8(b)] and induce vacancies [57, 58] to maintain charge neutrality [58].

The underlying electronic excitations in pristine and Cr-doped MgO are depicted in Figs. 11.8(c) and 11.8(d), respectively. For MgO, HHG can be understood within the two-band analysis in Sec. 11.2.3 above, because it has a full valence band (VB). Nevertheless, here we have to consider two conduction bands instead of one (three bands in total), to account for a secondary plateau in the experimental HHG spectrum [21]. We refer

[2] The effective electron mass significantly influences electron mobility in solids. Recent investigations have explored its energy-dependent behavior and its impact on attosecond electron transport [54–56].

[3] The pristine MgO crystal permits optical transitions only between states with equal crystal momentum within the first BZ.

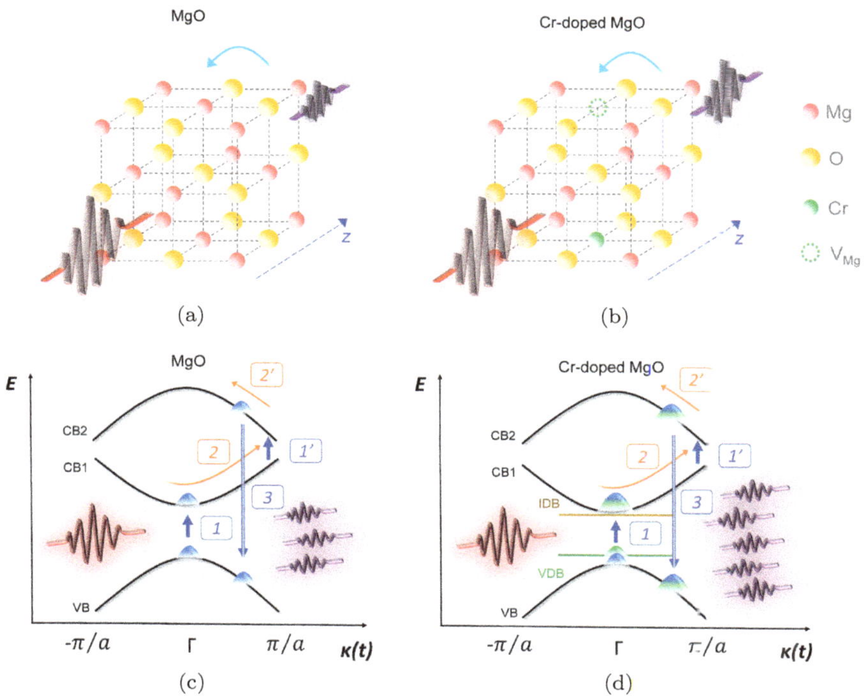


Fig. 11.8 Schematic representations of HHG in (a, c) MgO and (b, d) MgO:Cr in (a, b) coordinate space and (c, d) momentum space. In (a, b), the laser propagation axis is denoted as $z$, around which the crystals are rotated (indicated by the light blue arrow) during the anisotropy measurements [c.f., Figs. 11.9(a) and 11.9(b)]. In (b), a magnesium (Mg) vacancy replaces the nearest $Mg^{2+}$ ion. In (c, d) the laser electric field excites an electron wave packet from the valence band (VB) to the first conduction band (CB1) (1, indicated by the blue arrow). Subsequently, the electron experiences intraband acceleration (2, orange arrow). In certain cases, the electron wave packet may be excited to a higher conduction band (CB2) ($1'$, blue arrow), where intraband acceleration continues ($2'$, orange arrow). The excited electron wave packet eventually de-excites to the VB upon recombination with its residual hole (3, blue arrow), emitting harmonics with energies matching the instantaneous gap in the first BZ. In (d), the presence of the occupied impurity defect band (IDB) and vacancy defect band (VDB) in MgO:Cr leads to additional electronic transitions, resulting in an enhanced net high-order harmonic yield compared to MgO. Reproduced with permission from [21].

to the two conduction bands as CB1 and CB2. The overall HHG yield is expected to be dominated by the primary excitation between the valence and first conduction band, because tunneling from VB to CB1 is a necessary step for excitations from the first conduction band to higher bands. It can be described by Eq. (11.19). Even though from Eq. (11.19) it would be reasonable to expect that a reduction of the band gap enhances the yield, we recall that in this equation effective masses also play a role [33]. Thus, for MgO:Cr, it is the product of effective mass and band gap that needs to be reduced.

In the case of MgO:Cr, an impurity defect band (IDB) emerges near CB1, while an vacancy defect band (VDB) appears close to the VB. Adapting Eq. (11.19) to transitions originating from the VDB, we obtain an estimate for VDB contributions to the HHG yield,

$$Y_{\mathrm{MgO:Cr}}(\omega) \propto \exp\left[-\frac{\sqrt{8\, m^*_{cb1-vdb}}\,(\varepsilon_{cb1}-\varepsilon_{vdb})^{3/2}}{E_0}\right]. \tag{11.26}$$

In this equation, $m^*_{cb1-vdb}$ is the reduced effective mass [see Eq. (11.17)] between CB1 and VDB. Due to the substantial effective mass of the VDB, we can approximate $m^*_{vdb-cb1} \approx m^*_{cb1} > m^*_{vb-cb1}$. Consequently, the reduced effective mass is greater than that of the pristine sample. However, for the target used in this analysis, the reduction in the band gap induced by the vacancy results in a smaller numerator of the exponential term in Eq. (11.26) compared to Eq. (11.19). This suggests an enhancement in the HHG yield due to symmetry-allowed VDB-CB1 optical transitions.

### 11.4.4 *Comparison of Numerical and Experimental Results for Pristine and Doped Samples*

In this section, we compare the numerically modeled with the measured crystal orientation-dependent HHG response of MgO and MgO:Cr. The experimental results shown in this section are the first experimental confirmation of the yield enhancement in doped materials conducted by Nefedova *et al.* [21]. The authors investigated the HHG anisotropy as a function of crystal orientation relative to the laser polarization, as illustrated in Figs. 11.8(a, b).

Figures 11.9(a) and 11.9(b) present angle-dependent HHG spectra obtained from MgO and MgO:Cr. These spectra were recorded under conditions optimized for maximum HHG yields, specifically just below the damage threshold, with peak electric-field strengths of 1.23 V/Å (20 TW/cm$^2$) for MgO and 1.09 V/Å (15.7 TW/cm$^2$) for MgO:Cr. The observed four-fold symmetry corresponds to the face-centered cubic crystal structure of MgO [52]. Along the $\Gamma$-$X$ direction, the HHG yield is highest, while no signal is detected along the $\Gamma$-$K$ direction. Overall, the two anisotropy

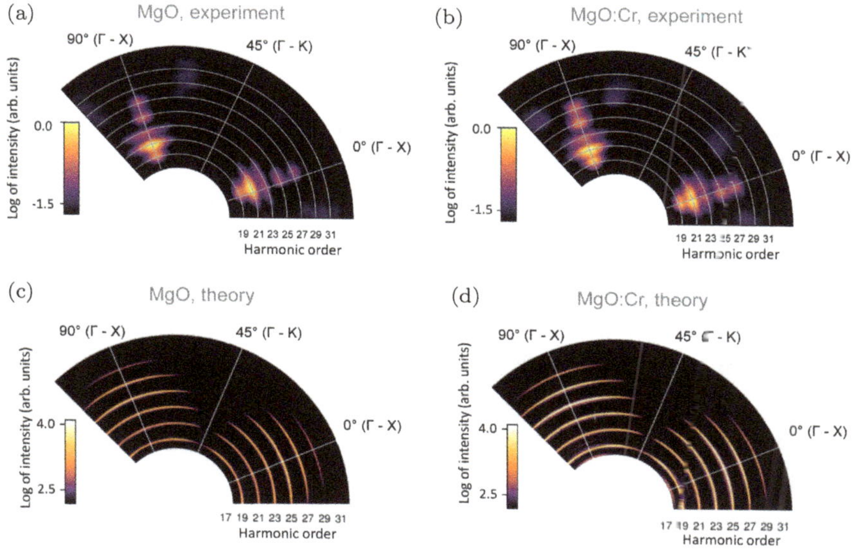


Fig. 11.9 Measured crystal orientation-dependent HHG spectra for (a, c) MgO and (b, d) MgO:Cr crystals. The crystals were rotated around the laser propagation direction in 3-degree increments, with the linear laser polarization direction held constant. Measured HHG spectra for (a) MgO, obtained at a peak laser field strength (in vacuum) of $1.23\,\text{V/Å}$ ($20\,\text{TW/cm}^2$) and (b) MgO:Cr, acquired at $1.09\,\text{V/Å}$ ($15.7\,\text{TW/cm}^2$). Due to the low doping concentration of 0.5%, the four-fold symmetry characteristic of the MgO cubic crystal structure remains practically unchanged in MgO:Cr. Calculated spectra for (c) MgO and (d) MgO:Cr at peak field strength $1.23\,\text{V/Å}$. Reproduced with permission from [21].

patterns are comparable, consistent with the preservation of crystal symmetry at our low dopant concentrations. Nefedova *et al.* [21] also observed that the harmonics emitted from MgO:Cr have a comparable yield at a lower driving-field intensities, as compared to harmonics from pure MgO. To analyze this behavior further, we calculated the angle-dependent HHG yield for MgO and MgO:Cr at the same peak vacuum electric-field strength of $1.23\,\text{V/Å}$ ($20\,\text{TW/cm}^2$). Our results are shown in Figs. 11.9(c) and 11.9(d). Although there are differences in the angular spread of the harmonics between theory and experiment, our numerical simulations reproduce maximal yields measured along the $\Gamma$-$X$ direction and the minimal yields along the $\Gamma$-$K$ direction. Across all crystal orientations, our simulations predict a larger HHG yield for MgO:Cr. These findings align well with the experimental results by Nefedova *et al.*

## 11.5 Conclusions

In this chapter we analyzed different aspects of the HHG process in solids. We showed how analytical semiclassical approximations promote the understanding of the underlying electron dynamics and help expose physical mechanisms at moderate numerical expense. We discussed how to numerically model HHG spectra, accounting for different targets and laser characteristics. In numerical calculation and semiclassical analytical approximations, we found it important to address contributions from initial states in the entire BZ for a more complete description of the observables. As compared to gas-phase HHG, the more complex HHG process in solids demands refined numerical calculations.

## Acknowledgements

We acknowledge support from Air Force Office of Scientific Research under award number FA9550-17-1-0369, NSF Grant No. 2110633 (theory for photoemission from surfaces) and, in part, the Chemical Sciences, Geosciences, and Biosciences Division, Office of Basic Energy Sciences, Office of Science, US Department of Energy Award DEFG02-86ER13491 (attosecond interferometry). Funding by the German Research Foundation, SFB 1477 "Light-Matter Interactions at Interfaces," Project No. 441234705, is gratefully acknowledged. We are indebted to V. E. Nefedova, S. Fröhlich, N. Tancogne-Dejean, D. Franz, A. Hamdou, S. Kaassamani, D. Gauthier, R. Nicolas, G. Jargot, M. Hanna, P. Georges, W. Boutu, and H. Merdji for a stimulating and productive collaboration on HHG from doped MgO crystals. M. F. C. acknowledges financial support from the Guangdong Province Science and Technology Major Project (Future functional materials under extreme conditions — 2021B0301030005) and the Guangdong Natural Science Foundation (General Program project No. 2023A1515010871).

## References

[1] A.-T. Le, R. R. Lucchese, S. Tonzani, T. Morishita and C. D. Lin, *Phys. Rev. A* **80**, p. 013401 (2009).
[2] F. Krausz and M. Ivanov, *Rev. Mod. Phys.* **81**, pp. 163–234 (2009).
[3] L. Plaja and L. Roso-Franco, *Phys. Rev. B* **45**, pp. 8334–8341 (1992).
[4] F. H. M. Faisal and J. Z. Kamiński, *Phys. Rev. A* **54**, pp. R1769–R1772 (1996).
[5] S. Ghimire, A. D. DiChiara, E. Sistrunk, P. Agostini, L. F. DiMauro and D. A. Reis, *Nat. Phys.* **7**, 2, pp. 138–141 (2011).

[6] G. Vampa, C. R. McDonald, G. Orlando, D. D. Klug, P. B. Corkum and T. Brabec, *Phys. Rev. Lett.* **113**, p. 073901 (2014).
[7] M. Wu, S. Ghimire, D. A. Reis, K. J. Schafer and M. B. Gaarde, *Phys. Rev. A* **91**, p. 043839 (2015).
[8] E. N. Osika, A. Chacón, L. Ortmann, N. Suárez, J. A. Pérez-Hernández, B. Szafran, M. F. Ciappina, F. Sols, A. S. Landsman and M. Lewenstein, *Phys. Rev. X* **7**, p. 021017 (2017).
[9] I. Floss, C. Lemell, G. Wachter, V. Smejkal, S. A. Sato, X.-M. Tong, K. Yabana and J. Burgdörfer, *Phys. Rev. A* **97**, p. 011401 (2018).
[10] K. K. Hansen, D. Bauer and L. B. Madsen, *Phys. Rev. A* **97**, p. 043424 (2018).
[11] F. Navarrete, M. Ciappina and U. Thumm, *Phys. Rev. A* **100**, 3 (2019).
[12] H. Jürß and D. Bauer, *Phys. Rev. B* **99**, 19 (2019).
[13] F. Navarrete and U. Thumm, *Phys. Rev. A* **102**, 6 (2020).
[14] L. Yue and M. B. Gaarde, *Phys. Rev. Lett.* **124**, 15, p. 153204 (2020).
[15] L. Li, P. Lan, X. Zhu and P. Lu, *Phys. Rev. Lett.* **127**, 22, p. 223201 (2021).
[16] G. Ndabashimiye, S. Ghimire, M. Wu, D. A. Browne, K. J. Schafer, M. B. Gaarde and D. A. Reis, *Nature* **534**, p. 520 (2016).
[17] Z. Chang, P. B. Corkum and S. R. Leone, *J. Opt. Soc. Am. B* **33**, p. 1081 (2016).
[18] T. T. Luu and H. J. Wörner, *Phys. Rev. A* **98**, p. 041802 (2018).
[19] A. J. Uzan, G. Orenstein, A. Jiménez-Galán, C. McDonald, R. E. F. Silva, B. D. Bruner, N. D. Klimkin, V. Blanchet, T. Arusi-Parpar, M. Krüger, A. N. Rubtsov, O. Smirnova, M. Ivanov, B. Yan, T. Brabec and N. Dudovich, *Nat. Photonics* **14**, p. 183 (2020).
[20] M. Lewenstein, M. F. Ciappina, E. Pisanty, J. Rivera-Dean, P. Stammer, T. Lamprou and P. Tzallas, *Nat. Phys.* **17**, 10, pp. 1104–1108 (2021).
[21] V. Nefedova, S. Fröhlich, F. Navarrete, N. Tancogne-Dejean, D. Franz, A. Hamdou, S. Kaassamani, D. Gauthier, R. Nicolas, G. Jargot, M. Hanna, P. Georges, M. Ciappina, U. Thumm, W. Boutu and H. Merdji, *Appl. Phys. Lett.* **118**, 20 (2021).
[22] J. Alcalà, U. Bhattacharya, J. Biegert, M. Ciappina, U. Elu, T. Graß, P. T. Grochowski, M. Lewenstein, A. Palau, T. P. H. Sidiropoulos, T. Steinle and I. Tyulnev, *Proc. Natl. Acad. Sci.* **119**, 40, p. e2207766119 (2022).
[23] C. Heide, Y. Kobayashi, D. R. Baykusheva, D. Jain, J. A. Sobota, M. Hashimoto, P. S. Kirchmann, S. Oh, T. F. Heinz, D. A. Reis and S. Ghimire, *Nat. Photonics* **16**, 9, pp. 620–624 (2022).
[24] K. Uchida, G. Mattoni, S. Yonezawa, F. Nakamura, Y. Maeno and K. Tanaka, *Phys. Rev. Lett.* **128**, 12, p. 127401 (2022).
[25] D. Garratt, L. Misiekis, D. Wood, E. W. Larsen, M. Matthews, O. Alexander, P. Ye, S. Jarosch, C. Ferchaud, C. Strüber *et al.*, *Nat. Commun.* **13**, 1, p. 3414 (2022).
[26] J. Freudenstein, M. Borsch, M. Meierhofer, D. Afanasiev, C. P. Schmid, F. Sandner, M. Liebich, A. Girnghuber, M. Knorr, M. Kira and R. Huber, *Nature* **610**, 7931, pp. 290–295 (2022).

[27] E. Goulielmakis and T. Brabec, *Nat. Photonics* **16**, 6, pp. 411–421 (2022).
[28] N. W. Ashcroft and N. D. Mermin, *Solid State Physics*. Halcourt College Publishers (1976).
[29] J. B. Krieger and G. J. Iafrate, *Phys. Rev. B* **33**, pp. 5494–5500 (1986).
[30] U. Lindefelt, H.-E. Nilsson and M. Hjelm, *Semicond. Sci. Technol.* **19**, 8, p. 1061 (2004).
[31] H. Haug and S. W. Koch, *Quantum Theory of the Optical and Electronic Properties of Semiconductors*. World Scientific Publishing Co. (2004).
[32] S. Ghimire *et al.*, *Nat. Phys.* **7**, p. 138 (2011).
[33] L. V. Keldysh, *JETP* **20**, p. 1307 (1965).
[34] G. Vampa, C. R. McDonald, G. Orlando, P. B. Corkum and T. Brabec, *Phys. Rev. B* **91**, p. 064302 (2015).
[35] Y.-N. Xu and W. Y. Ching, *Phys. Rev. B* **43**, pp. 4461–4472 (1991).
[36] T.-Y. Du and X.-B. Bian, *Opt. Express* **25**, 1, pp. 151–158 (2017).
[37] N. Moiseyev, *Phys. Rev. A* **91**, p. 053811 (2015).
[38] X. Song, S. Yang, R. Zuo, T. Meier and W. Yang, *Phys. Rev. A* **101**, p. 033410 (2020).
[39] R. W. Boyd, *Nonlinear Optics*. Academic Press (2003).
[40] D. Ray, F. He, S. De, W. Cao, H. Mashiko, P. Ranitovic, K. P. Singh, I. Znakovskaya, U. Thumm, G. G. Paulus, M. F. Kling, I. V. Litvinyuk and C. L. Cocke, *Phys. Rev. Lett.* **103**, p. 223201 (2009).
[41] M. Wu, D. A. Browne, K. J. Schafer and M. B. Gaarde, *Phys. Rev. A* **94**, p. 063403 (2016).
[42] S. R. Leone, C. W. McCurdy, J. Burgdörfer, L. S. Cederbaum, Z. Chang, N. Dudovich, J. Feist, C. H. Greene, M. Ivanov, R. Kienberger, U. Keller, M. F. Kling, Z.-H. Loh, T. Pfeifer, A. N. Pfeiffer, R. Santra, K. Schafer, A. Stolow, U. Thumm and M. J. J. Vrakking, *Nat. Photonics* **8**, pp. 162 (2014).
[43] U. Thumm, Q. Liao, E. M. Bothschafter, F. Süßmann, M. F. Kling and R. Kienberger, *Attosecond Physics: Attosecond Streaking Spectroscopy of Atoms and Solids*, chap. 13. John Wiley & Sons, pp. 387–441 (2015).
[44] T. Huang, X. Zhu, L. Li, X. Liu, P. Lan and P. Lu, *Phys. Rev. A* **96**, p. 043425 (2017).
[45] C. Yu, K. K. Hansen and L. B. Madsen, *Phys. Rev. A* **99**, p. 013435 (2019).
[46] H. Iravani, K. K. Hansen and L. B. Madsen, *Phys. Rev. Res.* **2**, p. 013204 (2020).
[47] A. Pattanayak, M. M. S. and G. Dixit, *Phys. Rev. A* **101**, p. 013404 (2020).
[48] H. Haug and S. Koch, *Theory of Transport Properties of Semiconductor Nanostructures*. Springer (1998).
[49] G. Vampa, C. McDonald, A. Fraser and T. Brabec, *IEEE Journal of Selected Topics in Quantum Electronics* **21**, 5, pp. 1–10 (2015).
[50] X. Liu, X. Zhu, P. Lan, X. Zhang, D. Wang, Q. Zhang and P. Lu, *Phys. Rev. A* **95**, p. 063419 (2017).
[51] R. Whited, C. J. Flaten and W. Walker, *Solid State Commun.* **13**, 11, pp. 1903–1905 (1973).

[52] Y. S. You, D. Reis and S. Ghimire, *Nat. Phys.* **13**, p. 345 (2017).
[53] S. Almalki, A. M. Parks, G. Bart, P. B. Corkum, T. Brabec and C. R. McDonald, *Phys. Rev. B* **98**, p. 144307 (2018).
[54] L. Kasmi, M. Lucchini, L. Castiglioni, P. Kliuiev, J. Osterwalder, M. Hengsberger, L. Gallmann, P. Krüger and U. Keller, *Optica* **4**, p. 1492 (2017).
[55] M. J. Ambrosio and U. Thumm, *Phys. Rev. A* **100**, p. 043412 (2019).
[56] Q. Liao, W. Cao, Q. Zhang, K. Liu, F. Wang, P. Lu and U. Thumm, *Phys. Rev. Lett.* **125**, p. 043201 (2020).
[57] F. Stavale, N. Nilius and H.-J. Freund, *New J. Phys.* **14**, p. 033006 (2012).
[58] E. Shablonin, A. I. Popov, A. Lushchik, A. Kotlov and S. Dolgova, *Physica B: Condensed Matter* **477**, p. 133 (2015).

https://doi.org/10.1142/9789811279560_0012

# Chapter 12

# High-Order Harmonic Generation from Unconventional Superconductors

Tobias Graß[1,2], Jordi Alcalà[3], Utso Bhattacharya[4], Jens Biegert[4,5], Marcelo F. Ciappina[6,7,8], Ugaitz Elu[4], Piotr T. Grochowski[9,10], Maciej Lewenstein[4,5], Anna Palau[3], Themistoklis P. H. Sidiropoulos[4], Tobias Steinle[4] and Igor Tyulnev[4]

[1]*DIPC — Donostia International Physics Center, Paseo Manuel de Lardizábal 4, 20018 San Sebastián, Spain*
[2]*IKERBASQUE, Basque Foundation for Science, Plaza Euskadi 5, 48009 Bilbao, Spain*
[3]*ICMAB-CSIC — Institut de Ciència de Materials de Barcelona, Campus UAB, 08193 Bellaterra, Spain*
[4]*ICFO — Institut de Ciències Fotòniques, The Barcelona Institute of Science and Technology, 08860 Castelldefels (Barcelona), Spain*
[5]*ICREA, Pg. Lluis Companys 23, ES-08010 Barcelona, Spain*
[6]*Guangdong Technion — Israel Institute of Technology, Shantou, 515063 Guangdong, China*
[7]*Technion — Israel Institute of Technology, 32000 Haifa, Israel*
[8]*Guangdong Provincial Key Laboratory of Materials and Technologies for Energy Conversion, Guangdong Technion — Israel Institute of Technology, 241 Daxue Road, Shantou, Guangdong, China, 515063*
[9]*Institute for Quantum Optics and Quantum Information, Austrian Academy of Sciences, A-6020 Innsbruck, Austria*
[10]*Institute for Theoretical Physics, University of Innsbruck, A-6020 Innsbruck, Austria*

## 12.1 Introduction

Despite decades of experimental and theoretical research, high-temperature superconductivity remains one of the unsolved problems in condensed matter physics. While conventional superconductivity is understood as an instability of the Fermi liquid due to electron pairing, the microscopic understanding of unconventional superconductivity is challenged by the fact that this phase is surrounded by strongly correlated, non-Fermi liquid phases, such as the pseudogap phases, strange metal phases, or even Mott insulators. Apart from the fundamental interest in this intriguing many-body behavior, understanding the mechanisms of high-temperature superconductivity will also have huge technological impact, potentially guiding us towards room-temperature superconductivity. Experimental techniques which have proven extremely insightful so far including angle-resolved photoemission spectroscopy [1–4], providing a window to the momentum-resolved electronic density of states, and resonant inelastic X-ray scattering [5, 6], which can be used to probe electron-phonon interactions. Non-linear optics has been used to probe the Higgs mode via third-harmonic generation [7] and nonlinear optical susceptibility [8]. Pump-probe experiments [9, 10] and responses to ultra-intense electromagnetic fields [11] have been used to explore the nonequilibrium dynamics of unconventional superconductors. The optical spectrum emitted by $YBa_2Cu_3O_{7-x}$ (YBCO), a primary representative of materials with unconventional superconductivity, after excitation with an ultrastrong mid-infrared light pulse has been studied in Ref. [11]. The high harmonic peaks, seen in the experimental data, have theoretically been reproduced by a semi-microscopic model of the electronic dynamics. This model combines the coherent optical drive of a Hubbard-like system with phenomenological decay processes. In this chapter, we review this simple yet powerful model.

## 12.2 Model

The degrees of freedom which are predominantly coupled to light are the electrons. Hence, in the microscopic description of the material, we restrict ourselves to electrons. As usual in a solid-state system, the electronic behavior is described by the band structure:

$$H_{\text{el}} = \sum_{\vec{k},\sigma,\nu} \varepsilon_{\vec{k},\nu} c^{\dagger}_{\vec{k},\sigma,\nu} c_{\vec{k},\sigma,\nu}, \tag{12.1}$$

where $\epsilon_{\vec{k},\nu}$ denote the energies in Bloch band $\nu$ at wave vector $\vec{k}$. Depending on the dimensionality of the system, $\vec{k}$ can be a scalar or a vector. The operators $c^{\dagger}_{\vec{k},\sigma,\nu}(c_{\vec{k},\sigma,\nu})$ create (annihilate) an electron of spin $\sigma$ at wave vector $\vec{k}$ in band $\nu$. The quadratic Hamiltonian in Eq. (12.1) describes electrons interacting only with an external potential. Additionally, the electrons interact amongst themselves, which gives rise to a quartic term in the Hamiltonian. Although there is little doubt that in a strongly correlated material, such as a high-temperature superconductor, the interaction term plays an important role for understanding the equilibrium phase diagram, we neglect it here. The practical reason behind this is the fact that the quartic term dramatically complicates the solution of the model. While modern methods, e.g., based on tensor network states and time-evolving block decimation, might provide tools to explicitly include electron-electron scattering, this would not automatically guarantee a more accurate description of the material. Other effects, such as interactions with lattice vibrations (phonons) or with defects in the lattice, may affect the electron dynamics on, possibly, similar time scales as bare electron-electron interactions. We therefore restrict ourselves to the simple quadratic form of the Hamiltonian, and account for effects beyond the model through phenomenological terms in the equation of motion (see Sec. 12.3.3).

We also note that a quadratic Hamiltonian like Eq. (12.1) may, to some extent, account for interactions. For instance, the band structure calculation can include a Hartree-Fock potential, which considers interactions on the mean-field level. In the model used in Ref. [11], the band structure was obtained through density functional theory (DFT), which allows for mapping the interacting electrons onto an efficient system of free electrons in the so-called Kohn-Sham orbitals.

### 12.2.1 *D-Wave Pairing Hamiltonian*

Nevertheless, the Hamiltonian in Eq. (12.1) is clearly not sufficient to capture the superconducting phase, as this requires a pairing interaction. On a mean-field level, this interaction is of the form $c^2$ or $c^{\dagger 2}$. We will derive such a pairing term from an assumed attractive interaction $U$ between electrons of opposite spins localized on neighboring sites $i$ and $j$:

$$H_{\mathrm{p}} = U \sum_{\langle\langle i,j \rangle\rangle} c^{\dagger}_{i\uparrow} c^{\dagger}_{j\downarrow} c_{j\downarrow} c_{i\uparrow}. \tag{12.2}$$

In this expression, the creation/annihilation operators are assumed to act in a basis of Wannier states of the Bloch band at the Fermi surface. For brevity, we have dropped the band index. We will re-introduce this index later after having derived the mean-field expression of the pairing term.

In cuprates, the CuO planes form a 2D square lattice, so a site at position $\vec{r}_j$ has four nearest neighbors at positions $\vec{r}_j \pm a\hat{x}$ and $\vec{r}_j \pm a\hat{y}$. With this, the Fourier-transformed pairing Hamiltonian reads

$$H_{\rm p} = U \sum_{\substack{\vec{k}_1,\vec{k}_2 \\ \vec{k}_3,\vec{k}_4}} \sum_j e^{i(\vec{k}_4+\vec{k}_3-\vec{k}_2-\vec{k}1)\cdot\vec{r}_j} [e^{i(k_{3x}-k_{2x})a} + e^{-i(k_{3x}-k_{2x})a} + e^{i(k_{3y}-k_{2y})a} + e^{-i(k_{3y}-k_{2y})a}] c^{\dagger}_{\vec{k}_1\uparrow}(t) c^{\dagger}_{\vec{k}_2\downarrow}(t) c_{\vec{k}_3\downarrow}(t) c_{\vec{k}_4\uparrow}(t). \tag{12.3}$$

We can now use the identity $\sum_j [e^{i(\vec{k}'-\vec{k})\cdot\vec{r}_j}] = \delta_{\vec{k},\vec{k}'}$, and separate into relative and center-of-mass coordinates,

$$\vec{q} = \frac{1}{2}(\vec{k}_1 - \vec{k}_2), \tag{12.4}$$

$$\vec{q}' = \frac{1}{2}(\vec{k}_4 - \vec{k}_3), \tag{12.5}$$

$$\vec{Q} = \vec{k}_1 + \vec{k}_2. \tag{12.6}$$

With this, the pairing Hamiltonian becomes

$$H_{\rm p} = \sum_{\vec{q},\vec{q}',\vec{Q}} U_{\vec{q}-\vec{q}'} c^{\dagger}_{\vec{q}+\frac{1}{2}\vec{Q}\uparrow} c^{\dagger}_{-\vec{q}+\frac{1}{2}\vec{Q}\downarrow} c_{-\vec{q}'+\frac{1}{2}\vec{Q}\downarrow} c_{\vec{q}'+\frac{1}{2}\vec{Q}\uparrow}, \tag{12.7}$$

where

$$U_{\vec{q}-\vec{q}'} = 2U[\cos(q_x a - q'_x a) + \cos(q_y a - q'_y a)]. \tag{12.8}$$

The possible gap symmetries can be inferred from the decomposition of $U_{\vec{q}-\vec{q}'}$ in irreducible representations of the point group symmetry of the square lattice. Thus, we get $U_{\vec{q}-\vec{q}'} = U^{(s)}_{\vec{q}-\vec{q}'} + U^{(p)}_{\vec{q}-\vec{q}'} + U^{(d)}_{\vec{q}-\vec{q}'}$ with

$$U^{(s)}_{\vec{q}-\vec{q}'} = U[\cos(q_x a) + \cos(q_y a)][\cos(q'_x a) + \cos(q'_y a)], \tag{12.9}$$

$$U^{(p)}_{\vec{q}-\vec{q}'} = U\{[\sin(q_x a) + i\sin(q_y a)][\sin(q'_x a) - i\sin(q'_y a)] + [\sin(q_x a) - i\sin(q_y a)][\sin(q'_x a) + i\sin(q'_y a)]\}, \tag{12.10}$$

$$U^{(d)}_{\vec{q}-\vec{q}'} = U[\cos(q_x a) - \cos(q_y a)][\cos(q'_x a) - \cos(q'_y a)]. \tag{12.11}$$

In the following, we retain only the d-type in $U_{\vec{q}-\vec{q}'}$ as for CuO it is known that the contribution of density of states for d-type is much higher than s-type or p-type. Defining $u_{\vec{q}} \equiv \cos(q_x a) - \cos(q_y a)$, the d-wave pairing Hamiltonian is written as

$$H_{\rm p} = U \sum_{\vec{q},\vec{q}',\vec{Q}} u_{\vec{q}} u_{\vec{q}'} c^{\dagger}_{\vec{q}+\frac{1}{2}\vec{Q}\uparrow} c^{\dagger}_{-\vec{q}+\frac{1}{2}\vec{Q}\downarrow} c_{-\vec{q}'+\frac{1}{2}\vec{Q}\downarrow} c_{\vec{q}'+\frac{1}{2}\vec{Q}\uparrow}. \tag{12.12}$$

As mentioned above, it is very hard to treat quartic interaction terms exactly, but by applying a mean field decoupling of the form $ab = \langle c\rangle b + a\langle b\rangle - \langle a\rangle\langle b\rangle$, we can reduce the pairing Hamiltonian to a quadratic form. Ignoring the constant term $(\langle a\rangle\langle b\rangle)$, the mean-field pairing Hamiltonian reads

$$H_{\rm mfp} = \sum_{\vec{Q},\vec{q}} \Delta_{\vec{Q},\vec{q}} c^{\dagger}_{\vec{q}+\frac{1}{2}\vec{Q}\uparrow} c^{\dagger}_{-\vec{q}+\frac{1}{2}\vec{Q}\downarrow} + \text{h.c.}, \tag{12.13}$$

where we have introduced the complex pairing field $\Delta_{\vec{Q},\vec{q}}$, defined as

$$\Delta_{\vec{Q},\vec{q}} = U u_{\vec{q}} \left[ \sum_{\vec{q}'} u_{\vec{q}'} \langle c_{-\vec{q}'+\frac{1}{2}\vec{Q}\downarrow} c_{\vec{q}'+\frac{1}{2}\vec{Q}\uparrow} \rangle \right]. \tag{12.14}$$

Next, we will assume that the center-of-mass momentum of the pairing vanishes, $\Delta_{\vec{Q},\vec{q}} = \Delta_{\vec{q}} \delta_{\vec{Q},\vec{0}}$.

### 12.2.2 *Two-Band Model*

Let us now return to the band structure described by Eq. (12.1). In a minimal model, we assume that bands below the Fermi surface are inert, and the dynamics is described by considering a band at the Fermi surface, which we will denote by the letter 'F', and an empty band into which electrons can be excited, denoted by the letter 'E'. The system Hamiltonian of this two-band system reads

$$H_{\rm sys}(t) = \sum_{\substack{\vec{k},\sigma \\ \nu=\{\rm F,E\}}} \varepsilon_{\vec{k},\nu} c^{\dagger}_{\vec{k},\sigma,\nu} c_{\vec{k},\sigma,\nu} + \sum_{\vec{k}} \Delta^{*}_{\vec{k}}(t) c_{-\vec{k},\downarrow,\rm F} c_{\vec{k},\uparrow,\rm F} + \text{h.c.} \tag{12.15}$$

The Hamiltonian above appears with an explicit time dependence, since, if the system is not in equilibrium, the pairing field becomes time-dependent. In the context of an optical experiment, such a non-equilibrium situation is produced by quenching the system Hamiltonian with a light-matter interaction Hamiltonian $H_{\rm l-m}(t)$.

### 12.2.3 *Light-Matter Interaction*

To describe the light-matter interaction we need to make a gauge choice. In the so-called *velocity gauge*, the vector potential $\vec{A}(\vec{r},t)$ of the electromagnetic field couples to the phase of the wave function by replacing the momentum operator $(-i\hbar\nabla)$ by $(-i\hbar\nabla - e\vec{A}(\vec{r},t))$. Alternatively, in the *length gauge*, the electric field of the light, $\vec{E}(\vec{r},t)$, interacts with the charge distribution $\rho(\vec{r})$ in the material through the term $\int \mathrm{d}V \rho(\vec{r})\vec{r}\cdot\vec{E}(\vec{r},t)$. In the dipole approximation, one essentially assumes a homogeneous electric field $\vec{E}(\vec{r},t) = \vec{E}(t)$, justified by the fact that spatial variations of light occur on length scales given by the wavelength of the light (thousands of nanometers), whereas spatial variations of the electronic wave function occur on length scales on the order of a unit cell (fraction of a nanometer). The integral over the electron distribution is calculated using the Bloch theorem, splitting the electronic wave function into a lattice-periodic Bloch function $u_{\vec{k},\nu}(\vec{r}) = u_{\vec{k},\nu}(\vec{r}+\vec{R})$, where $\vec{R}$ is an arbitrary lattice vector, and a plane wave $e^{i\vec{k}\cdot\vec{r}}$:

$$\begin{aligned}\int \mathrm{d}\vec{r}\rho(\vec{r})\vec{r} &= e\sum_{\vec{k},\vec{k}'}\sum_{\nu,\nu',\sigma} c^{\dagger}_{\vec{k}',\sigma,\nu'}c_{\vec{k},\sigma,\nu}\int \mathrm{d}V\ u^{*}_{\vec{k}',\nu'}(\vec{r})e^{i\vec{k}'\cdot\vec{r}}\ \vec{r}\ u_{\vec{k},\nu}(\vec{r})e^{-i\vec{k}\cdot\vec{r}} \\ &= e\sum_{\vec{k},\vec{k}'}\sum_{\nu,\nu',\sigma} c^{\dagger}_{\vec{k}',\sigma,\nu'}c_{\vec{k},\sigma,\nu}\sum_{\vec{R}} e^{i(\vec{k}'-\vec{k})\cdot\vec{R}} \\ &\quad\times\int_{\text{cell}} \mathrm{d}V\ u^{*}_{\vec{k}',\nu'}(\vec{r})(i\nabla_{\vec{k}})u_{\vec{k},\nu}(\vec{r}).\end{aligned} \tag{12.16}$$

Here, we have split the spatial integral into a sum over lattice vectors $\vec{R}$ and an integral over the unit cell. We have anticipated that summation yields a Kronecker delta in $\vec{k}$, such that the plane waves drop out of the integral. We have also used the identity $\vec{r} = i\nabla_{\vec{k}}$, such that we can apply the product rule. Exploiting also the orthonormality of Bloch functions we arrive at

$$\begin{aligned}H_{\mathrm{l-m}}(t) &= \vec{E}(t)\cdot\int \mathrm{d}V\rho(\vec{r})\vec{r} = e\vec{E}(t)\cdot\sum_{\vec{k},\vec{k}'}\sum_{\nu,\nu',\sigma}\vec{r}_{\vec{k},\nu;\vec{k}',\nu'}c^{\dagger}_{\vec{k}',\sigma,\nu'}c_{\vec{k},\sigma,\nu} \\ &= e\vec{E}(t)\cdot\sum_{\vec{k}}\sum_{\nu,\nu',\sigma}\left[\delta_{\nu',\nu}(i\nabla_{\vec{k}}) + \vec{d}_{\vec{k},\nu',\nu}\right]c^{\dagger}_{\vec{k}',\sigma,\nu'}c_{\vec{k},\sigma,\nu}.\end{aligned} \tag{12.17}$$

In the second line, we have introduced $\vec{r}_{\vec{k},\nu;\vec{k}',\nu'}$ as the matrix element of the position operator in a basis of Bloch bands. In the third line, we have introduced $\vec{d}_{\vec{k},\nu',\nu} = i\int \mathrm{d}\vec{r}\, u^{*}_{\vec{k}',\nu'}(\vec{r})\nabla_{\vec{k}} u_{\vec{k}',\nu'}(\vec{r})$, which is the interband dipole moment, if $\nu \neq \nu'$, or the Berry connection of the Bloch band $\nu$, if $\nu = \nu'$.

## 12.3 Dynamics of an Optically Driven Superconductor

From now on we are going to concentrate on a two-band model as given by $H_{\mathrm{sys}}(t)$ in Eq. (12.15), and its dynamical response to the presence of light-matter coupling as given by $H_{\mathrm{l-m}}(t)$ in Eq. (12.17). We assume that initially the system is in equilibrium at a given temperature $T$.

### 12.3.1 *Equilibrium*

The Hamiltonian $H_{\mathrm{sys}}(t)$ of Eq. (12.15) is diagonal only in the block which corresponds to the excited band, but not in the lower band due to the presence of pairing interactions. To diagonalize also this part, we apply a Bogoliubov transformation, that is, we express the creation and annihilation operators in the lower band in terms of quasiparticle operators $\gamma_{\vec{k},\sigma}$:

$$c_{\vec{k},\uparrow,\mathrm{F}} = a_{\vec{k}}\gamma_{\vec{k},\uparrow} + b_{\vec{k}}\gamma^{\dagger}_{-\vec{k},\downarrow} \tag{12.18}$$

$$c^{\dagger}_{\vec{k},\uparrow,\mathrm{F}} = a^{*}_{\vec{k}}\gamma^{\dagger}_{\vec{k},\uparrow} + b^{*}_{\vec{k}}\gamma_{-\vec{k},\downarrow} \tag{12.19}$$

$$c^{\dagger}_{-\vec{k},\downarrow,\mathrm{F}} = -b^{*}_{\vec{k}}\gamma_{\vec{k},\uparrow} + a^{*}_{\vec{k}}\gamma^{\dagger}_{-\vec{k},\downarrow} \tag{12.20}$$

$$c_{-\vec{k},\downarrow,\mathrm{F}} = -b_{\vec{k}}\gamma^{\dagger}_{\vec{k},\uparrow} + a_{\vec{k}}\gamma_{-\vec{k},\downarrow}. \tag{12.21}$$

The coefficients $a_{\vec{k}}$ and $b_{\vec{k}}$ are defined by the system parameters as

$$|a_{\vec{k}}|^2 - |b_{\vec{k}}|^2 = \frac{\varepsilon_{\vec{k},\mathrm{F}}}{\sqrt{(\varepsilon_{\vec{k},\mathrm{F}})^2 + |\Delta_{\vec{k}}|^2}}, \tag{12.22}$$

$$|a_{\vec{k}}|^2 = \frac{1}{2}\left(1 + \frac{\varepsilon_{\vec{k},\mathrm{F}}}{\sqrt{(\varepsilon_{\vec{k},\mathrm{F}})^2 + |\Delta_{\vec{k}}|^2}}\right), \tag{12.23}$$

$$|b_{\vec{k}}|^2 = \frac{1}{2}\left(1 - \frac{\varepsilon_{\vec{k},\mathrm{F}}}{\sqrt{(\varepsilon_{\vec{k},\mathrm{F}})^2 + |\Delta_{\vec{k}}|^2}}\right), \tag{12.24}$$

$$a_{\vec{k}}b_{\vec{k}} = \frac{\Delta_{\vec{k}}}{2\sqrt{(\varepsilon_{\vec{k},\mathrm{F}})^2 + |\Delta_{\vec{k}}|^2}}. \tag{12.25}$$

In this Bogoliubov basis, the system Hamiltonian thus reads

$$H_{\rm sys} = \sum_{\vec{k},\sigma} E_{\vec{k},\rm E} c^{\dagger}_{\vec{k},\sigma,\rm E} c_{\vec{k},\sigma,\rm E} + \sum_{\vec{k},\sigma} E_{\vec{k},\rm F} \gamma^{\dagger}_{\vec{k},\sigma} \gamma_{\vec{k},\sigma} + \text{const.} \tag{12.26}$$

where $E_{\vec{k},\rm F} = \sqrt{(\varepsilon_{\vec{k},\rm F})^2 + |\Delta_{\vec{k}}|^2} - \mu$, $E_{\vec{k},\rm e} = \varepsilon_{\vec{k},\rm F} - \mu$. Here, we have explicitly introduced a chemical potential $\mu$, which allows us to adjust the Fermi surface to the desired filling, which in the experiment is controlled by the doping and/or gate voltage.

The thermal equilibrium for the system can now easily be written down. As the electrons, the Bogoliubov quasiparticles are fermions, and hence, the equilibrium occupation of the fermionic modes is given by the Fermi-Dirac distribution:

$$\left\langle \gamma^{\dagger}_{\vec{k},\uparrow} \gamma_{\vec{k},\uparrow} \right\rangle = n_{\rm FD}(E_{\vec{k},\rm F}, T), \tag{12.27}$$

$$\left\langle \gamma_{-\vec{k},\downarrow} \gamma^{\dagger}_{-\vec{k},\downarrow} \right\rangle = 1 - \left\langle \gamma^{\dagger}_{-\vec{k},\downarrow} \gamma_{-\vec{k},\downarrow} \right\rangle = 1 - n_{\rm FD}(E_{\vec{k},\rm F}, T), \tag{12.28}$$

$$\left\langle c^{\dagger}_{\vec{k},\uparrow,\rm E} c_{\vec{k},\uparrow,\rm E} \right\rangle = \left\langle c^{\dagger}_{\vec{k},\downarrow,\rm E} c_{\vec{k},\downarrow,\rm E} \right\rangle = n_{\rm FD}(E_{\vec{k},\rm E}, T). \tag{12.29}$$

Here, $n_{\rm FD}(E,T) = (1 + e^{E/(k_B T)})^{-1}$ is the Fermi-Dirac distribution.

In equilibrium, there are no correlations between the eigenmodes of the system. This means that correlators of the form $(\gamma)^2$, $(\gamma^{\dagger})^2$, $c_{\rm E}^2$, $(c_{\rm E}^{\dagger})^2$, $c_{\rm E}\gamma$, $c_{\rm E}^{\dagger}\gamma$, $c_{\rm E}\gamma^{\dagger}$ or $c_{\rm E}^{\dagger}\gamma^{\dagger}$ are all zero. However, if we transform back from the Bogoliubov quasiparticles to the original electrons, we will find pairing correlations between electrons in the lower band F. Considering the correlations in terms of the original electron operators is useful also from the perspective of dynamics, since the light-matter Hamiltonian, Eq. (12.17), has a simple form in terms of electron operators rather than Bogoliubov operators. We introduce all quadratic operator products in the electron basis in Table 12.1.

From the population of the Bogoliubov eigenmodes, Eq. (12.27), we find for the equilibrium values of the electronic operator products:

$$\begin{aligned} n_{\vec{k},\uparrow,\rm F} &= \left\langle c^{\dagger}_{\vec{k},\uparrow,\rm F} c_{\vec{k},\uparrow,\rm F} \right\rangle = \left\langle \left( a^*_{\vec{k}} \gamma^{\dagger}_{\vec{k},\uparrow} + b^*_{\vec{k}} \gamma_{-\vec{k},\downarrow} \right) \left( a_{\vec{k}} \gamma_{\vec{k},\uparrow} + b_{\vec{k}} \gamma^{\dagger}_{-\vec{k},\downarrow} \right) \right\rangle \\ &= |a_{\vec{k}}|^2 n_{\rm FD}(E_{\vec{k},\rm F}, T) + |b_{\vec{k}}|^2 [1 - n_{\rm FD}(E_{\vec{k},\rm F}, T)] = n_{-\vec{k},\uparrow,\rm F}, \end{aligned} \tag{12.30}$$

$$\begin{aligned} n_{-\vec{k},\downarrow,\rm F} &= \left\langle c^{\dagger}_{\vec{k},\downarrow,\rm F} c_{\vec{k},\downarrow,\rm F} \right\rangle = \left\langle \left( -b^*_{\vec{k}} \gamma_{\vec{k},\uparrow} + a^*_{\vec{k}} \gamma^{\dagger}_{-\vec{k},\downarrow} \right) \left( -b_{\vec{k}} \gamma^{\dagger}_{\vec{k},\uparrow} + a_{\vec{k}} \gamma_{-\vec{k},\downarrow} \right) \right\rangle \\ &= |a_{\vec{k}}|^2 n_{\rm FD}(E_{\vec{k},\rm F}, T) + |b_{\vec{k}}|^2 [1 - n_{\rm FD}(E_{\vec{k},\rm F}, T)] = n_{\vec{k},\downarrow,\rm F}, \end{aligned} \tag{12.31}$$

Table 12.1 All possible operator products of two-electron operators.

| | |
|---|---|
| $n_{\vec{k},\uparrow,\mathrm{E}} = \left\langle c^{\dagger}_{\vec{k},\uparrow,\mathrm{E}} c_{\vec{k},\uparrow,\mathrm{E}} \right\rangle$ | Population (upper band, spin-up) |
| $n_{\vec{k},\downarrow,\mathrm{E}} = \left\langle c^{\dagger}_{\vec{k},\downarrow,\mathrm{E}} c_{\vec{k},\downarrow,\mathrm{E}} \right\rangle$ | Population (upper band, spin-down) |
| $n_{\vec{k},\uparrow,\mathrm{F}} = \left\langle c^{\dagger}_{\vec{k},\uparrow,\mathrm{F}} c_{\vec{k},\uparrow,\mathrm{F}} \right\rangle$ | Population (lower band, spin-up) |
| $n_{\vec{k},\downarrow,\mathrm{F}} = \left\langle c^{\dagger}_{\vec{k},\downarrow,\mathrm{F}} c_{\vec{k},\downarrow,\mathrm{F}} \right\rangle$ | Population (lower band, spin-down) |
| $V_{\vec{k}} = \left\langle c^{\dagger}_{\vec{k},\uparrow,\mathrm{F}} c^{\dagger}_{-\vec{k},\downarrow,\mathrm{F}} \right\rangle$ | Pairing correlation (lower band) |
| $C_{\vec{k}} = \left\langle c^{\dagger}_{\vec{k},\uparrow,\mathrm{E}} c^{\dagger}_{-\vec{k},\downarrow,\mathrm{E}} \right\rangle$ | Pairing correlation (upper band) |
| $P^{*}_{\vec{k},\uparrow} = \left\langle c^{\dagger}_{\vec{k},\uparrow,\mathrm{E}} c_{\vec{k},\uparrow,\mathrm{F}} \right\rangle$ | Interband polarization (spin-up) |
| $P^{*}_{\vec{k},\downarrow} = \left\langle c^{\dagger}_{\vec{k},\downarrow,\mathrm{E}} c_{\vec{k},\downarrow,\mathrm{F}} \right\rangle$ | Interband polarization (spin-down) |
| $A_{\vec{k},\uparrow\downarrow} = \left\langle c^{\dagger}_{\vec{k},\uparrow,\mathrm{E}} c^{\dagger}_{-\vec{k},\downarrow,\mathrm{F}} \right\rangle$ | Interband pairing correlation 1 |
| $A_{\vec{k},\downarrow\uparrow} = \left\langle c^{\dagger}_{\vec{k},\downarrow,\mathrm{E}} c^{\dagger}_{-\vec{k},\uparrow,\mathrm{F}} \right\rangle$ | Interband pairing correlation 2 |

$$\begin{aligned} V_{\vec{k}} &= \left\langle c^{\dagger}_{\vec{k},\uparrow,\mathrm{F}} c^{\dagger}_{-\vec{k},\downarrow,\mathrm{F}} \right\rangle = \left\langle \left(a^{*}_{\vec{k}} \gamma^{\dagger}_{\vec{k},\uparrow} + b^{*}_{\vec{k}} \gamma_{-\vec{k},\downarrow}\right) \left(-b^{*}_{\vec{k}} \gamma_{\vec{k},\uparrow} + a^{*}_{\vec{k}} \gamma^{\dagger}_{-\vec{k},\downarrow}\right) \right\rangle \\ &= a^{*}_{\vec{k}} b^{*}_{\vec{k}} [1 - 2n_{\mathrm{FD}}(E_{\vec{k},\mathrm{F}}, T)], \end{aligned} \tag{12.32}$$

$$C_{\vec{k}} = P^{*}_{\vec{k},\uparrow} = P^{*}_{\vec{k},\downarrow} = A_{\vec{k},\uparrow\downarrow} = A_{\vec{k},\downarrow\uparrow} = 0. \tag{12.33}$$

We note that through Eq. (12.14), the pairing field $\Delta_{\vec{k}}$ is related to the correlators $V_{\vec{k}'}$, which themselves, through the equations Eq. (12.25) and Eq. (12.32), are related to the fields $\Delta_{\vec{k}'}$. This leads to a famous self-consistency equation, also known as gap equation:

$$\Delta_{\vec{k}} = U u_{\vec{k}} \sum_{\vec{k}'} u_{\vec{k}'} \frac{\Delta_{\vec{k}'}}{2\sqrt{\varepsilon^{2}_{\vec{k}',\mathrm{F}} + |\Delta_{\vec{k}'}|^{2}}} [1 - 2n_{\mathrm{FD}}(E_{\vec{k}',\mathrm{F}}, T)]. \tag{12.34}$$

Above a critical temperature, which depends on the strength of the pairing interaction $U$, the only solution to the gap equation is the trivial one, $\Delta_{\vec{k}} = 0$. Hence, knowing the critical temperature, we can fix the model parameter $U$. We plot in Fig. 12.1(a) the band structure of the material, calculated using DFT, joint with the solution of Eq. (12.34) (Fig. 12.1(b)).

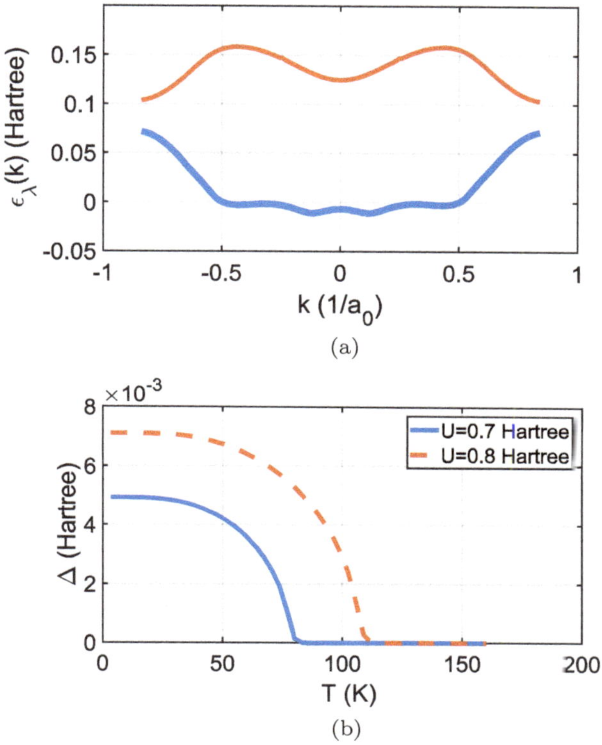


Fig. 12.1 (a) Band structure, obtained from DFT calculation, which has been used in the two-band model. (b) The pairing field $\Delta$, summed over $k$, self-consistently obtained from the gap equation, Eq. (12.34), in the two-band model as a function of temperature. Knowing that the critical temperature of YBCO is around 90 K, we are able to choose the parameter $U$ of the pairing interaction.

### 12.3.2 *Superconductor Bloch Equations*

To describe the dynamical evolution of the material once an optical field has brought it out of equilibrium, we use the formalism of optical Bloch equations [12]. The formalism is based on the Heisenberg equation of motion for an operator $O(t)$:

$$i\frac{\partial}{\partial t}O(t) = -[H(t), O(t)]. \tag{12.35}$$

The Hamiltonian shall now include also the driving field, $H(t) = H_{\mathrm{sys}}(t) + H_{\mathrm{l-m}}(t)$. Replacing the operators on both sides by their expectation values, we obtain an equation of motion for $\langle O\rangle(t)$. As operators $O$, we consider any of the quadratic operator products listed in Table 12.1. Since the Hamiltonian is quadratic, all the terms on the right-hand side can be reduced to quadratic operator products. Hence, considering all choices for quadratic operator products, we obtain a closed set of equations. Moreover, the Hamiltonian mixes, at most, operators at wave vector $\vec{k}$ with operators at $-\vec{k}$. Hence, a set of 20 coupled differential equations, two for each of the 10 operators products listed in Table 12.1, at wave vectors $\vec{k}$ and $-\vec{k}$ has to be solved. Since these equations are straightforwardly derived from Eq. (12.35) by merely using some commutator algebra, we do not explicitly write them down.

We note that, in the absence of a pairing term, this set of coupled equations is further reduced. In this case, the Hamiltonian would be diagonal in $\vec{k}$, and only the populations $n_{\vec{k},\sigma,\nu}$ and the polarizations $P_{\vec{k},\sigma}$ would be non-zero. In the literature the closed set of equations of motion for band populations and interband polarization is known as *semiconductor Bloch equations*. Our analogous set of equations, extended by the pairing correlations and anomalous polarizations, might therefore be considered *superconductor Bloch equations*. We emphasize that despite choosing a system Hamiltonian without pairing interactions in the upper band and/or between the bands, the pairing correlations in the upper band, $C_{\vec{k}}$, and the intraband pairing correlations, $A_{\vec{k},\uparrow\downarrow}$ and $A_{\vec{k},\downarrow\uparrow}$, acquire non-zero values in the course of the dynamical evolution, if pairing correlations are initially present in the lower band, i.e., if $C_{\vec{k}} \neq 0$. This effect can be seen as some form of light-induced superconductivity.

### 12.3.3 *Phenomenological Model*

While the microscopic model discussed so far might capture the essential light-induced excitation processes in a superconducting material, it is of course too simple to describe the dynamical evolution of a real sample. This might be cured by adding more degrees of freedom, such as phonons or additional bands, and/or adding more terms, such as explicit couplings beyond the mean-field description. However, this would quickly lead to a model which is too complicated to solve. Instead, we choose to augment our microscopic model by additional phenomenological terms. Specifically, we argue that, after exciting the system with a short pulse, it decays back to equilibrium within typical time scales $\tau_1$, for excitations in the populations, and $\tau_2$, for populations in the correlations.

Such a decay mechanism is easily included in the equations of motion, Eq. (12.35), by adding a damping term

$$i\frac{\partial}{\partial t}O(t) = -[H(t), O(t)] - i\frac{O(t) - O(0)}{T}. \tag{12.36}$$

The damping time $T$ is $T = \tau_1$, if $O$ is a band population, or $T = \tau_2$ if $O$ is a correlation function. These times $\tau_{1,2}$ are assumed to account phenomenologically for all the processes which have not been explicitly included in the model. As such, these times are temperature-dependent, and we expect them to drop as temperature increases. However, we note that these times and their temperature dependence are not necessarily identical with the ones in linear response. Hence, they are not directly related to resistivity, and they are *a priori* unknown. Obviously, this reduces the predictive power of our description. Nevertheless, as we will show below, in combination with experimental data, some interesting insights into the material can be obtained from our theoretical model.

## 12.4 Results

We conclude by briefly reviewing the results which are obtained when the theoretical model, as detailed above in this chapter, is applied to the experimental data from a high harmonic spectroscopy experiment applied to an YBCO sample. The results have originally been published in Ref. [11].

### 12.4.1 *Electric Current*

For a direct comparison between theoretical and experimental data, we need to calculate the electric current as a function of time. Its Fourier transform into frequency space can then be compared with the experimentally obtained emission spectrum. To find a theoretical expression of the current in terms of the operator products given in Table 12.1, we first note that the velocity operator $\vec{v}$ in a non-relativistic system is given by $\vec{v} = \frac{1}{i\hbar}[\vec{r}, H_{\text{sys}}]$. Thus, its matrix elements in a basis of Bloch bands is given by

$$\vec{v}_{\vec{k},\nu;\vec{k}',\nu'} = \frac{1}{i\hbar}(\vec{r}_{\vec{k},\nu;\vec{k}',\nu'}\varepsilon_{\vec{k}',\nu'} - \varepsilon_{\vec{k},\nu}\vec{r}_{\vec{k},\nu;\vec{k}',\nu'}). \tag{12.37}$$

Here, $\vec{r}_{\vec{k},\nu;\vec{k}',\nu'}$ are the matrix elements of the position operator which have been introduced before in the derivation of the light-matter coupling, see Eq. (12.17). We get

$$\vec{v}_{\vec{k},\nu;\vec{k}',\nu'} = \frac{1}{\hbar}\delta_{\vec{k},\vec{k}'}[\delta_{\nu,\nu'}\nabla_{\vec{k}}\varepsilon_{\vec{k},\nu} - id_{\vec{k},\nu,\nu'}(\varepsilon_{\vec{k},\nu'} - \varepsilon_{\vec{k},\nu})]. \tag{12.38}$$

Therefore, the operator for the current is given by

$$\vec{J} = \sum_{\sigma} \sum_{\nu,\nu'} \sum_{\vec{k},\vec{k}'} \vec{v}_{\vec{k},\nu;\vec{k}',\nu'} c^{\dagger}_{\vec{k},\sigma,\nu} c_{\vec{k}',\sigma,\nu'}. \tag{12.39}$$

Accordingly, the expectation value of the current is obtained in terms of band populations and interband polarizations:

$$\langle \vec{J} \rangle = \frac{1}{\hbar} \sum_{\sigma,\vec{k}} \left[ \sum_{\nu} \nabla_{\vec{k}} \varepsilon_{\vec{k},\nu} n_{\vec{k},\sigma,\nu} - 2\vec{d}_{\vec{k},\mathrm{F},\mathrm{E}} (\varepsilon_{\vec{k},\mathrm{E}} - \varepsilon_{\vec{k},\mathrm{F}}) \mathrm{Im}(P_{\vec{k},\sigma}) \right]. \tag{12.40}$$

### 12.4.2 *High-Order Harmonics*

We are interested in the currents generated by a short, intense monochromatic light pulse. Specifically, for the laser electric field we take

$$E(t) = E_0 \sin^2\left(\frac{\omega t}{2n_{\mathrm{cyc}}}\right) \sin(\omega t), \tag{12.41}$$

where $n_{\mathrm{cyc}} = 8$ is the number of cycles, $\omega = 2\pi \times 94$ THz is the frequency of the light (corresponding to a laser wavelength $\lambda = 3200$ nm), and $E_0 = 0.083\mathrm{V}/\mathrm{Å}$ is the electric field strength. After Fourier transforming the current into frequency space, we find that it exhibits peaks at odd multiples of the incoming frequency $\omega$. The height of the peaks decreases with the harmonic order, but peaks remain clearly visible up to the 7th order. The spectrum of the calculated current matches qualitatively well with the experimentally obtained power spectrum of the emission, especially after including a filtering procedure into the Fourier transform, see Fig. 12.2(a–c).

In order to match the relative heights of the different peaks quantitatively, the relaxation times $\tau_1$ and $\tau_2$ can be used as two temperature-dependent fit parameters. The result is shown in Fig. 12.2(d), over a range of temperatures from 80 K to 300 K. For the fit, we enforce monotonic decay of the relaxation times as a function of temperature, leading to the choice plotted in Fig. 12.2(e). Strikingly, three different temperature regimes are obtained in which the scattering times exhibit different functional behavior. These regimes agree well with the expected three phases, superconducting phase, pseudogap phase, and strange metallic phase. This demonstrates that high harmonic spectroscopy, combined with the semi-microscopic model presented in this chapter, is able to discern between different strongly correlated phases of matter in a high-temperature superconductor.

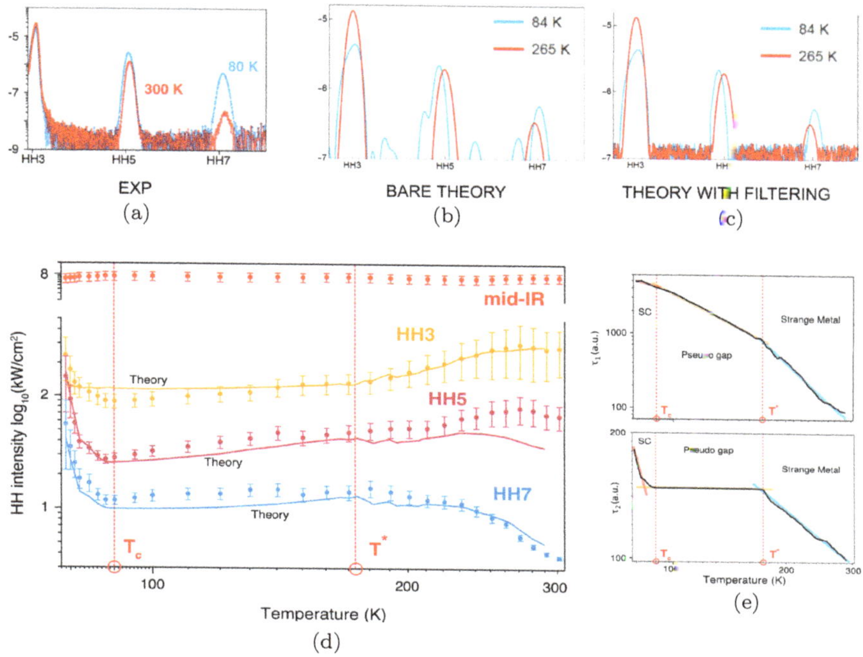


Fig. 12.2 (a) Experimentally obtained emission spectrum of YBCO at two different temperatures, at $T = 80\,\mathrm{K}$ (superconducting phase) and $T = 300\,\mathrm{K}$ (strange metal phase). Clearly visible are peaks at odd high harmonic (HH) frequencies (HH3 $= 3\omega$, HH5 $= 5\omega$, and HH7 $= 7\omega$). (b, c) The Fourier components of the theoretically calculated current, with or without filtering, also exhibit peaks at odd frequencies, and similarly to the experimental data, the height of the peaks decays with its harmonic order. (d) The height of odd peaks as a function of temperature, experimental and theoretical data. (e) The scattering times $\tau_1$ and $\tau_2$ have been used as temperature-dependent fit parameters, as plotted here, in order to match experimental and theoretical data in panel (d). Reproduced with permission from [11].

### 12.4.3 *Conclusions*

Our empirical and theoretical results present an initial instance of how high harmonic spectroscopy can effectively differentiate highly correlated phases in superconductors and open avenues toward a clearer comprehension of the intriguing physical mechanisms at play in a high-temperature superconductor. High harmonic spectroscopy, though straightforward in its application, necessitates a theoretical framework to offer microscopic insights. We have introduced a robust quasi-Hubbard model under strong fields, linking the measurable optical spectrum to the microscopic dynamics of carriers. This model elucidates the underlying physics behind the conversion of composite bosons from fermion spin-1/2 particles through Cooper pairing and the electron hopping processes among the material's orbitals. The synergy between experiment and theory facilitates the derivation of phenomenological scattering times, distinctly pinpointing the quantum critical points. It is noteworthy that, in the linear response range, these phenomenological amplitudes and phase decay times directly correspond to the material's conductivity. Our Bardeen-Cooper-Schrieffer-type model replicates the observed high harmonic spectroscopy sensitivity to all quantum critical points in a reasonable way. Our strong-field analysis of a high-temperature superconductor and the measurement of its quantum phase transitions allows us to better understand the electron dynamics of a superconductor driven by a strong laser field. The approach unequivocally illustrates its effectiveness in identifying various quantum phases, such as the pseudogap phase, the strange metal, and the superconducting phase, along with their associated quantum critical points. We are confident that applying attosecond technology to quantum materials affords insights into the multi-body physics of strongly correlated materials, enabling the probing of Cooper pair dynamics. This understanding will advance our comprehension of strong-field and light-field control over light–matter hybrids, potentially influencing competing mechanisms in quantum matter. We anticipate harnessing strong-field control to either facilitate or impede superconductivity and manipulate quantum phases at light frequencies. This bears implications for developing novel approaches to energy-efficient information processing and optical quantum computation.

## Acknowledgments

T. G. acknowledges funding by Gipuzkoa Provincial Council (QUAN-000021-01), by the Department of Education of the Basque Government through the IKUR strategy and through the project PIBA_2023_1_0021 (TENINT), by the Agencia Estatal de Investigación (AEI) through Proyectos de Generación de Conocimiento PID2022-142308NA-I00 (EXQUSMI), and by the BBVA Foundation (Beca Leonardo a Investigadores en Física 2023). The BBVA Foundation is not responsible for the opinions, comments and contents included in the project and/or the results derived therefrom, which are the total and absolute responsibility of the authors. U. B. acknowledges the project that gave rise to these results, and received the support of a fellowship (funded from the European

Union's Horizon 2020 research and innovation programme under the Marie Sklodowska-Curie grant agreement No 847648) from "la Caixa" Foundation (ID 100010434). The fellowship code is "LCF/BQ/PR23/11980043". The authors acknowledge financial support from the Spanish Ministry of Science and Innovation MCIN/AEI/10.13039/501100011033/ through HTSUPERFUN PID2021-124680OB-I00 co-financed by ERDF A way of making Europe. J. B. acknowledges financial support from the European Research Council for ERC Advanced Grant "TRANSFORMER" (788218), ERC Proof of Concept Grant "miniX" (840010), FET-OPEN "PETACom" (829153), FET-OPEN "OPTOlogic" (899794), FET-OPEN "TwistedNano" (101046424), Laserlab-Europe (871124), MINECO for Plan Nacional PID2020–112664 GB-I00; AGAUR for 2017 SGR 1639, MINECO for "Severo Ochoa" (CEX2019-000910-S), Fundació Cellex Barcelona, the CERCA Programme/Generalitat de Catalunya, and the Alexander von Humboldt Foundation for the Friedrich Wilhelm Bessel Prize. I. T. and J. B. acknowledge support from Marie Skłodowska-Curie ITN "smart-X" (860553). M. F. C. acknowledges support by the National Key Research and Development Program of China (Grant No. 2023YFA1407100), Guangdong Province Science and Technology Major Project (Future functional materials under extreme conditions — 2021B0301030005) and the Guangdong Natural Science Foundation (General Program project No. 2023A1515010871). M. L. acknowledges support from ERC AdG NOQIA; MCIN/AEI (PGC2018-0910.13039/501100011033, CEX2019-000910-S/10.13039/501100011033, Plan National FIDEUA PID2019-106901GB-I00, Plan National STAMEENA PID2022-139099NB-I00 project funded by MCIN/AEI/10.13039/501100011033 and by the European Union NextGenerationEU (PRTR-C17.I1), FPI); QUANTERA MAQS PCI2019-111828-2; QUANTERA DYNAMITE PCI2022-132919 (QuantERA II Programme co-funded by European Union's Horizon 2020 program under Grant Agreement No 101017733), Ministry of Economic Affairs and Digital Transformation of the Spanish Government through the QUANTUM ENIA project call — Quantum Spain project, and by the European Union through the Recovery, Transformation, and Resilience Plan — NextGenerationEU within the framework of the Digital Spain 2026 Agenda; Fundació Cellex; Fundació Mir-Puig; Generalitat de Catalunya (European Social Fund FEDER and CERCA program, AGAUR Grant No. 2021 SGR 01452, QuantumCAT U16-011424, co-funded by ERDF Operational Program of Catalonia 2014–2020); Barcelona Supercomputing Center MareNostrum (FI-2023-1-0013); EU Quantum Flagship (PASQuanS2.1, 101113690); EU

Horizon 2020 FET-OPEN OPTOlogic (Grant No 899794); EU Horizon Europe Program (Grant Agreement 101080086 — NeQST), ICFO Internal "QuantumGaudi" project; European Union's Horizon 2020 program under the Marie Sklodowska-Curie grant agreement No 847648 "La Caixa" Junior Leaders fellowships, La Caixa" Foundation (ID 100010434): CF/BQ/PR23/11980043. Views and opinions expressed are, however, those of the author(s) only and do not necessarily reflect those of the European Union, European Commission, European Climate, Infrastructure and Environment Executive Agency (CINEA), or any other granting authority. Neither the European Union nor any granting authority can be held responsible for them.

## References

[1] A. Damascelli, Z. Hussain and Z.-X. Shen, Angle-resolved photoemission studies of the cuprate superconductors, *Reviews of Modern Physics* **75**, pp. 473–541 (2003).

[2] M. Hashimoto, I. M. Vishik, R.-H. He, D. T. P. and Z.-X. Shen, Energy gaps in high-transition-temperature cuprate superconductors, *Nature Physics* **10**, p. 483 (2014).

[3] J. M. Bok, J. J. Bae, H.-Y. Choi, C. M. Varma, W. Zhang, J. He, Y. Zhang, L. Yu and X. J. Zhou, Quantitative determination of pairing interactions for high-temperature superconductivity in cuprates, *Science Advances* **2**, 3, p. e1501329 (2016).

[4] T. Yu, C. E. Matt, F. Bisti, X. Wang, T. Schmitt, J. Chang, H. Eisaki, F. Donglai and V. N. Strocov, The relevance of ARPES to high-Tc superconductivity in cuprates, *npj Quantum Materials* **5**, p. 46 (2020).

[5] L. Chaix, G. Ghiringhelli, Y. Y. Peng, M. Hashimoto, B. Moritz, K. Kummer, N. B. Brookes, Y. He, S. Chen, S. Ishida, Y. Yoshida, H. Eisaki, M. Salluzzo, L. Braicovich, Z.-X. Shen, T. P. Devereaux and W.-S. Lee, Dispersive charge density wave excitations in $Bi_2Sr_2CaCu_2O_8^{+\delta}$, *Nature Physics* **13**, 10, pp. 952–956 (2017).

[6] W. S. Lee, K.-J. Zhou, M. Hepting, J. Li, A. Nag, A. C. Walters, M. Garcia-Fernandez, H. C. Robarts, M. Hashimoto, H. Lu, B. Nosarzewski, D. Song, H. Eisaki, Z. X. Shen, B. Moritz, J. Zaanen and T. P. Devereaux, Spectroscopic fingerprint of charge order melting driven by quantum fluctuations in a cuprate, *Nature Physics* **17**, 1, pp. 53–57 (2021).

[7] H. Chu, M.-J. Kim, K. Katsumi, S. Kovalev, R. D. Dawson, L. Schwarz, N. Yoshikawa, G. Kim, D. Putzky, Z. Z. Li, H. Raffy, S. Germanskiy, J.-C. Deinert, N. Awari, I. Ilyakov, B. Green, M. Chen, M. Bawatna, G. Cristiani, G. Logvenov, Y. Gallais, A. V. Boris, B. Keimer, A. P. Schnyder, D. Manske, M. Gensch, Z. Wang, R. Shimano and S. Kaiser, Phase-resolved Higgs

response in superconducting cuprates, *Nature Communications* **11**, 1, p. 1793 (2020).

[8] K. Katsumi, N. Tsuji, Y. I. Hamada, R. Matsunaga, J. Schneeloch, R. D. Zhong, G. D. Gu, H. Aoki, Y. Gallais and R. Shimano, Higgs mode in the $d$-wave superconductor $Bi_2Sr_2CaCu_2O_{8+x}$ driven by an intense terahertz pulse, *Physical Review Letters* **120**, p. 117001 (2018).

[9] L. Perfetti, B. Sciolla, G. Biroli, C. J. van der Beek, C. Piovera, M. Wolf and T. Kampfrath, Ultrafast dynamics of fluctuations in high-temperature superconductors far from equilibrium, *Physical Review Letters* **114**, p. 067003 (2015).

[10] F. Giusti, A. Marciniak, F. Randi, G. Sparapassi, F. Boschini, H. Eisaki, M. Greven, A. Damascelli, A. Avella and D. Fausti, Signatures of enhanced superconducting phase coherence in optimally doped $Bi_2Sr_2Y_{0.08}Ca_{0.92}Cu_2O_{8+\delta}$ driven by midinfrared pulse excitations, *Physical Review Letters* **122**, p. 067002 (2019).

[11] J. Alcalà, U. Bhattacharya, J. Biegert, M. Ciappina, U. Elu, T. Graß, P. T. Grochowski, M. Lewenstein, A. Palau, T. P. H. Sidiropoulos T. Steinle and I. Tyulnev, High-harmonic spectroscopy of quantum phase transitions in a high-Tc superconductor, *Proceedings of the National Academy of Sciences* **119**, 40, p. e2207766119 (2022).

[12] H. Haug and S. W. Koch, *Quantum Theory of the Optical and Electronic Properties of Semiconductors*. World Scientific (2004).

https://doi.org/10.1142/9789811279560_0013

# Chapter 13

# Interference Effects in High-Order Harmonic Generation in Solids

Guang-Rui Jia[1], Tian-Jiao Shao[2], Tao-Yuan Du[3,4], Ling-Jie Lü[5], Xin-Qiang Wang[6] and Xue-Bin Bian[5]

[1] *School of Physics, Henan Normal University, Xinxiang 453007, China*
[2] *School of Information Science and Engineering, NingboTech University, Ningbo 315100, China*
[3] *School of Mathematics and Physics, China University of Geosciences, Wuhan 430074, China*
[4] *Key Laboratory of Computational Physical Sciences (MOE), Fudan University, Shanghai 200433, China*
[5] *State Key Laboratory of Magnetic Resonance and Atomic and Molecular Physics, Wuhan Institute of Physics and Mathematics, Innovation Academy for Precision Measurement Science and Technology, Chinese Academy of Sciences, Wuhan 430071, China*
[6] *College of Physics and Optoelectronic Engineering, Shenzhen University, ShenZhen 518060, China*

## 13.1 Introduction

Interference effect is one of the most important phenomena in physical sciences. For high-order harmonic generation (HHG) in gases, there are various types of interference in HHG processes, such as interference between different atomic centers within molecular targets [1,2], interference between different orbits in molecules [3–5], and interference between long and short trajectories [6]. From 2011, HHG in solids has become a hot research direction [7]. Rich interference effects in crystal HHG have been revealed by theoretical and experimental studies recently. From the energy band theory [8], high-order harmonics mainly come from the interband polarization and intraband oscillation in crystals. In this chapter, we will review the multi-electron interference in intraband currents, the interference between long and short trajectories in interband currents, the interference between intraband and interband transitions, multichannel interference, and subcycle interference.

## 13.2 Interference of $k$-Resolved Electrons in Intraband HHG

Driven by ultrafast intense lasers, crystal HHG is an ultrafast process involving many-body electron and hole dynamics. Multiple energy bands in semiconductors may participate in HHG processes, and only by considering all the electrons in a band can we get the concepts of electron and hole which play an important role in HHG. In this section, we will demonstrate that the collective motion of electrons leads to an interference in the intraband HHG.

### 13.2.1 *$k$-Resolved Intraband Harmonic Model*

In practical simulations, all initial electrons with different lattice momentum $k$ in the Brillouin zone (BZ) should be considered. Valence electrons with different lattice momentum have different dispersion, thus they have different nonlinear responses to the driving lasers. In order to fully understand the HHG contributed by all electrons in BZ and optimize the crystal HHG, the following part introduces a semiclassical model of $k$-resolved intraband HHG.

Driven by an intense laser field $F(t)$, the crystal momentum satisfies the acceleration theorem,

$$dk/dt = -F. \tag{13.1}$$

By neglecting the envelope, we simplify the driving field into a cosine electric field $F(t) = -F_p \cos(\omega t)$. According to Eq. (13.1),

$$k(t) = k_0 + A_0 \sin(\omega t), \tag{13.2}$$

where $k_0$ is the initial momentum, and $A_0 = F_p/\omega$ is the peak value of the vector potential. Here, we expand the energy bands by the Taylor series:

$$\varepsilon_m(k) = c_{m,0} + c_{m,1}(k - k_0) + c_{m,2}(k - k_0)^2 + \cdots . \tag{13.3}$$

According to Eqs. (13.2) and (13.3), the group velocity $v_m(k) = \frac{\partial}{\partial k}\varepsilon_m(k)$, and some trigonometric formulas, we can obtain the acceleration,

$$\begin{aligned}\dot{v}_m(k_0, A_0, t) &= Y_{m,1}\cos(\omega t) + Y_{m,2}\cos(2\omega t - \pi/2) + \cdots \\ &\quad + Y_{m,n}\cos[n\omega t + (1-n)\pi/2],\end{aligned} \tag{13.4}$$

where

$$Y_{m,n} = n\omega H_{m,n}. \tag{13.5}$$

$n = 1, 2, 3, \ldots$ represents the harmonic order, and $H_{m,n}$ is a polynomial function of $c_{m,n}$, which can be obtained by the empirical rule introduced in Ref. [9]. Harmonic amplitudes are proportional to the accelerations of carriers. Thus, $Y_{m,n}$ represents the harmonic amplitude as functions of $k_0$ and $A_0$. Considering the difference of only one coefficient $n\omega$, harmonic amplitudes can also be expressed by $H_{m,n}$. In the two-band model, the total harmonic amplitudes produced by both the electron in the conduction band and the hole in the valence band can be written as $Y_{t,n} = Y_{h,n} - Y_{e,n}$.

Equation (13.5) gives the analytical relationship between the band structure $\varepsilon_m(k)$, laser $A_0$, and intraband harmonic $Y_{m,n}$, which is lattice momentum $k_0$ resolved. To graphically demonstrate this relationship, Fig. 13.1 shows the intraband harmonic amplitudes for $SiO_2$ driven by a laser with $A_0 = 0.42$ a.u. in the two-band model. As can be seen from the figure, the even-order and odd-order harmonics have different distribution characteristics. The even-order harmonic amplitudes are odd functions, so they will disappear after integration, and only the odd-order harmonics are retained at the end. In addition, the distributions of harmonics with different orders are different, which makes it possible to optimize harmonics selectively. It is worth noting that the point $k_0 = 0$ is not necessarily the point whose contribution to the harmonics is the strongest. It is also important to note that the distributions of harmonic amplitudes shown

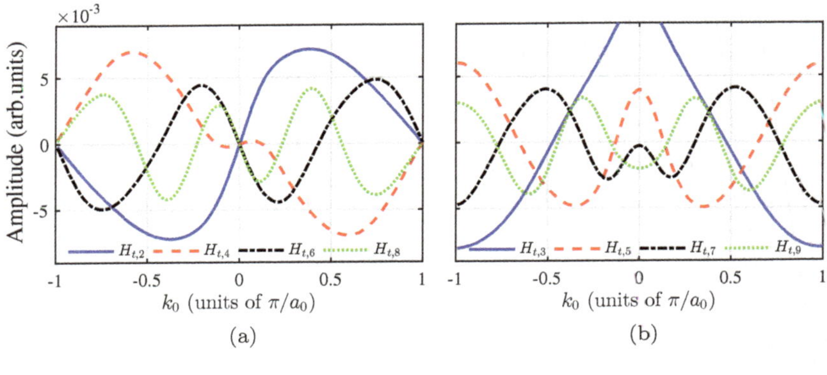


Fig. 13.1 The even-order (a) and odd-order (b) $k$-resolved intraband harmonic amplitudes jointly contributed by electrons and holes in the bands of $SiO_2$, where $A_0 = 0.42$ a.u. Reproduced with permission from [9].

in the figure are positive in some intervals and negative in others. The difference between positive and negative harmonic amplitudes implies a harmonic phase difference $\pi$, which may lead to destructive interference. Therefore, it provides a guideline to optimize harmonics, that is, avoiding the carrier distribution in the interval with different symbols can reduce destructive interference and enhance the harmonic yield.

### 13.2.2 *Selective Enhancement of Intraband Harmonics*

As can be seen from Fig. 13.1, the amplitudes of harmonics with different orders are distributed differently in the BZ, so we can adjust the distribution of electrons and holes in the BZ by using a pump-probe scheme to selectively enhance some harmonics. In this scheme, there is a short wavelength pump pulse in front of the probe light, and its single photon energy is equal to a certain band gap, which is used to excite electron-hole pairs at a specified position. When enough carriers are excited by the pump pulse, the probe light drives electrons and holes to move in the energy bands, radiating intraband harmonics.

We simulate this process by calculating the semiconductor Bloch equation (SBE). In the two-band model of $SiO_2$, the band gap varies from 8.20 to 10.28 eV. We change the energy of the pump pulse in this range, so as to excite electrons at different positions in the BZ. The wavelength and the peak electric field strength of the probe light are 3500 nm and $5.5\times10^{-3}$ a.u., thus the peak value of its vector potential is the same as that in Fig. 13.1, about 0.42 a.u. Figure 13.2 shows the variation of the 3rd, 5th, 7th, and 9th harmonic intensities with the pump light energy. For the convenience of comparison, the intensities of the harmonics without pump are marked by horizontal lines. It can be seen that the curves of harmonic intensities are similar to that in Fig. 13.1(b), and the pump light can not only enhance but also suppress harmonics. This is caused by the interference between the intraband harmonics generated by the electrons and holes excited by the pump pulse and the probe light. Specifically, when the phase of harmonics generated by the pumped electrons coincides with the phase of harmonics generated by the probe light, the harmonic is enhanced. When the phase is opposite, the harmonic is weakened. Because the phase distributions of different harmonics are different, the intensity of harmonics is selectively enhanced in this scheme. For example, when the pump light of 8.20 eV is used, the 3rd and 9th harmonics are enhanced, while other harmonics are suppressed.

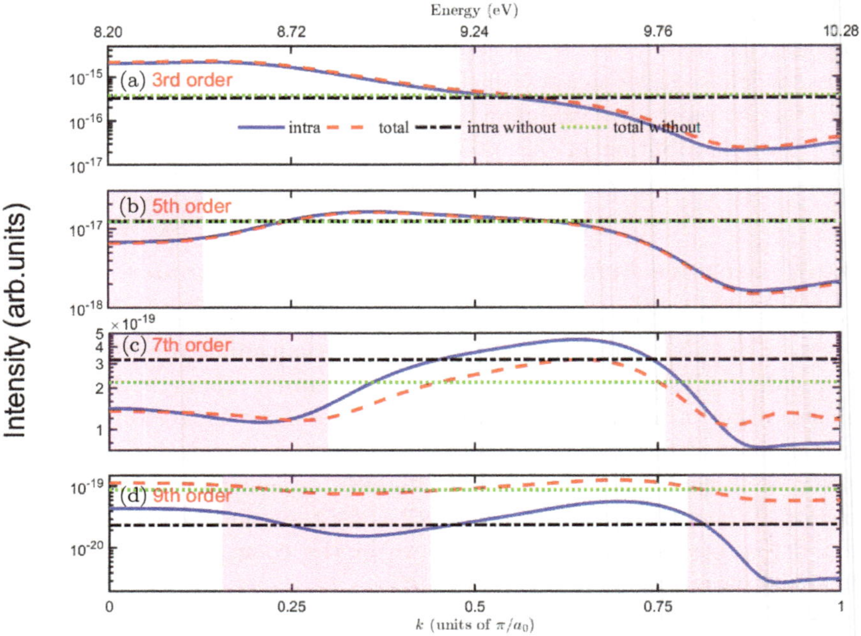


Fig. 13.2 The $k$-resolved harmonic intensities. (a)–(d) represent the yield of the 3rd, 5th, 7th, and 9th harmonics, respectively. For clarity, the energy of the pre-excitation pulse which is the top abscissa is equal to the band gap at $k$ which is the bottom abscissa in this figure. Horizontal lines represent harmonic intensities without the pump light. In the shadings, the electrons excited by the pump pulse enhance destructive interference. The situation is opposite outside the shadings. In fact, the $k$-resolved harmonic intensities reproduce the $k$-resolved harmonic amplitudes in Fig. 13.1(b). Reproduced with permission from [9].

### 13.2.3 *Field Strength-Dependent Intraband Harmonic Interference*

In the process of crystal HHG, due to the high laser intensity and long wavelength, electrons are generally excited non-resonantly. In this case, the distribution of electrons and holes in BZ is relatively wide, spanning multiple intervals with positive and negative harmonic amplitudes. As illustrated in Fig. 13.1, the harmonics of electrons at different positions in the BZ can interfere constructively or destructively. Therefore, for some laser parameters, the positive amplitude provided by the carrier is equal

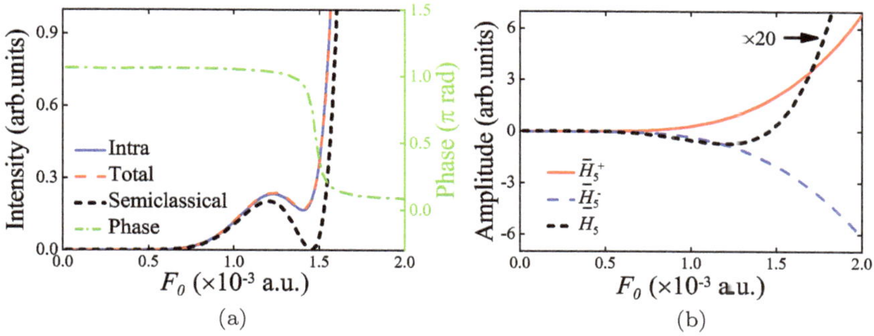


Fig. 13.3 (a) The solid blue, dashed red and short-dash black lines represent the 5th harmonic intensity varying with $F_0$ calculated by SBE and semiclassical model, respectively. The dashed-dotted green line represents the harmonic phase. (b) The solid red and dashed blue lines represent the positive and negative amplitudes of the 5th harmonic, respectively. The short-dash black line represents the 5th harmonic amplitude generated by all carriers in the first BZ. Reproduced with permission from [9].

to the negative amplitude, and the destructive interference will result in a minimum value of harmonic yield.

An example of such harmonic minimum is shown in Fig. 13.3(a). The curves show the relationship between numerically calculated harmonic intensity and the driving laser amplitude. The black dotted line is calculated by the semiclassical intraband model introduced in this section, and it is the same as the results calculated by the SBE. Near the minimum value of harmonic intensity, the phase of harmonics changes by $\pi$, which means that there is a destructive interference between the harmonics of electrons at different $k$ points. The data in Fig. 13.3(b) further verify this interference mechanism. The red solid line and the blue dotted line represent the positive and negative amplitudes of the 5th harmonic respectively, and the short, black dotted line is the sum of them, that is, the total 5th harmonic contributed by all carriers in BZ. It can be seen that the positive amplitude exceeds the negative amplitude when $F_0 = 1.4 \times 10^{-3}$ a.u., which results in the minimum value of harmonic intensity.

In this section, we explain the intraband harmonic interference mechanism from $k$-resolved electrons in the pump-probe scheme and the dependence of driving light intensity, respectively. This could be utilized to optimize the crystal HHG. It is also helpful for the research of dynamic Bloch oscillation detection and energy band reconstruction [10].

## 13.3 Interference between Long and Short Trajectories in Interband Solid HHG

HHG in gases is understood by the three-step model [11]: (1) ionization, (2) acceleration, and (3) recombination. The motion of ionized electrons in the continuum can be regarded as nearly free electrons and described classically based on Newton's law. For non-perturbative harmonics in the plateau, each of them is mainly contributed by two independent trajectories in a half optical cycle. According to the recombination time $t_r$ and ionization time $t_i$ of the electron, the trajectories are divided into long trajectories with a later ionization time and an earlier recombination time, and short trajectories with an earlier ionization time and a later recombination time. The interference between the long and short trajectories in atomic HHG processes leads to the modulation of the spectra, which has been verified by experimental measurements and theoretical calculations [6, 12].

The interband transition in crystal HHG can also be explained by the three-step model [13, 14] picture as illustrated in Fig. 13.4, similar to HHG in gases. Driven by the laser field, the valence band electrons are firstly excited to the conduction band through tunneling and holes are formed in the valence band. Then, the electrons and holes do intraband motions in the conduction and valence bands driven by the laser field, respectively. Since the laser field changes direction in each half optical period, electrons and holes will recombine and emit high-energy photons. The interband transition is similar to the recollision picture of HHG in gases, which also contains the short and long trajectories depending on their excursion time. In this section, we use the density matrix formulation [13, 15] and the saddle point method to explain the modulation of the HHG spectra caused by the interference between long and short trajectories in interband transitions.

### 13.3.1 *Simulation of the Density Matrix Formulation*

The density matrix formulations (DMFs) in the reciprocal $k$-space [13, 15] can be used to describe the laser-induced electron-hole pair dynamics in crystals. The polarization strength $\pi(K, t)$ between the conduction band and the valence band and carrier population $n_b(K, t)$ within the conduction ($b \to C$) and valence bands ($b \to V$) can be obtained by solving the

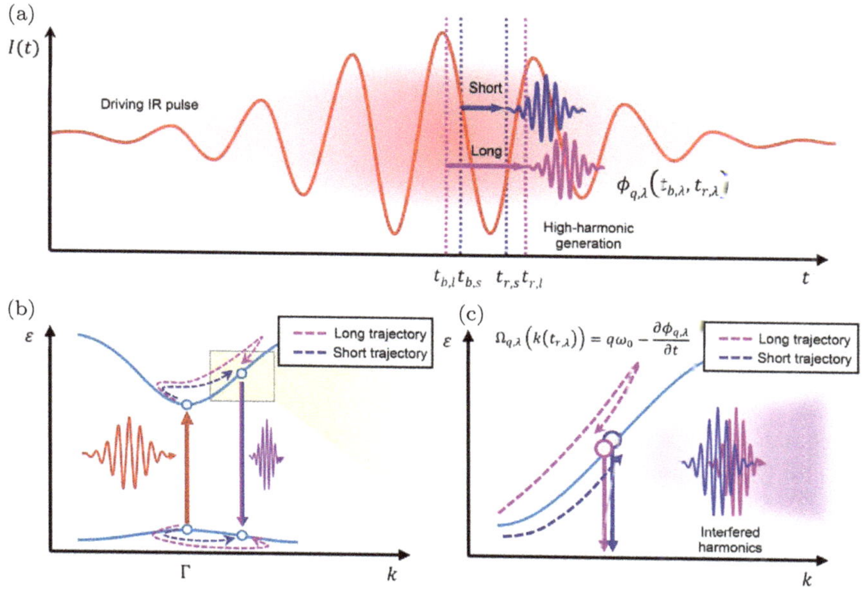


Fig. 13.4 Long and short trajectories in solids. (a) Long (l) and short (s) trajectories depicted relative to the driving laser field in the temporal domain. (b) Long and short trajectories traced within a crystal band structure. (c) Enlarged view of the recombination process followed by interference between the long and short trajectories. $\Omega_{q,\lambda}$ is the chirped center frequency of the $q$-th harmonic. Reproduced with permission from [16].

two-band DMF as follows,

$$\dot{\pi}(K,t) = -\frac{\pi(K,t)}{T_2} - i\Omega(K,t)w(K,t)e^{-iS(K,t)}, \tag{13.6}$$

$$\dot{n}_b(K,t) = is_b\Omega^*(K,t)\pi(K,t)e^{iS(K,t)} + c.c., \tag{13.7}$$

where $T_2$ is the dephasing time used to describe the coherence between the electrons and holes. $w(K,t) = n_C(K,t) - n_V(K,t)$ denotes the difference between the population of the conduction band and the valence band. $S(K,t) = \int_{-\infty}^{t} \varepsilon_g(K + A(t'))dt'$ is the classical action which describes the accumulated phase of the carrier in intraband motion driven by the laser field. $\varepsilon_g$ is the band gap between the conduction and valence bands. $s_b = 1, -1$ for $(b \to C)$ and $(b \to V)$, respectively. $\Omega(K,t) = d(K + A(t))E(t)$ is the Rabi frequency.

The DMFs can be solved by the Euler method or Runge-Kutta method, and then the interband current $j_{inter}$ and intraband current $j_{intra}$ can be calculated as follows,

$$j_{inter}(t) = \frac{d}{dt}\int_{BZ} p(K,t)d^3K, \tag{13.8}$$

$$j_{intra}(t) = \sum_{b=c,v}\int_{\overline{BZ}} v_b[K+A(t)]n_b(K,t)d^3K \tag{13.9}$$

where $p(K,t) = d(K+A(t))\pi(K,t)\mathrm{e}^{iS(K,t)} + c.c.$ is the interband polarization and $d(K+A(t))$ is the transition dipole moment. $v_b[K+A(t)]$ denotes the band velocities. The band structure, the transition dipole moment of the solid target, and laser parameters such as wavelength greatly influence the interband and intraband currents. The integral in the above formulas contain all $k$-points in the BZ. HHG spectra contributed by the interband and intraband currents can be obtained by Fourier transform (FT) of $j_{inter}(t)$ and $j_{intra}(t)$ as $|FT\{j_{inter}(t)\}|^2$ and $|FT\{j_{intra}(t)\}|^2$, respectively. The total HHG spectra can be obtained by Fourier transform of the total current $j_{total}(t) = j_{inter}(t) + j_{intra}(t)$.

In the case where the excitation yield is low, by using the Keldysh approximation $w(K,t) \approx 1$, the interband current can be calculated as,

$$j_{inter}(t) = \frac{d}{dt}\int_{BZ} d^3Kd(K)\int_{-\infty}^{t} dt'E\left(t'\right)d^*\left(K+A\left(t'\right)\right) \times e^{-iS\left(K,t',t\right)-\frac{t-t'}{T_2}} + c.c., \tag{13.10}$$

then in the frequency domain, the HHG contributed by the interband current can be calculated as,

$$j_{inter}(\omega) = \omega\int_{\mathrm{BZ}} d^3Kd(K)\int_{-\infty}^{\infty} dte^{-i\omega t}\int_{-\infty}^{t} dt'E\left(t'\right)d^*\left(K+A(t')\right) \times e^{-iS\left(K,t',t\right)-\left(t-t'\right)/T_2} + \mathrm{c.c.}, \tag{13.11}$$

where the accumulated phase of the carrier is expressed as,

$$-i\phi = -iS\left(K,t',t\right) - \frac{t-t'}{T_2}. \tag{13.12}$$

To distinguish between the short and long trajectories in the interband transition, the window function $F(t-t')$ is added to select the trajectory,

$$j_{inter}(\omega) = \omega\int_{\mathrm{BZ}} d^3Kd(K)\int_{-\infty}^{\infty} dte^{-i\omega t}\int_{-\infty}^{t} dt'E\left(t'\right)d^*\left(K+A(t')\right) \times e^{-iS\left(K,t',t\right)-\left(t-t'\right)/T_2}F(t-t') + \mathrm{c.c.} \tag{13.13}$$

We set $F(t-t') = 1$ in the interval $t - t' < 0.65T_0$ to select the short trajectory, where $T_0$ is the optical cycle of the driving field. To select the long trajectory, we set $F(t-t') = 1$ when $0.65T_0 < t - t' < T_0$. $\tau_{cutoff} = (t-t') \approx 0.65T_0$ is the excursion time of the electron-hole pair contributing to the cutoff energy. $\tau_{cutoff}$ can be estimated by calculating the classical trajectories obtained by the saddle-point approximation:

$$\nabla_K \phi = \int_{t'}^{t} \Delta \mathbf{v} \left(K + A\left(t''\right)\right) dt'' = 0, \tag{13.14}$$

$$\frac{d\phi}{dt'} = \varepsilon_{\rm g} \left[K + A\left(t'\right)\right] - \frac{i}{T_2} = 0, \tag{13.15}$$

$$\frac{d\phi}{dt} = \varepsilon_{\rm g}(K + A(t)) - \omega + \frac{i}{T_2} = 0. \tag{13.16}$$

### 13.3.2 *Intensity-Dependent Spectral Modulation in Bulk Sapphire*

Figure 13.5 illustrates the dependence of the HHG spectra on the field strength from bulk sapphire target obtained from experimental measurements and theoretical simulations [16]. The two-dimensional figure shows the optical spectrum of harmonic photons from the 7th order (114 nm) to the 11th order (72.7 nm) in the field strength range of 1.75–2.26 V/nm. A clear splitting pattern was observed near the 7th harmonic. In addition to the 114 nm peak at the normal integer peak position, a red-shifted peak at 118 nm was also observed. When the driving laser field strength is weak, the splitting of the 7th harmonic can be ignored [Fig. 13.5(c) and Fig. 13.5(d)], and only a single peak is observed. When the field strength of the driving laser increases, the 7th harmonic will be broadened, and a second red-shifted peak emerges gradually. When the intensity of the redshift peak increases, the original integer-order 114 nm peak shrinks. When the field strength of the driving laser is strong, the intensity dependence of the two peaks is completely opposite [Fig. 13.5(e) and Fig. 13.5(f)]. The simulation results well reproduce the experimental observations of anomalous field strength dependence. When we continue increasing the driving light field strength, the red-shifted peak (118 nm) gradually increases until it exceeds the normal peak at 114 nm. It is found that there are slight differences in the relative intensity of the normal peak at 114 nm and the red-shifted peak at 118 nm between the experimental data and the calculated results. This might be caused by the neglect of the effect of macroscopic phase matching

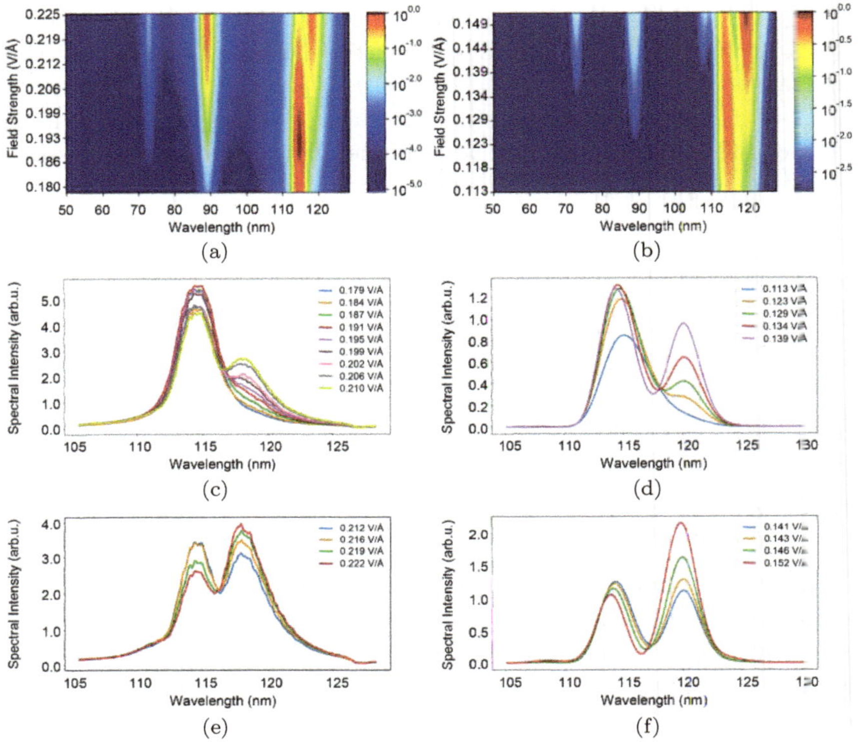

Fig. 13.5 Harmonic spectrum as a function of laser field strength. The experimental results are on the left, while the theoretical simulations are on the right. (c)–(f) are the 7th harmonic with varying field strengths. Reproduced with permission from [16].

and the contribution of multiple bands in the simulation. The propagation effects like plasma defocusing and self-phase modulation are assumed to be small during the experiment. This is because a tight focusing geometry is adopted which results in a relatively short laser-medium interaction length.

The theoretical simulations using the DMF and saddle-point approximation are shown in Fig. 13.6. The band structure of bulk sapphire is calculated by the density functional theory with Vienna Ab initio Simulation Package [17]. In the theoretical calculation, the computation parameters are selected to accord with the actual experimental conditions. From Fig. 13.6(b), one may find that the intraband harmonic yields are about 1–3 orders of magnitude lower than the interband harmonics. Therefore,

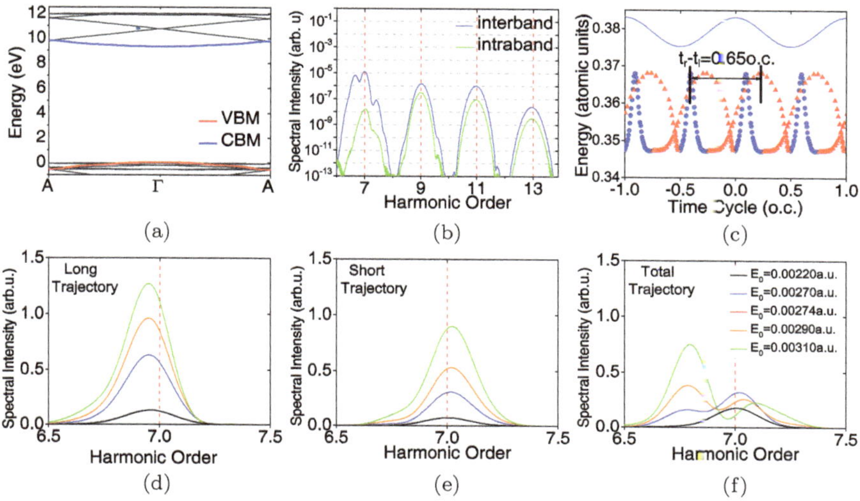


Fig. 13.6 Long and short trajectory contributions. (a) Band structure of sapphire along the Γ-A direction. VBM and CBM are the valence and conduction bands finally selected for simulations. (b) Expected interband and intraband contributions ($E_0 = 1.39\,\mathrm{V/nm}$). (c) Harmonic photon energy as a function of ionization (blue line) and recombination (red line) times. The solid blue line is the driving laser field. (d, e, f) Calculated harmonic spectra with varying laser intensity for long, short, and coherently combined trajectories, respectively. The dephasing time is set as $T_2 = 2.67\,\mathrm{fs}$ (1.0 o.c.). Reproduced with permission from [16].

we only consider the contribution of the interband transitions. By adding a window function in the time domain and changing the range of time-domain integration in the calculation using saddle-point approximation, the 7th harmonic contributed by the short and long trajectories in the time domain can be separated. The calculation results in Figs. 13.6(d) and 13.6(e) show the red-shifted harmonic peak contributed by the long trajectory and the slightly blue-shifted harmonic peak contributed by the short trajectory under different driving laser field strengths. Compared with the slight blue shift of the short trajectory contribution, the long trajectory contribution has a larger magnitude of red shift. The shift of spectral peak energy towards low energy (red shift) or high energy (blue shift) mainly depends on the intensity gradient of the rising and falling edges of the laser pulse with time. Thus the magnitude of the frequency shift should increase with the increment of the amplitude of the driving laser field. When the contributions of the short and the long trajectories are coherently superposed [Fig. 13.6(f)], the generated spectrum shows a clear broadening, and the secondary peak at 118 nm increases with the driving field strength, as well as the peak reversal. These results prove that the splitting and broadening observed in the experiment of HHG in bulk sapphire come from the interference of the long and short trajectories. Since our discussion is not confined to a specific harmonic order or the shape of the band structure, this kind of interference should also appear in other harmonic orders and other materials. However, the precondition is that the dephasing time should be long enough.

To sum up, the interference of long and short trajectories in interband transitions may lead to anomalous spectral modulation, broadening, splitting, and shift in HHG. This interference is also observed in the transmission and reflection HHG in bulk ZnO in monochromatic and two-color laser fields [18]. The trajectory may be selected by adjusting the phase difference in the two-color scheme [19]. Theoretical calculation of HHG in MgO shows that the interference could also be used to reconstruct the energy band [19].

## 13.4 Interference between Interband and Intraband Currents

The HHG in solids is usually separated into intraband and interband contributions according to the laser-induced electron dynamics. They correspond to the Bloch wave packets moving at the group and phase velocities, respectively [20]. Intraband and interband currents can interfere with each other [21]. The total HHG can be calculated by the Fourier transform of the total current

$$\begin{aligned} S_{\text{total}}(\omega) &= \left| \int j_{\text{total}}(t) \exp(-i\omega t) dt \right|^2 \\ &= S_{\text{inter}}(\omega) + S_{\text{intra}}(\omega) + S_{\text{interference}}(\omega), \end{aligned} \tag{13.17}$$

where the interference term is

$$\begin{aligned} S_{\text{interference}}(\omega) = &\int j^*_{\text{inter}}(t) \exp(i\omega t) dt \int j_{\text{intra}}(t) \exp(-i\omega t) dt \\ &+ \int j^*_{\text{intra}}(t) \exp(i\omega t) dt \int j_{\text{inter}}(t) \exp(-i\omega t) dt. \end{aligned}$$

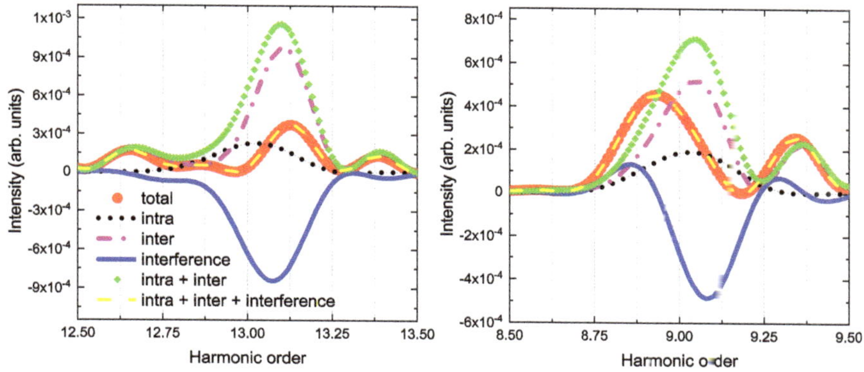


(a) (b)

Fig. 13.7 Fine structures of harmonics in ZnO caused by the interference between intraband and interband currents. The wavelength and intensity of the laser pulse is $3.2545\,\mu$m and $1.9920 \times 10^{12}$ W/cm$^2$, respectively. (a) The 13th order harmonic. Linear scale harmonic spectra are presented: total (red solid circle marker line) intraband (black dotted line), interband (magenta dot-dashed line), interference (blue solid line), addition of intraband and interband spectra (green diamond marker line), and sum of intraband and interband spectra and their interference (yellow dashed line). (b) is the same as (a) but for the 9th order harmonic. Reproduced with permission from [21].

We take the interaction between ZnO and strong laser pulses as an example to investigate this kind of interference. DMF is solved numerically [13, 21, 22]. The laser pulse is linearly polarized along the $\Gamma - M$ direction in ZnO with a trapezoidal-like pulse envelope. The initial state of ZnO is selected to be the whole valence band. The results are illustrated in Fig. 13.7.

The incoherent superposition of intraband and interband harmonic radiation only produces a single strong peak near $13.1\omega_0$, which is shown by the green diamond marker line in Fig. 13.7(a). The interference term between intraband and interband contributions significantly offsets the intraband and interband radiation, leading to the reduction of harmonic intensity. It also makes the bandwidth narrow. In Fig. 13.7(b), the peaks of both intraband and interband harmonic spectra are on the right side of $9\omega_0$ (blue shift), but the total harmonic peak appears on the left side of $9\omega_0$ (red shift). There are two peaks and one valley in the interference spectrum (blue solid line). When the interference term is added to the sum of intraband and interband harmonic spectra, the red shift and the oscillation of the total harmonic peak can be well explained.

In summary, the intraband current and interband current do not oscillate synchronously. Their interference may lead to obvious fine structures in harmonic peaks and the suppression of harmonic yield.

## 13.5 Multichannel Interference in HHG in Crystals

The interference effects discussed above are mainly restricted to the electron-hole dynamics in the first BZ as shown in Fig. 13.8(a). However, when the laser vector potential $A_0$ is big enough, the electron and hole will surpass the edge of the first BZ, leading to multiple channels in HHG. We take the initial excitation at $k = 0$ in Fig. 13.8(b) as an example to show the scheme in single-active-electron approximation. i) The electron on the top of the valance band (VB) is pumped into the first conduction band (CB1) via tunneling at the peak of the laser field, where the band gap between them is minimal. ii) The electron oscillates on CB1 and is driven to the edge of the first BZ. iii) It experiences a Bragg reflection in the same band and Zener tunneling to a higher conduction band, i.e., Bloch-Zener oscillation (BZO). In other words, the electron crosses the boundary, then it may stay in CB1 in the second BZ, or it may be pumped to CB2 in the second BZ. iv) When the laser reverses its direction, the electron on CB2 may stay in CB2 or transit to CB1 in the first BZ. The electron on CB1 may stay in CB1 or tunnel to CB2. The step-by-step population [14] in

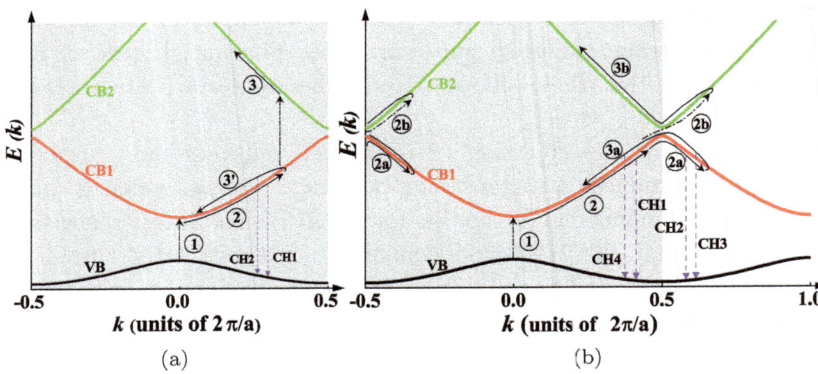


Fig. 13.8 Schematic of Bloch-Zener oscillations. (a) and (b) show two scenarios, in which electrons are restricted within the first BZ and extended to the second BZ, respectively. The electronic dynamics in the second BZ can be equivalent to entering the first BZ from the inverse edge of BZ. Dotted purple arrows indicate the emission channels (CH1–CH4) of one harmonic energy. Reproduced with permission from [23].

higher conduction bands results in the multi-plateau structure in the HHG spectra.

For a given harmonic $N\omega$ from CB1, it is usually emitted twice within each half cycle in Fig. 13.8(a), i.e., the short (CH1) and long (CH2) trajectories respectively. However, BZO plays a significant role in the HHG. The intrinsic boundary of the BZ will regulate the ultrafast electron dynamics. As presented in Fig. 13.8(b), one may find another two channels CH3 and CH4. These channels have different phases, and their interference will produce complex HHG spectra.

To illustrate the spectral features in the multiple-channel case involving the BZO effect, we calculate the HHG spectra involving the moderate and stronger intensities of laser fields shown in Fig. 13.9. For the convenience of discussion, Fig. 13.9(a) presents the top valence band and the lowest two conduction bands of a periodic potential model [23]. In Fig. 13.9(b), the HHG spectrum obtained from the moderate intensity of the laser fields, i.e., $0.488\,\mathrm{TW/cm^2}$ and wavelength $4.8\,\mu\mathrm{m}$, has a specific cutoff energy, which shows a good agreement with the energy band gap between CB1 and VB bands. When the BZO effect is included under a stronger driving laser field with the intensity $1.25\,\mathrm{TW/cm^2}$, the cutoff energy of this plateau is extended in Fig. 13.9(c). The extension of cutoff energy is denoted as $\Delta E_{\mathrm{cutoff}}$. Due to the fact that the electrons can arrive at the top of CB1 under higher field intensity, the electrons of CB1 can be easily pumped into CB2 via the BZO effect. Thus the HHG plateau yield is also enhanced about two orders of magnitude. To understand this enhancement of yield, we also resolve the intraband and interband contributions shown in Figs. 13.9(b) and 13.9(c). One can observe that the yield enhancement $\Delta Z_2$ is attributed to the collective contributions of the intraband and interband dynamics involving the BZO process. As per our prediction, the multiple channels with different phases make the spectra complicated. Considering the phase matching effect, this interference may not be easily observed in the experimental measurement.

Note that the scenarios discussed above just involve the excitation of valence electrons around the $\Gamma$ point of the system. When one takes into account all the valence electrons involved in full BZ, the multichannel interference of concern here is still applicable since the tunnel excitation mainly occurs around $k = 0$ with a minimum band gap.

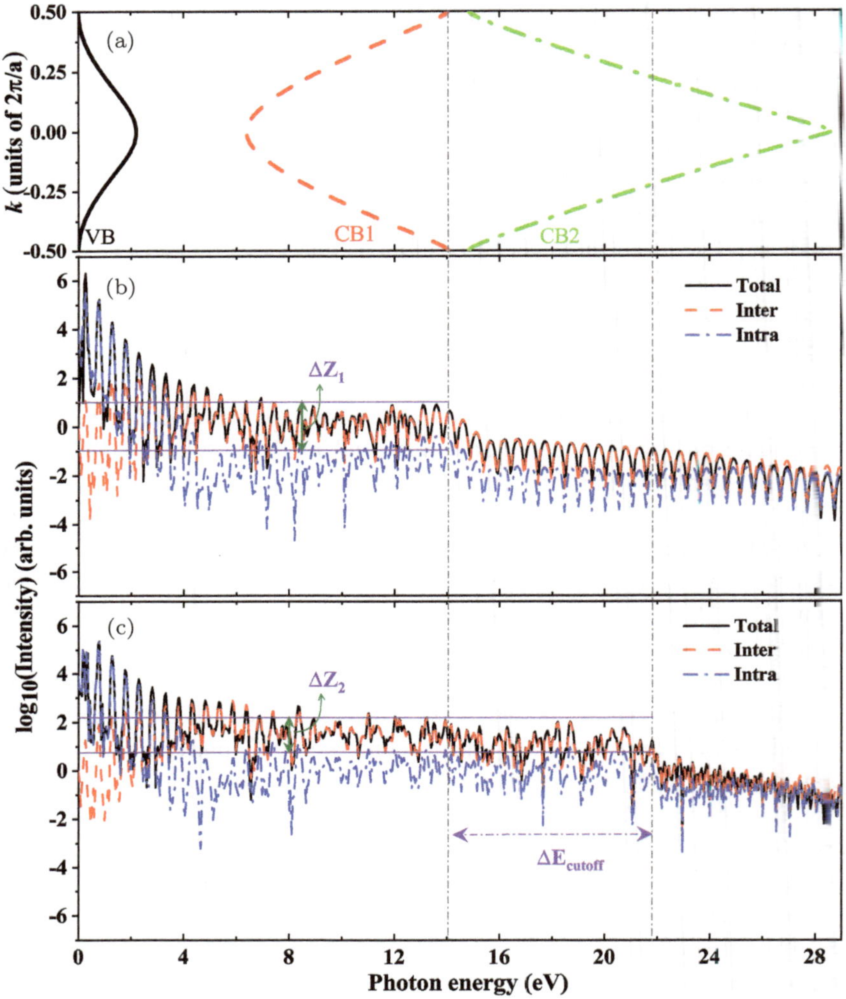


Fig. 13.9 Role of Bloch-Zener oscillation effect in HHG spectrum. (a) Energy band dispersion of the periodic potential model. (b) and (c) respectively display the HHG spectra without and with BZO effect. The variation of the relative contribution between intraband and interband currents caused by the BZO effect is described by $\Delta Z_1$ and $\Delta Z_2$. Here the $\Delta E_{\text{cutoff}}$ denotes the extension of cutoff energy. Reproduced with permission from [23].

## 13.6 Subcycle Interference in HHG in Solids

Different from HHG in gases, we found that the yield of HHG in solids exhibits unexpected modulations as a function of the driving laser intensity and wavelength. Its mechanism can be unraveled by the interference between currents induced by two adjacent Zener tunneling events. In Fig. 13.10, we show the schematic diagram of subcycle interference in solid-state HHG. It is well-known that the electronic wave packet is pumped into the conduction band via the minimum band gap at the peaks of the laser field, which is indicated by the arrows in Fig. 13.10(a). The subsequent electronic dynamics is presented in Fig. 13.10(b). One can see that the electronic wave packets, excited at the two adjacent peaks of the laser field, have different phases. They will interfere with each other.

Then in Fig. 13.11(a), we show the signature of subcycle interference as a function of the laser intensity, in which the dephasing time is set to be 10 fs. For the sake of analyses, we also show that the integral yields within HHG plateau zone (i.e., 3.3–12 eV) display distinct modulations, as illustrated by the blue solid curve in Fig. 13.11(b). In contrast, this yield modulation in the HHG spectra can be destroyed under the situation of a shorter dephasing time, as shown by the blue dashed-dotted curve in Fig. 13.11(b).

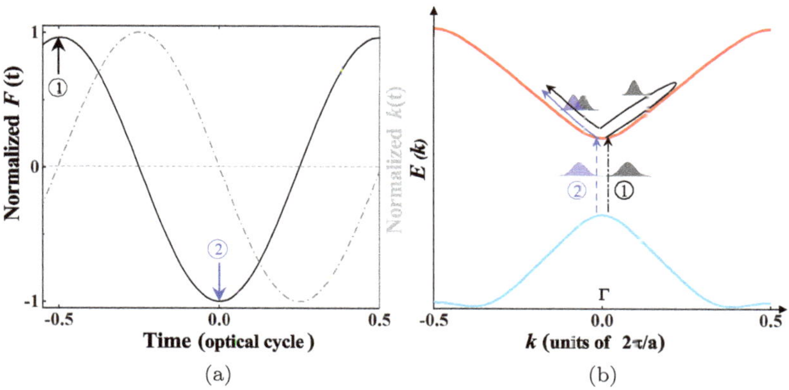


Fig. 13.10 Schematic of the subcycle interference. At the adjacent peaks, two electronic wave packets are produced by Zener transition. Temporal evolution of the electronic wave packets follows the vector potential, i.e., $k(t) = k_0 + A(t)$, as shown in (a). The subcycle dynamics in (b) depicts the quantum interference between two electronic wave packets produced at two adjacent peak times. Reproduced with permission from [24].

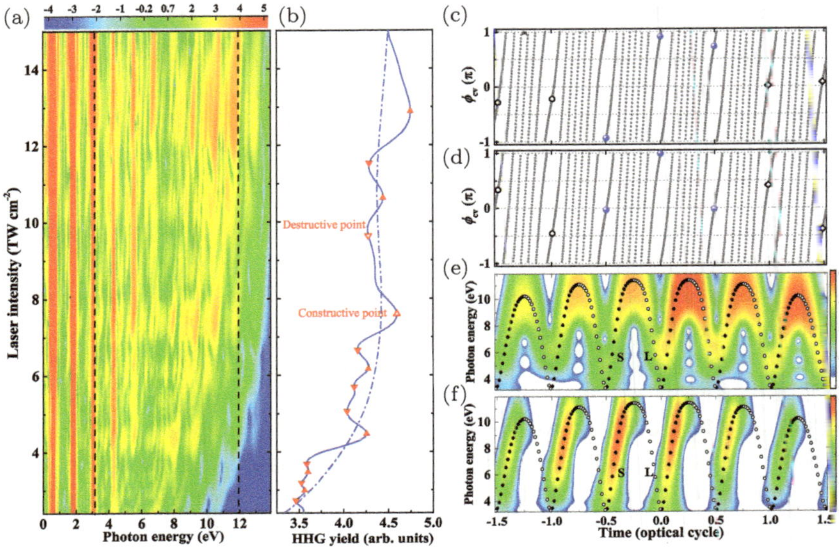


Fig. 13.11 (a) HHG spectra as a function of the laser intensity. The HHG plateau in (a) is marked by two vertical dashed lines. Its integral yields under $T_2 = 10$ fs and a shorter dephasing time are respectively shown by the solid and dashed-dotted curves in (b). The phase differences of two characteristic cases in (c) and (d) verify the subcycle interference processes. Role of dephasing time in subcycle interference has been illustrated by comparison of the time-frequency analyses in (e) and (f). Reproduced with permission from [24].

To verify this subcycle interference mechanism, we take two cases as the representative examples, which are defined as the constructive and destructive points in Fig. 13.11(b). These two extreme points in Fig. 13.11(b) are also predicted by the dynamical phase difference between the two wave packets ionized at the two adjacent peaks of the laser field, in which a $\pi$-phase shift of the transition phase is involved because of the opposite laser polarization directions [24]. Thus the destructive and constructive points in Fig. 13.11(b) are clarified by the phase difference $\phi_{cv}$ shown in Figs. 13.11(c) and 13.11(d), respectively. To assess the role of the dephasing time in the subcycle interference, we show the time-frequency analyses of HHG spectra with $T_2 = 10$ fs and 3 fs in Figs. 13.11(e) and 13.11(f), respectively. Actually, the results of time-frequency analyses show that the subcycle interference channels are closed under the situation of a short dephasing time.

The simulation in Fig. 13.11 is also performed in the single-active electron approximation. It should be applicable to the real solid system since Zener tunneling occurs molstly around the Γ point. This interference model is also applicable in both the intraband and interband transitions. The modulation of harmonic yield as a function of laser intensity can be found in the experimental results in Refs. [25, 26]. Our model predicts that this kind of modulation will also be present in the driving laser wavelength scaling [24].

## 13.7 Discussion

The appearance of interference in HHG spectra sensitively depends on the laser conditions and band structures. When the driving laser wavelength is long, the intraband harmonics usually dominate the spectra, then one may observe the multi-electron interference in intraband transitions. At the same time, laser vector potential $A_0$ may be bigger than $\pi/a$, i.e., electrons and holes may travel beyond the edge of the first BZ, and then multichannel interference may occur. If the interband and intraband transitions are comparable, the interference between them may be measured. When the driving laser wavelength is short and the interband transtion dominates, the interference between long and short trajectories will be present. Subcycle interference requires a bigger dephasing time, and may exist in both the intraband and interband harmonics. These interference effects may be used to understand the origin of rich spectral structures in HHG, reconstruct band structure, and probe ultrafast electron-hole dynamics in solids.

## Acknowledgments

This work is supported by the National Natural Science Foundation of China (NSFC) (Grants No. 12174090, No. 12204492, and No. 12104395), Zhejiang Provincial Natural Science Foundation of China (No. LQ22A040004), and Cooperative Project of Henan Normal University with Fuhuide Trade Co., Ltd of Henan Province (Grant No. XH2019005).

## References

[1] M. Lein, N. Hay, R. Velotta, J. P. Marangos and P. Knight, Role of the intramolecular phase in high-harmonic generation, *Physical Review Letters* **88**, 18, p. 183903 (2002).

[2] T. Kanai, S. Minemoto and H. Sakai, Quantum interference during high-order harmonic generation from aligned molecules, *Nature* **435**, 7041, pp. 470–474 (2005).
[3] B. K. McFarland, J. P. Farrell, P. H. Bucksbaum and M. Gühr, High harmonic generation from multiple orbitals in N2, *Science* **322**, 5905, pp. 1232–1235 (2008).
[4] O. Smirnova, Y. Mairesse, S. Patchkovskii, N. Dudovich, D. Villeneuve, P. Corkum and M. Y. Ivanov, High harmonic interferometry of multi-electron dynamics in molecules, *Nature* **460**, 7258, pp. 972–977 (2009).
[5] H. J. Wörner, J. B. Bertrand, P. Hockett, P. B. Corkum and D. M. Villeneuve, Controlling the interference of multiple molecular orbitals in high-harmonic generation, *Physical Review Letters* **104**, 23, p. 233904 (2010).
[6] A. Zaïr, M. Holler, A. Guandalini, F. Schapper, J. Biegert, L. Gallmann, U. Keller, A. S. Wyatt, A. Monmayrant, I. A. Walmsley, E. Cormier, T. Auguste, J. P. Caumes and P. Salières, Quantum path interferences in high-order harmonic generation, *Physical Review Letters* **100**, p. 143902 (2008).
[7] S. Ghimire, A. D. DiChiara, E. Sistrunk, P. Agostini, L. F. DiMauro and D. A. Reis, Observation of high-order harmonic generation in a bulk crystal, *Nature Physics* **7**, 2, pp. 138–141 (2011).
[8] H. Haug and S. W. Koch, *Quantum Theory of the Optical and Electronic Properties of Semiconductors*, 5th edition. World Scientific (2009).
[9] L.-J. Lü and X.-B. Bian, Multielectron interference of intraband harmonics in solids, *Physical Review B* **100**, p. 214312 (2019).
[10] L.-J. Lü and X.-B. Bian, Ultrafast intraband electron dynamics of preexcited $SiO_2$, *Optics Express* **28**, 9, pp. 13432–13442 (2020).
[11] P. B. Corkum, Plasma perspective on strong field multiphoton ionization, *Physical Review Letters* **71**, 13, p. 1994 (1993).
[12] C. M. Heyl, J. Güdde, U. Höfer and A. L'Huillier, Spectrally resolved maker fringes in high-order harmonic generation, *Physical Review Letters* **107**, p. 033903 (2011).
[13] G. Vampa, C. R. McDonald, G. Orlando, D. D. Klug, P. B. Corkum and T. Brabec, Theoretical analysis of high-harmonic generation in solids, *Physical Review Letters* **113**, p. 073901 (2014).
[14] T.-Y. Du and X.-B. Bian, Quasi-classical analysis of the dynamics of the high-order harmonic generation from solids, *Optics Express* **25**, 1, pp. 151–158 (2017).
[15] G. Vampa, T. J. Hammond, N. Thiré, B. E. Schmidt, F. Légaré, C. R. McDonald, T. Brabec, D. D. Klug and P. B. Corkum, All-optical reconstruction of crystal band structure, *Physical Review Letters* **115**, p. 193603 (2015).
[16] Y. W. Kim, T.-J. Shao, H. Kim, S. Han, S. Kim, M. Ciappina, X.-B. Bian and S.-W. Kim, Spectral interference in high harmonic generation from solids, *ACS Photonics* **6**, 4, pp. 851–857 (2019).

[17] G. Kresse and J. Hafner, Ab initio molecular dynamics for liquid metals, *Physical Review B* **47**, pp. 558–561 (1993).
[18] Y. Wang, T. Shao, X. Li, Y. Liu, P. Jiang, W. Zheng, L. Zhang, X.-B. Bian, Y. Liu, Q. Gong and C. Wu, Trajectory-controlled high-order harmonic generation in ZnO crystals, *Optics Express* **31**, 2, pp. 3379–3389 (2023).
[19] T.-J. Shao, L.-J. Lü, J.-Q. Liu and X.-B. Bian, Quantum path interferences and selection in interband solid high-order harmonic generation in mgo crystals, *Physical Review A* **101**, p. 053421 (2020).
[20] T.-Y. Du, X.-H. Huang and X.-B. Bian, High-order-harmonic generation from solids: The contributions of the Bloch wave packets moving at the group and phase velocities, *Physical Review A* **97**, p. 013403 (2018).
[21] X.-Q. Wang, Y. Xu, X.-H. Huang and X.-B. Bian, Interference between inter- and intraband currents in high-order harmonic generation in solids, *Physical Review A* **98**, p. 023427 (2018).
[22] G. Vampa, C. R. McDonald, G. Orlando, P. B. Corkum and T. Brabec, Semiclassical analysis of high harmonic generation in bulk crystals, *Physical Review B* **91**, p. 064302 (2015).
[23] T.-Y. Du, D. Tang, X.-H. Huang and X.-B. Bian, Multichannel high-order harmonic generation from solids, *Physical Review A* **97**, p. 043413 (2018).
[24] T.-Y. Du, D. Tang and X.-B. Bian, Subcycle interference in high-order harmonic generation from solids, *Physical Review A* **98**, p. 063416 (2018).
[25] G. Ndabashimiye, S. Ghimire, M. Wu, D. A. Browne, K. J. Schafer, M. B. Gaarde and D. A. Reis, Solid-state harmonics beyond the atomic limit, *Nature* **534**, 7608, pp. 520–523 (2016).
[26] A. A. Lanin, E. A. Stepanov, A. B. Fedotov and A. M. Zheltikov, Mapping the electron band structure by intraband high-harmonic generation in solids, *Optica* **4**, 5, pp. 516–519 (2017).

https://doi.org/10.1142/9789811279560_bmatter

# Index

Milton Keynes UK
Ingram Content Group UK Ltd.
UKHW051933270624
444728UK00036B/4

9 789811 279553